Nanochitosan-Based Enhancement of Fisheries and Aquaculture

Patrick Omoregie Isibor
Aina Olukukola Adeogun • Alex Ajeh Enuneku
Editors

Nanochitosan-Based Enhancement of Fisheries and Aquaculture

Aligning with Sustainable Development Goal 14 – Life Below Water

Springer

Editors
Patrick Omoregie Isibor
Department of Biological Sciences,
College of Science and Technology
Covenant University
Ota, Ogun State, Nigeria

Aina Olukukola Adeogun
Zoology
University of Ibadan
Ibadan, Nigeria

Alex Ajeh Enuneku
Environmental Management and Toxicology
University of Benin
Benin City, Nigeria

ISBN 978-3-031-52260-4 ISBN 978-3-031-52261-1 (eBook)
https://doi.org/10.1007/978-3-031-52261-1

This Springer imprint is published by the registered company Springer Nature Switzerland AG
The registered company address is: Gewerbestrasse 11, 6330 Cham, Switzerland

Paper in this product is recyclable.

The book is dedicated to God Almighty, the giver of life and inspirations. It is also dedicated to the impoverished fishermen in the Niger Delta areas of Nigeria who are impacted by polluted water bodies and loss of aquatic organisms.

Preface

Nanochitosan-Based Enhancement of Fisheries and Aquaculture presents a comprehensive analysis detailing the application of novel nanochitosan in the management of aquatic resources. This meticulous examination sheds light on the transformative potential of nanochitosan, an advanced nanomaterial poised to revolutionize the current challenges in fisheries and aquaculture.

This scholarly work is tailored to cater to a diverse readership encompassing researchers, practitioners, and students, elucidating intricate scientific concepts with precision and clarity. Its focal point revolves around sustainability, elucidating nanochitosan's exceptional capability to enhance water quality, enhance spawning and breeding, promote detoxification, bolster aquatic well-being, and redefine disease control and feed quality in aquaculture.

Each chapter harmoniously intertwines, culminating in a schematic concept centered on the pivotal theme of fisheries and aquaculture enhancement for attainment of biodiversity conservation. With scrupulous attention to detail, the narrative portrays nanochitosan as a novel biopolymer, underscoring its unparalleled efficacy in ensuring the safety of seafood products by mitigating contaminants and pathogens. Conservation is the bedrock ethos permeating the architecture of this academic discourse. Nanochitosan assumes a central role as a transformative agent steering aquatic resource management toward a more harmonious and ecologically attuned paradigm. Its transformative potential in water purification, nurturing aquatic life, and revolutionizing disease control and feed quality within aquaculture is accentuated.

This compendium bridges the realms of theory and application, harnessing interdisciplinary perspectives to underscore the profound impact and promise of nanochitosan in reshaping our comprehension and utilization of aquatic resources.

Through pragmatic applications, multidisciplinary insights, and an unwavering focus on food safety and aquatic resource conservation, readers acquire a multifaceted comprehension of the evolutionary trajectory of this field. As it delineates future prospects and challenges in integrating nanotechnology, this book stands as an indispensable roadmap for envisioning a balanced and prosperous future in aquatic resource management.

Ota, Nigeria Patrick Omoregie Isibor
Ibadan, Nigeria Aina Olukukola Adeogun
Benin City, Nigeria Alex Ajeh Enuneku

Acknowledgments

Covenant University is acknowledged for providing uninterrupted power supply, reliable Internet bandwidth, and auspicious environment that enabled the seamless execution of this work.

About the Book

Nanochitosan-Based Enhancement of Fisheries and Aquaculture explores the amalgamation of nanotechnology with aquatic resource management. It delineates the capabilities of nanochitosan, an advanced nanomaterial, in potentially revolutionizing the practices within fisheries and aquaculture. The book caters to a diverse audience, including researchers, practitioners, and students, elucidating intricate scientific concepts with a focus on sustainability.

This comprehensive work highlights nanochitosan's potential to enhance water quality, improve aquatic health, and transform disease control and feed quality in aquaculture. It particularly emphasizes the role of nanochitosan in ensuring food safety by mitigating contaminants and pathogens, portraying it as a safeguard for the edibility of aquacultural products. Central to its narrative is the alignment with Sustainable Development Goal 14, aimed at life underwater, which permeates the book's structure. Nanochitosan emerges as a crucial component in promoting a more environmentally conscious approach to sustainable aquatic resource management. Its ability to convert ordinary water sources into conducive habitats, nurture aquatic life, and address disease control and feed quality in aquaculture is underscored.

By bridging theoretical concepts with practical applications and employing interdisciplinary perspectives, the book elucidates the potential and impact of nanochitosan in redefining the utilization of aquatic resources. It serves as a guide to comprehend the advancements in the field and envisions a balanced and prosperous future for aquatic resource management by integrating nanotechnology.

Contents

About the Editors

Patrick Omoregie Isibor, Ph.D., is a specialist in Pollution Studies and Ecotoxicology and a lecturer and researcher in the Department of Biological Sciences at Covenant University. He received his Ph.D. and M.Sc. in Environmental Quality Management from the University of Benin and a B.Sc. in Zoology from Ambrose Alli University. Dr. Isibor's research interests include ecotoxicology, hydrobiology, bioaccumulation, biosequestration, biodiversity conservation, and aquatic ecology. He is a member of the Association for Environmental Impact Assessment of Nigeria (AEIAN), the International Association of Risk and Compliance Professionals (IARCP), and the African Society for Toxicological Sciences (ASTS). Dr. Isibor is co-editor of the book *Biotechnological Approaches to Sustainable Development Goals* (Springer, 2023), an editor for the *African Journal of Health, Safety, and Environment*, and a reviewer for several reputable international journals.

Aina Olukukola Adeogun, Ph.D., is a professor of Aquatic Toxicology in the Department of Zoology, University of Ibadan, Nigeria. She has pioneered research on endocrine disruption and developed protocols for molecular and cellular toxicology in aquatic organisms toward water and food safety, especially for the vulnerable subpopulation (women and children) in tropical environments dependent on natural fish food as preferred protein sources. She received the Society of Toxicology Global Senior Scholar Award in 2018 and has recently published on the detection of microplastics in fish gut in a tropical municipal lake and showed potential human health effects due to the ability of MPs to be contaminant vectors. She has published over 80 articles in peer-reviewed journals, international conferences, as well as over 20 first-authored articles. She is the Past President of Toxicologists of African Origin (TAO) and chaired the TAO Informational Session on the COVID-19 pandemic in Africa, highlighting Africa's current status toward developing global best practices for curtailing this pandemic.

Alex Ajeh Enuneku, Ph.D., is a specialist in Pollution Studies and Ecotoxicology and Head of the Department of Environmental Management and Toxicology, Faculty of Life Sciences, University of Benin. He obtained his Ph.D. and M.Sc. in Pollution Studies and Ecotoxicology and a B.Sc. in Zoology from the University of Benin. He was certified as a chemical hazard control expert in 2016 by Hokkaido University, Sapporo. Professor Enuneku's research interests include ecotoxicology, hydrobiology, pollution studies, endocrine disruption, health risk assessment, environmental impact assessment, environmental management, remote sensing, and geographic information systems. He is a member of the Society for Environmental Chemistry and Toxicology (SETAC), the Nigerian Environmental Society, and the Waste Management Association of Nigeria. He is a reviewer for several reputable international journals.

Introduction to Application of Nanochitosan in Aquaculture

Patrick Omoregie Isibor and Ifeoluwa Ihotu Kayode-Edwards

Contents

P. O. Isibor (✉) · I. I. Kayode-Edwards
Department of Biological Sciences, College of Science and Technology, Covenant University, Ota, Ogun State, Nigeria
e-mail: patrick.isibor@covenantuniversity.edu.ng

P. O. Isibor et al. (eds.), *Nanochitosan-Based Enhancement of Fisheries and Aquaculture*, https://doi.org/10.1007/978-3-031-52261-1_1

1 The Significance of Fisheries and Aquaculture

The integral role of fisheries and aquaculture in global food security is a cornerstone in addressing the dietary needs of a burgeoning world population. This exploration aims to dissect the interplay between these industries, scrutinize their ecological footprints, and unearth the prospects for sustainable methodologies, ultimately highlighting their significance in nourishing an expanding global populace.

As the world population escalates and challenges of impoverishment and malnutrition persist, the imperative to satiate the nutritional requirements of global denizens remains pivotal. Fisheries and aquaculture stand as pivotal contributors to global food security, supplying a rich and accessible reservoir of protein. Fish, renowned for its abundance in indispensable nutrients such as Omega-3 fatty acids, vitamins, and minerals, constitutes a crucial component for human well-being (Sarojnalini & Hei, 2019; Hei, 2020; Kwasek et al., 2020). Particularly in coastal and developing regions, fish serves as an economically viable and widely attainable source of sustenance, especially for marginalized communities (Sharma, 2011; Batista et al., 2014). The management of fisheries in a sustainable manner plays a pivotal role in ensuring the perpetuity of fish stocks, averting the menace of overexploitation, and safeguarding livelihoods reliant on fish resources. This approach not only mitigates the risk of food scarcity but also contributes to poverty alleviation by securing steady access to food sources. Furthermore, the ascent of aquaculture, the cultivation of aquatic organisms, assumes a progressively pivotal role in global food production (Little et al., 2016). This methodological shift offers an avenue for augmenting food supplies while relieving the pressures exerted on natural fish stocks.

Aquaculture presents a viable solution to relieve strain on natural fish populations while accommodating the escalating demand for fish-based commodities (Engle & van Senten, 2022). Its controlled cultivation of fish offers a dependable supply chain, less susceptible to environmental fluctuations, catering to burgeoning requirements (Brye, 2023). Nevertheless, concerns persist regarding the ecological ramifications of both fisheries and aquaculture, necessitating a contemplation of sustainable practices. Issues like overfishing, habitat degradation, and pollution loom large, imperiling the enduring viability of these industries (Sumaila & Tai, 2020). Depletion of fish stocks due to overfishing poses a threat to marine ecosystems and livelihoods dependent on fishing activities. Strategic interventions encompassing catch restrictions, establishment of marine sanctuaries, and community-involved management frameworks stand as effective measures to ameliorate this predicament and restore fish populations for sustained fisheries (Cooke et al., 2023). Conversely, aquaculture encounters its own set of environmental challenges including nutrient discharge, disease outbreaks, and habitat deterioration (Islam & Yasmin, 2017; Olaussen, 2018). However, integrating sustainable methodologies such as efficient feed management, proficient treatment of wastewater, and prudent site selection can assuage these impacts (Greenberg, 2014). Prioritizing eco-friendly feed sources and diminishing reliance on wild fish stocks for feedstuff further augments the sustainability quotient within aquaculture practices.

The potential for sustainable contributions from fisheries and aquaculture to global food security, despite facing environmental hurdles, underscores the need for conscientious practices within these sectors. Efficient resource utilization and the advocacy of sustainable techniques in both fishing and farming operations stand as pivotal factors in realizing this objective. The implementation of effective governance structures, scientific inquiry, and collaborative international alliances assumes paramount importance in steering fisheries and aquaculture toward sustainability (Msomphora, 2018; Roderick, 2020). Active involvement of local communities and small-scale fishers in decision-making processes is imperative, enhancing the efficacy and equity of management strategies (Charles, 2017; March & Failler, 2022). Encouraging community fish farmers to actively participate in decision-making processes can significantly benefit the sustainability and productivity of their endeavours. This can be achieved through programs and strategies such as training and education initiatives; demonstration farms; access to information; community-based organizations; financial support and access to resources; extension services; inclusive decision-making platforms; and monitoring and evaluation can effectively engage and empower the grassroot farmers. Developing programs that offer comprehensive training sessions and workshops on various aspects of fish farming (Sumaila & Tai, 2020). These sessions can cover topics such as modern farming techniques, aquaculture best practices, environmental sustainability, financial management, and market analysis. Providing educational resources empowers farmers to make informed decisions. Establishing demonstration farms within communities allows farmers to witness firsthand the implementation of new techniques or technologies (Roderick, 2020). These farms serve as practical learning platforms, enabling farmers to observe successful practices and their outcomes, fostering confidence in decision-making. Providing timely and relevant information through various channels such as mobile apps, community meetings, newsletters, or local radio programs. Accessible information on market trends, weather forecasts, disease outbreaks, and technological advancements aids farmers in making informed decisions about their farming practices. Supporting or creating community-based organizations focused on fish farming can facilitate collective decision-making. These groups enable farmers to collaborate, share experiences, pool resources, and collectively address challenges, ultimately enhancing their decision-making capacity. Offering financial assistance, subsidies, or access to microcredit programs can empower farmers to invest in improved infrastructure, equipment, or higher quality fish breeds. Access to resources enhances their ability to make choices that positively impact their productivity and sustainability. Employing extension officers or agents who regularly engage with farmers, providing guidance, troubleshooting assistance, and continuous support. These professionals can act as mentors, advising farmers on various aspects of fish farming and aiding in decision-making processes. Creating forums or platforms where community fish farmers can voice their opinions, share insights, and actively participate in decision-making concerning policies, regulations, or community initiatives related to fish farming (Charles, 2017). Involving them in discussions and decision-making processes fosters a sense of ownership and empowerment. Implementing systems to track and evaluate the impact of decisions made within the community. Regular assessments help identify successful practices and areas needing improvement, enabling adaptive decision-making based on empirical data. By employing these programs and strategies, community fish farmers can be empowered to actively engage in decision-making processes, fostering a more sustainable and prosperous aquaculture industry within their communities.

Investment in research and technology remains a linchpin in advancing sustainable practices within aquaculture (Moehl et al., 2006; Niner et al., 2022). Innovations spanning feed formulation, disease control methodologies, and monitoring systems hold promise in reducing the industries' ecological impact and enhancing their operational efficiency (Al-Emran & Griffy-Brown, 2023). Encouraging eco-certification and sustainable labeling practices can establish market incentives, fostering widespread adoption of sustainable approaches throughout the industry (Olopade & Dienye, 2017). Fisheries and aquaculture serve as indispensable sources of protein and nutrients for millions globally (Pradeepkiran, 2019), despite persistent challenges like overfishing and environmental consequences. The potential lies in sustainable management practices, robust governance frameworks, and technological innovations, offering avenues to address these pressing concerns. A comprehensive approach is necessary, balancing nutritional requirements, ecosystem conservation, and livelihood sustenance for communities reliant on these industries. Prioritizing sustainability in both sectors is pivotal, ensuring the realization of their full potential and substantial strides toward global food security.

Fish represents a crucial reservoir of superior-quality protein, essential fatty acids, and an array of vitamins (such as vitamin D) and minerals (including iodine, zinc, iron, and selenium) imperative for human physiological well-being (Pakkiam Muniyasamy, 2023). Particularly in coastal and developing regions worldwide, fish stands as a principal source of animal protein. The Food and Agriculture Organization (FAO) reported a noteworthy surge in global per capita fish consumption, surpassing 20 kg annually for the first time in 2016. Shahbandeh's (2023) findings further support this trajectory, estimating a per capita fish consumption of approximately 20.9 kg globally in 2019. This consumption accounts for 6.7% of the total protein intake in the human diet.

Recent trends in aquatic food provisioning indicate a transformative shift in its sourcing dynamics: while input from capture fisheries has remained relatively stable since the late 1980s, aquaculture production has experienced rapid expansion. In 1974, aquaculture contributed merely 7% of fish for human consumption, a figure that ascended to 26% by 1994 and escalated to 50% by 2013. A prognostication from the World Bank anticipates that by 2030, aquaculture will cater to 60% of fish (including finfish, molluscs, and crustaceans) directly consumed by humans. Aquaculture emerges as the fastest-growing primary production sector, displaying an average annual growth rate exceeding 8% over the past three decades, surpassing the pace of human population expansion (Norman et al., 2019).

2 Nutritional Value of Fish

Fisheries and aquaculture play pivotal roles in ensuring global food security, contributing significantly to human nutrition. The nutritional value of fish extends beyond sustenance, encompassing a diverse array of essential nutrients vital for overall health. Fish stands out as an exceptional source of high-quality protein, providing all essential amino acids necessary for human growth and tissue repair (Ghaly et al., 2013; Weinert et al., 2014; Li et al., 2021). Its low-fat content distinguishes it as a healthier alternative to other animal protein sources (de Boer et al., 2020). Furthermore, fish is abundant in omega-3 fatty acids, which are crucial for brain development, cardiovascular health, and mitigating the risk of chronic diseases such as heart ailments and stroke (Punia et al., 2019; Balta et al., 2021). In addition to its protein and healthy fat content, fish offers a rich array of vitamins and minerals. Notably, it serves as a significant source of vitamin D, essential for bone health and immune function, alongside substantial amounts of vitamins A and B12. Various minerals present in fish, including iodine, zinc, iron, and selenium, play pivotal roles in bodily functions such as thyroid regulation, immune support, and red blood cell formation (Lall & Kaushik, 2021). Especially in coastal areas and developing countries, fish constitutes a staple in daily diets, serving as a primary source of animal protein. Its accessibility and affordability make it indispensable in addressing malnutrition and deficiencies, particularly among vulnerable populations like children and pregnant women.

Different fish species, from oily varieties like salmon and mackerel to lean types such as tilapia and cod, offer distinct nutritional profiles, contributing to a well-rounded and diverse diet. Promoting responsible fishing practices, ethical management of aquaculture operations, and conservation of marine ecosystems are imperative for ensuring sustained access to this nutritional resource. Recognizing and harnessing the nutritional contributions of fisheries and aquaculture can lead to a more secure and nourished global population.

2.1 Macronutrients in Fish

The high-quality protein content found in fish is a key nutritional attribute that distinguishes it as an important dietary component. Fish is considered a complete protein source as it contains all essential amino acids required by the human body (Ghaly et al., 2013; Weinert et al., 2014; Li et al., 2021). These amino acids play crucial roles in various bodily functions, including muscle growth, repair of tissues, and immune function. The protein in fish is highly bioavailable, meaning the body can efficiently absorb and utilize these proteins. This quality makes fish an excellent option for individuals looking to meet their protein needs effectively.

Fish often contains lower levels of saturated fats compared to other animal protein sources, making it a leaner protein option (Zhubi-Bakija et al., 2021). This is particularly true for many white fish varieties like cod or haddock. Fish proteins are generally easy to digest, which can be beneficial for individuals with digestive sensitivities or those who have difficulty digesting certain protein-rich foods (Dallas et al., 2017). The digestibility is characterized by the lower levels of connective tissue, which contributes to their tenderness and easier digestion compared to some land animals (Purslow, 2005). The amino acids obtained from fish protein are crucial for muscle health and repair (Papadopoulou, 2020). They are especially valued among athletes or individuals engaged in physical activities due to their role in muscle recovery and growth. Protein, in general, contributes to feelings of fullness and satiety (Morell & Fiszman, 2017). Including fish in the diet can aid in managing appetite and supporting weight management goals.

Incorporating fish as a source of high-quality protein in the diet can be beneficial for individuals aiming to diversify their protein sources, particularly those seeking lean protein options with significant nutritional benefits. Omega-3 fatty acids, notably EPA (eicosapentaenoic acid) and DHA (docosahexaenoic acid), found abundantly in fish, offer numerous health benefits, particularly for heart and brain health (Khalid et al., 2022; Chauhan et al., 2023). They have been linked to a reduced risk of heart disease. They help lower triglycerides, reduce blood clotting, and decrease inflammation, all of which contribute to a healthier cardiovascular system. Regular consumption of omega-3s, often found in fatty fish like salmon, mackerel, and sardines, may help in lowering blood pressure, thus reducing the risk of hypertension and related heart problems. Omega-3s can increase high-density lipoproteins (HDL) which are good cholesterol, while reducing low-density lipoproteins (LDL) which are bad cholesterol and triglycerides, promoting a healthier lipid profile (Djuricic & Calder, 2021). Amino acids in fish may also play a role in improving metabolic health by influencing insulin sensitivity and glucose regulation, potentially benefiting individuals with conditions like diabetes or metabolic syndrome (Abachi et al., 2023).

DHA, in particular, is crucial for brain development in infants and children (Cohen Kadosh et al., 2021). It is a key component of the brain and retina and is associated with cognitive function and visual acuity (Lafuente et al., 2021). Omega-3 fatty acids have shown promise in reducing the risk of cognitive decline in older adults. They may help protect against conditions like Alzheimer's disease and age-related cognitive impairments (Cohen Kadosh et al., 2021). Some studies suggest that omega-3s could potentially help in mood regulation and reducing the risk of depression and anxiety (DiNicolantonio & O'Keefe, 2020).

Omega-3 fatty acids have anti-inflammatory properties, which can benefit various inflammatory conditions, including arthritis, by reducing joint pain and stiffness (Simonetto et al., 2019). Adequate intake of omega-3s during pregnancy is crucial for the development of the baby's brain and eyes (Cohen Kadosh et al., 2021). It is often recommended for pregnant women to consume sufficient omega-3s for fetal development (Archibong, 2023).

Incorporating fish rich in omega-3 fatty acids into the diet, whether through fresh fish or supplements, can provide these essential nutrients, promoting heart health, brain function, and overall well-being. However, it is essential to consider sustainable sourcing and be mindful of potential contaminants in certain fish species, especially for individuals consuming fish regularly or relying on supplements. However, the benefits of fisheries and aquaculture can be derived through sustainable aquaculture practices aimed at minimizing environmental impact while ensuring the responsible production of seafood. It is important to select aquaculture sites that minimize environmental impact and account for factors like water quality, currents, and ecosystem health. Regular monitoring of water quality, waste management, and disease control to maintain a healthy environment for farmed species is a sustainable practice (Abachi et al., 2023). Another strategy to attain a sustainable aquaculture practice is developing and using feeds that reduce waste and minimize environmental impact, such as utilizing alternative protein sources. Recirculating aquaculture systems (RAS) are also required. These are systems that recycle water, minimizing waste and environmental contamination. Preventing farmed species from escaping into the wild, which can disrupt local ecosystems is also required. Other strategies include implementation of integrated multitrophic aquaculture (IMTA) systems that cultivate various species together, utilizing waste from one species as nutrients for another. Limiting the use of antibiotics and chemicals to prevent water pollution and reduce risks of developing antibiotic-resistant bacteria and implementing measures like vaccination and proper husbandry practices to minimize disease outbreaks are also imperative (Lafuente et al., 2021). It is essential for farmers to seek certification from regulatory organizations to ensure adherence to sustainability standards. Involving local communities and stakeholders in decision-making processes through community-based programs ensure the social and economic well-being of the area. By integrating these practices, aquaculture can become more sustainable, ensuring the production of seafood while minimizing environmental impacts and preserving aquatic ecosystems for future generations.

2.2 Micronutrients in Fish

2.2.1 Vitamins

Fish contains various vitamins, each contributing to its nutritional value. Fish is one of the few natural food sources rich in vitamin D, particularly fatty fish like salmon, mackerel, and tuna (Schmid & Walther, 2013). Vitamin D is essential for calcium absorption, promoting bone health and supporting immune function. Some fish, especially oily varieties, contain vitamin A in the form of retinol (Mohanty et al., 2013). Vitamin A is vital for vision, immune function, and skin health.

Fish, especially shellfish and certain types like salmon and trout, are excellent sources of vitamin B12 (Watanabe, 2007). This vitamin is crucial for nerve function, red blood cell production, and DNA synthesis. Fish is a good source of several B vitamins, including vitamin B6 (pyridoxine) and niacin (vitamin B3) (Shabbir et al., 2020). These vitamins are involved in energy metabolism, nerve function, and the production of red blood cells. Some fish, such as tuna and salmon, contain folate, an essential B vitamin especially important for pregnant women as it helps prevent certain birth defects (de Seymour et al., 2022).

While not as abundant as in some other foods, certain fish, like trout and mackerel, contain vitamin E, an antioxidant that supports immune function and skin health (Merdzhanova et al., 2013). Including a variety of fish in the diet can provide a diverse range of vitamins, contributing to overall health and supporting various bodily functions. Additionally, the combination of these vitamins with other nutrients in fish, like omega-3 fatty acids and minerals, enhances their overall nutritional value.

2.2.2 Minerals

Fish is a rich source of various minerals, each playing essential roles in numerous bodily functions. While not as abundant in fish as in dairy products, certain types like canned sardines or salmon with bones contain notable amounts of calcium (Singh et al., 2021). Calcium is vital for bone and teeth health, nerve transmission, muscle function, and blood clotting. Fish is also a good source of phosphorus, a mineral important for bone health, energy production (as part of ATP molecules), and maintaining proper pH balance in the body (Sarojnalini & Hei, 2019). Phosphorous is found in various fish species, magnesium is involved in hundreds of biochemical reactions in the body, including energy production, muscle function, nerve transmission, and bone health. Magnesium is available many fish species and it is involved many functions in human, including energy production, muscle function, nerve transmission, and bone health. Magnesium is an essential mineral that plays a fundamental role in various metabolic functions within the human body. Its functions are diverse and crucial for maintaining overall health. Magnesium is a cofactor for more than 300 enzymes involved in diverse biochemical reactions (Colombo and Mazal, 2020). These enzymes are pivotal for energy metabolism, DNA synthesis, protein synthesis, and cellular signaling pathways. ATP (adenosine triphosphate) is the primary energy currency of cells. Magnesium is essential for ATP synthesis, contributing significantly to energy production within cells. Magnesium is crucial for muscle contraction and relaxation. It plays a role in the regulation of neuromuscular signals and is involved in the transport of calcium and potassium ions across cell membranes, influencing muscle function. It participates in neurotransmitter release and regulation, affecting nerve function and transmission (Venugopal & Gopakumar, 2017). Magnesium also plays a role in maintaining a balanced nervous system, contributing to relaxation and stress reduction. Alongside calcium and vitamin D, magnesium is vital for maintaining bone health. It helps regulate calcium levels, impacting bone density and structure. Magnesium also plays a role in maintaining a steady heartbeat by influencing the heart's electrical activity (Pal et al., 2018). It helps regulate electrolyte balance, supporting cardiovascular function. Magnesium influences insulin secretion and the function of insulin receptors, contributing to the regulation of blood sugar levels. Magnesium is involved in the synthesis and stability of DNA and RNA, as well as in the synthesis of proteins, contributing to various cellular functions and repair mechanisms.

Fish, particularly in species like salmon, contains potassium (Colombo & Mazal, 2020). This mineral helps regulate fluid balance, nerve signals, and muscle contractions, contributing to heart health and overall bodily function. Many types of fish are also excellent sources of selenium (Pal et al., 2018). Selenium is an antioxidant that supports immune function, thyroid health, and helps prevent cellular damage caused by free radicals. While fish is not as high in iron as some other animal protein sources, it still contains this essential mineral (Khalili Tilami & Sampels, 2018). Iron is crucial for making hemoglobin, which carries oxygen in the blood, and for various enzymatic reactions in the body. Fish, particularly shellfish like oysters and crab, contains zinc (Venugopal & Gopakumar, 2017). Zinc is important for immune function, wound healing, DNA synthesis, and supports the senses of taste and smell.

Seafood, especially seaweed and fish from iodine-rich waters, is a significant source of iodine (Pal et al., 2018). Iodine is crucial for thyroid hormone production, which regulates metabolism and supports growth and development. These minerals in fish, when incorporated into a balanced diet, contribute to overall health and support various bodily functions, including bone health, immune function, energy metabolism, and the proper functioning of nerves and muscles.

Incorporating fish in human diet thus serves notable nutritional and health benefits. Table 1 breaks down and summarizes the catalogue of nutritional value of fish at a glance.

Table 1 Nutritional benefits of fish

Nutrient	Source	Importance
Protein	Found abundantly in most fish and seafood	Providing essential amino acids necessary for growth and repair
Omega-3 fatty acids	Particularly prevalent in oily fish like salmon, mackerel, sardines, and trout	These fatty acids are essential for heart health, brain function, and reducing inflammation
Vitamin D	Fish, especially fatty fish like salmon and tuna, is an excellent natural source of vitamin D	Crucial for bone health and immune function
Vitamin A	Commonly found in fish liver oils and some fish species like herring, mackerel, and salmon	Vitamin A is essential for vision, immune function, and skin health
Vitamin B12	Abundant in most fish, particularly shellfish like clams, mussels, and crab	Vitamin B12 is critical for nerve function and the production of DNA and red blood cells
Selenium	Present in many types of fish, particularly in shellfish like shrimp and crab	Selenium acts as an antioxidant and supports thyroid function
Iodine	Abundant in seafood, especially in seaweed, shrimp, cod, and tuna	Iodine is crucial for thyroid function and the production of thyroid hormones
Zinc	Found in various seafood such as oysters, crab, lobster, and fish like salmon and sardines	Zinc supports immune function, wound healing, and DNA synthesis
Iron	Predominantly in shellfish like clams, mussels, and oysters, as well as in some fish like sardines and tuna	Iron is essential for oxygen transport in the blood
Magnesium	Present in fish like salmon, mackerel, and halibut	Magnesium supports nerve function, muscle health, and bone strength
Calcium	Particularly in small fish consumed with bones, like sardines and anchovies	Calcium is vital for bone health and muscle function

3 Health Benefits of Fish

While fish consumption offers numerous health advantages, it is essential to consider factors like sustainability, mercury levels (especially in larger predatory fish), and individual dietary needs when incorporating fish into a regular diet. Mercury toxicity in fish is a concerning issue due to the bioaccumulation of mercury in aquatic ecosystems. Mercury exists in various forms, with methylmercury being the most common and toxic form found in water bodies (Chauhan et al., 2023). It's produced through natural processes like volcanic eruptions and industrial activities like coal burning and mining, entering water bodies and eventually accumulating in fish. Fish, especially predatory species higher up the food chain, tend to accumulate mercury over time. Small organisms in aquatic ecosystems absorb mercury from water and sediments. When larger fish eat these smaller organisms, they take in the mercury. As these larger fish continue to consume smaller ones, mercury accumulates in their tissues. Predatory fish at the top of the food chain, like shark, swordfish, or king mackerel, tend to accumulate higher levels of mercury because they consume many smaller fish containing mercury (Djuricic & Calder, 2021). Mercury concentrations increase as you move up the food chain, leading to higher levels in top predators. Consuming fish with high mercury levels can pose health risks to humans. Mercury is a neurotoxin that affects the nervous system, especially in developing fetuses and young children. Chronic exposure to high levels of mercury can lead to neurological problems, impaired cognitive function, and developmental issues. Monitoring mercury levels in water bodies and fish is crucial. Efforts to control industrial discharges and minimize mercury release into the environment help reduce overall exposure. Educating the public about the risks of consuming fish high in mercury and promoting awareness about safer options is essential. Balancing the nutritional benefits of consuming fish with the risks of mercury exposure remains a challenge in managing the health impact of mercury-contaminated fish. Consulting with a healthcare professional or nutritionist can provide personalized guidance on the consumption of fish for optimal health benefits.

3.1 Cardiovascular Health

Omega-3 fatty acids, notably EPA (eicosapentaenoic acid) and DHA (docosahexaenoic acid), wield profound effects in mitigating the risk of heart disease through multifaceted mechanisms that impact various facets of cardiovascular health (Chauhan et al., 2023). One pivotal role of omega-3s in cardiovascular health is their ability to modulate triglyceride levels in the bloodstream (Djuricic & Calder, 2021). Elevated triglycerides pose a significant risk for heart disease, and the capacity of omega-3 fatty acids to lower these lipid levels contributes substantially to reducing this risk.

Another vital cardiovascular benefit stems from the potential of omega-3s to regulate blood pressure (von Schacky, 2020). Their incorporation into dietary patterns demonstrates an ability to aid in maintaining optimal blood pressure levels, a pivotal factor in averting cardiovascular conditions such as heart attacks and strokes.

The anti-inflammatory properties of omega-3 fatty acids hold immense significance in cardiovascular health (Simonetto et al., 2019). Chronic inflammation is intricately linked to the progression and development of heart disease. Omega-3s' capacity to attenuate this inflammation within the body serves as a protective mechanism against cardiovascular ailments (Balta et al., 2021). Furthermore, these fatty acids play a pivotal role in enhancing endothelial function, crucial for the integrity of blood vessel linings (Mallick & Duttaroy, 2022). Improved endothelial function translates to enhanced blood flow and diminished clot formation, thereby reducing the risk of heart disease by ensuring optimal cardiovascular health.

DHA, in particular, assumes a critical role in the maintenance of normal heart rhythm (Albert et al., 2020). Its influence in preventing arrhythmias, characterized by irregular heartbeats, serves as a protective measure against sudden cardiac events, thus further bolstering the cardioprotective effects of omega-3 fatty acids. Numerous robust studies, encompassing randomized controlled trials and observational research, underscore the substantial benefits of omega-3 fatty acids in mitigating the risk of cardiovascular events (Mozaffarian & Wu, 2011; Siscovick et al., 2017; Khan et al., 2021). However, it is imperative to acknowledge that while omega-3s confer noteworthy cardiovascular advantages, they should complement rather than replace other essential heart-healthy practices. Adopting a balanced diet, regular exercise regimen, abstaining from smoking, and effectively managing additional risk factors like high cholesterol and diabetes remain integral components of comprehensive cardiovascular health.

The American Heart Association (2023) underscores the importance of consuming fatty fish such as salmon, mackerel, sardines, and trout at least twice a week to ensure an adequate intake of omega-3 fatty acids. For individuals grappling with heart disease or elevated triglyceride levels, additional omega-3 supplements may be recommended under the guidance and supervision of healthcare professionals to augment their cardiovascular health strategies. Fish, especially fatty fish like salmon, mackerel, and sardines, can have a positive impact on cholesterol levels, primarily due to their high content of omega-3 fatty acids.

Omega-3 fatty acids found abundantly in fish can significantly lower triglyceride levels in the blood (Djuricic & Calder, 2021). Elevated triglycerides are often associated with higher levels of "bad" LDL cholesterol and lower levels of "good" HDL cholesterol, increasing the risk of heart disease. By reducing triglycerides, fish consumption indirectly supports a healthier cholesterol profile. Omega-3s may also affect the size and composition of LDL cholesterol particles, making them larger and less likely to contribute to arterial plaque buildup (DePace et al., 2019). Larger LDL particles are considered less harmful than smaller, denser ones.

While the impact of fish on HDL cholesterol (the "good" cholesterol) is not as pronounced as on triglycerides or LDL cholesterol, some studies suggest that omega-3s can help maintain or slightly increase HDL levels (Li & Palmer-Keenan, 2016), which is beneficial for overall cardiovascular health. Choosing fish as a protein source over red meat or processed meats, which are high in saturated fats, can positively impact cholesterol levels. Fish is generally lower in saturated fats and transfats, making it a heart-healthy alternative (Astrup et al., 2020).

However, the impact of fish on cholesterol levels can vary among individuals. Factors such as genetics, overall diet, lifestyle habits, and the method of fish preparation can influence its effects on cholesterol (Gioia et al., 2020). It is important to note that while fish can contribute to a healthier cholesterol profile, it is just one component of a heart-healthy diet. A balanced diet that includes a variety of nutrient-dense foods, along with regular physical activity and healthy lifestyle choices, plays a crucial role in managing cholesterol levels and overall cardiovascular health.

Consulting with a healthcare professional or a registered dietitian can provide personalized guidance on incorporating fish and other heart-healthy foods into a cholesterol-lowering diet plan.

3.2 Brain Health

Omega-3 fatty acids, particularly EPA (eicosapentaenoic acid) and DHA (docosahexaenoic acid) found abundantly in fish, play a crucial role in supporting cognitive function and brain health in several ways (Khalid et al., 2022). Structural Component of the Brain: DHA, in particular, is a major structural component of brain tissue. It constitutes a significant portion of the brain's cell membranes, aiding in the fluidity and functionality of brain cells (Balakrishnan et al., 2021). Adequate intake of DHA, primarily sourced from fish, is essential for optimal brain development in infants and supports cognitive function throughout life (Cohen Kadosh et al., 2021).

Maternal consumption of omega-3-rich fish during pregnancy and breastfeeding is associated with improved cognitive development in infants (Archibong, 2023). DHA, passed from the mother to the fetus or infant through the placenta or breast milk, respectively, supports brain development during these critical stages (Zeng et al., 2019; Khalid et al., 2022). Maintenance of Cognitive Function: Omega-3s, especially DHA, contribute to maintaining cognitive function in adults and the elderly (Singh, 2020). Research suggests that regular consumption of fish or omega-3 supplements may help preserve cognitive abilities, such as memory, attention, and problem-solving skills, and reduce the risk of cognitive decline related to aging (Archibong, 2023).

Omega-3 fatty acids possess anti-inflammatory and antioxidant properties (Oppedisano et al., 2020). Chronic inflammation and oxidative stress are implicated in cognitive decline and neurological disorders. The anti-inflammatory effects of omega-3s may help mitigate these processes, thereby supporting brain health (Stefaniak et al., 2022). Neuroprotective Effects: DHA, in particular, exhibits neuroprotective properties, potentially reducing the risk of neurodegenerative diseases like Alzheimer's disease (Thomas et al., 2015). It may help maintain the integrity of nerve cells and support their optimal function, contributing to a lower risk of cognitive impairment.

While the evidence regarding the positive impact of omega-3s on cognitive function is promising, further research is ongoing to explore their precise mechanisms and potential benefits for various aspects of brain health. Incorporating fish, especially fatty fish like salmon, mackerel, and sardines, into a balanced diet is a recommended way to obtain omega-3 fatty acids. However, individual dietary needs, potential mercury content in certain fish species, and other factors should be considered. Consulting with a healthcare professional or a registered dietitian can offer personalized guidance on incorporating omega-3s into one's diet to support cognitive function and overall brain health.

Omega-3 fatty acids, particularly DHA (docosahexaenoic acid) found in fish, have shown promise in potentially reducing the risk of neurodegenerative diseases such as Parkinson's diseases (Vega & Cepeda, 2021). DHA is a key structural component of brain tissue and plays a crucial role in maintaining neuronal structure and function. Adequate levels of DHA may contribute to preserving brain health as individuals age, potentially reducing the risk of cognitive decline and neurodegeneration.

Chronic inflammation and oxidative stress are implicated in the development and progression of neurodegenerative diseases. Omega-3 fatty acids, particularly DHA, possess anti-inflammatory and antioxidant properties, which may help mitigate these processes, potentially lowering the risk of neurodegeneration (Vega & Cepeda, 2021). DHA supports the integrity of nerve cells and their optimal functioning. It is involved in various aspects of neuronal communication and synaptic transmission, which are vital for cognitive processes (Suvarna & Singh, 2019). Adequate intake of DHA may help maintain neuronal function and connectivity, potentially reducing the risk of neurodegenerative disorders.

Some studies suggest that higher consumption of omega-3 fatty acids, particularly DHA, is associated with a reduced risk of developing Alzheimer's disease (Thomas et al., 2015). While further research is needed to establish a definitive causal relationship, these findings indicate a potential protective effect of omega-3s against this neurodegenerative condition. Omega-3s, through their various mechanisms such as supporting neuronal structure, reducing inflammation, and acting as antioxidants, may exert neuroprotective effects, potentially slowing down the progression of neurodegenerative diseases (Stefaniak et al., 2022; Zhou et al., 2022). It is important to note that while omega-3 fatty acids show promise in potentially reducing the risk of neurodegenerative diseases, more research is needed to establish clear causal relationships and determine specific dosage recommendations or therapeutic interventions.

Incorporating fish, particularly fatty fish rich in omega-3s, into a balanced diet is a recommended approach to obtain these beneficial fatty acids. However, it is crucial to consider individual dietary needs and potential mercury content in certain fish species and consult with healthcare professionals for personalized advice on diet and potential supplementation to support brain health and reduce the risk of neurodegenerative diseases.

4 Economic Importance of Fish

Fisheries and aquaculture play a pivotal role in global livelihoods and economies, constituting vital contributors to food security and economic prosperity worldwide. These sectors encompass a broad spectrum of activities, from fishing to processing, marketing, and distribution, thereby offering diverse employment opportunities to millions of people.

The significance of fisheries and aquaculture transcends mere sustenance, wielding substantial economic influence that extends well beyond local waters. These sectors serve as robust engines for employment and income generation, engaging a vast workforce directly or indirectly associated with their operations. In coastal regions and developing nations particularly, these activities foster economic stability and resilience within communities. The economic impact of fisheries and aquaculture resonates deeply in both local and international markets, significantly contributing to global trade (Bjørndal et al., 2024). The trade in fish and fishery products commands a substantial market value annually, reinforcing the economic importance of these sectors. Nations heavily reliant on fishery exports benefit from strengthened economies, facilitating economic growth, foreign exchange earnings, and investments across related industries.

Beyond economic considerations, fisheries and aquaculture hold paramount significance in ensuring food access and affordability, particularly in regions where alternative protein sources are limited or costly (Waite et al., 2014; Bjørndal et al., 2024). The availability of fish, as a comparatively affordable and nutrient-rich protein source, plays a crucial role in enhancing food security. This provision of a cost-effective dietary option helps alleviate hunger and malnutrition among diverse populations, thus contributing to improved public health outcomes. The economic influence exerted by fisheries and aquaculture extends far beyond their immediate activities, engendering a broader impact on interconnected industries and supportive services. The development of infrastructure, technological innovations, as well as research and advancements within these sectors, initiates a cascading effect that stimulates growth in associated domains. This ripple effect contributes to the amplification of economic activities and the generation of employment opportunities across diverse sectors.

The sustained prosperity of fisheries and aquaculture hinges upon the implementation of sustainable management practices, the infusion of investments in cutting-edge technology, and the establishment of robust regulatory frameworks. These elements are pivotal in unlocking the full economic potential of these sectors while safeguarding their longevity and environmental integrity. By acknowledging and augmenting their economic significance, nations can strategically harness these sectors as potent instruments for catalyzing economic development, mitigating poverty, and fortifying global food security. This strategic approach allows nations to reap the multifaceted benefits offered by fisheries and aquaculture, thereby positioning these sectors as integral components of sustainable socio-economic progress.

5 Challenges Faced by the Fisheries and Aquaculture Industry

The expansion of aquaculture scope has sparked various concerns, encompassing pollution, feeding methods, disease control, antibiotic usage, habitat utilization, introduction of non-native species, food safety, fraudulent practices, animal welfare, influence on conventional wild fisheries, water and space accessibility, market competitiveness, genetic considerations, overfishing and depleting fish stocks, illegal, unreported, and unregulated fishing, environmental impact and habitat degradation, climate change and ocean acidification, governance and regulatory issues, socioeconomic issues, and disease (Bjørndal et al., 2024). Addressing these challenges requires a multi-faceted approach, including sustainable fishing practices, effective regulations and well-designed policies, improved aquaculture technologies, conservation of habitats, international cooperation, and support for small-scale fishers to ensure the long-term sustainability and resilience of fisheries and aquaculture industries. Addressing these challenges requires also collaborative efforts involving governments, stakeholders, scientific communities, and international organizations. Implementing science-based management practices, promoting sustainable fishing methods, improving monitoring and compliance, and investing in community development are key strategies toward achieving sustainable fisheries and aquaculture practices while preserving marine ecosystems for future generations.

5.1 Overfishing and Depleting Fish Stocks

Overfishing refers to the practice of catching too many fish at a rate that outpaces their ability to reproduce and replenish their populations. This issue has surged due to multiple factors, chiefly the rising global demand for seafood driven by population growth, changing dietary habits, and the expansion of commercial fishing operations. Unsustainable fishing methods like bottom trawling, gillnetting, and longlining exacerbate this problem. These methods often result in the unintended capture of non-target species, known as bycatch, which includes dolphins, turtles, and other marine life (Gray & Kennelly, 2018). Additionally, destructive fishing practices can harm the seafloor and other habitats, impacting the broader marine ecosystem. The consequences of overfishing are far-reaching. Depleted fish stocks not only affect the species being targeted but also disrupt the intricate balance within marine ecosystems. When key species decline or disappear, it can trigger a cascade effect, impacting the food web and leading to imbalances in predator–prey relationships (Eisenberg, 2013). As a consequence, numerous species face decline or extinction, disrupting food chains and jeopardizing marine biodiversity. Some notable examples of fish species facing decline or extinction due to overfishing include Atlantic bluefin tuna (*Thunnus thynnus*) and Pacific Bluefin Tuna (*Thunnus orientalis*), which are highly sought after for sushi and sashimi; *Orange roughy* (*Hoplostethus atlanticus*); Atlantic cod (Gadus morhua); Grouper species (*Epinephelus* spp. and *Mycteroperca* spp.); Haddock (*Melanogrammus aeglefinus*); Atlantic salmon (*Salmo salar*); and European eel (*Anguilla Anguilla*).

The implications for human populations relying on fisheries are profound. Coastal communities and economies heavily reliant on fishing suffer as fish stocks dwindle. Jobs are lost, incomes decrease, and food security becomes a pressing concern. Furthermore, small-scale and artisanal fishers often face the brunt of these impacts, despite contributing less to the overfishing problem compared to large-scale commercial operations (Auld, 2021; Okafor-Yarwood et al., 2022). To combat overfishing and promote sustainable practices, various measures have been proposed and implemented. These include establishing marine protected areas, setting catch limits, implementing quotas, and promoting more selective and eco-friendly fishing techniques. Sustainable aquaculture and fish farming also offer alternatives to alleviate pressure on wild fish stocks.

Addressing overfishing requires a multi-faceted approach involving governments, fishing industries, conservation organizations, and consumers. Sustainable fishing practices, coupled with responsible consumption habits and informed policies, are crucial to ensure the long-term health of our oceans, the preservation of marine biodiversity, and the sustenance of fishing communities worldwide.

5.2 Illegal, Unreported, and Unregulated (IUU) Fishing

Illegal, unreported, and unregulated (IUU) fishing practices represent a significant threat to global fisheries and marine ecosystems. These practices occur outside the bounds of established regulations and agreements, circumventing laws meant to manage and conserve fish stocks. IUU fishing contributes to stock depletion, damages ecosystems, and disrupts the fair competition that legitimate fisheries rely on (Chen et al., 2023). One of the most damaging aspects of IUU fishing is its impact on fish populations. IUU operations often target species without regard for quotas or conservation measures, leading to the rapid decline of vulnerable fish stocks. These activities not only jeopardize the sustainability of fisheries but also compromise the ability of ecosystems to function properly. Moreover, IUU operators frequently use destructive fishing methods, such as bottom trawling or illegal gear, causing habitat destruction and further harming non-target species (Gray & Kennelly, 2018). The damage caused by IUU fishing extends beyond environmental concerns. Legitimate fishing industries that adhere to regulations face unfair competition from those engaging in illegal practices (Telesetsky, 2014). This unfair competition can lead to economic losses for law-abiding fishers and exacerbate social and economic disparities within communities dependent on fishing.

Addressing IUU fishing poses a complex challenge. The clandestine nature of these operations makes them difficult to detect and regulate. Furthermore, IUU fishing often occurs in international waters or in regions with limited monitoring and enforcement capacity, complicating efforts to combat these practices effectively (Telesetsky, 2014). Effective solutions to tackle IUU fishing require collaboration and coordination among governments, international organizations, law enforcement agencies, and the fishing industry. Strengthening monitoring and surveillance measures, employing technology such as satellite tracking and vessel monitoring systems, and enhancing cross-border cooperation are crucial steps to detect and deter IUU activities.

Implementing robust legal frameworks and sanctions, along with strict enforcement mechanisms, is vital to dissuade and penalize those engaged in illegal fishing practices (Kuemlangan et al., 2023). Improving transparency and traceability throughout the seafood supply chain can also help identify and eliminate illegally caught fish from entering markets (Lewis & Boyle, 2017). Engaging local communities and empowering them to participate in sustainable fishing practices can serve as a deterrent against IUU activities while promoting conservation efforts.

Ultimately, combating IUU fishing requires a comprehensive and concerted effort at local, regional, and global levels. By addressing the root causes, enhancing regulations, and enforcing stringent measures, it is possible to mitigate the detrimental impacts of IUU fishing and ensure the long-term sustainability of fisheries and marine ecosystems.

5.3 Environmental Impact and Habitat Degradation

While aquaculture and fishing practices can help meet the increasing global demand for seafood, they are not without their environmental consequences. Aquaculture, often hailed as a solution to overfishing, has its own set of challenges that can negatively impact the environment.

Habitat destruction is a significant concern associated with aquaculture. The construction of aquaculture facilities such as shrimp farms or fish ponds often involves altering coastal areas, mangroves, or other ecosystems to create suitable conditions for farming (Bosma et al., 2020). This alteration can lead to the loss of crucial habitats for various species, disrupting the natural balance and biodiversity of these areas. Furthermore, aquaculture operations can generate pollution through the release of excess feed, antibiotics, and waste products into surrounding waters (Mavraganis et al., 2020; González-Gaya et al., 2022). Uneaten feed and excrement from farmed fish can contribute to nutrient imbalances, leading to water-quality degradation and harmful algal blooms (Musalia et al., 2020). These blooms can deplete oxygen levels in water, causing "dead zones" where marine life struggles to survive.

The introduction of non-native species is another concern. In some cases, farmed species escape into the wild, potentially outcompeting native species or introducing diseases for which local species have no natural defenses (Haubrock et al., 2021; Kang et al., 2023). This disrupts ecosystems and can lead to the decline or extinction of indigenous species, further exacerbating biodiversity loss. Certain fishing practices, such as bottom trawling or the use of destructive gear, also contribute to environmental degradation (Carneiro & Martins, 2021; Willer et al., 2022). Bottom trawling involves dragging heavy nets along the seabed, damaging fragile habitats like coral reefs and disrupting the seabed's structure. This practice not only destroys essential habitats but also impacts non-target species and alters the ecosystem dynamics.

Reducing the environmental impacts of aquaculture and fishing practices requires implementing sustainable approaches. This involves adopting more eco-friendly aquaculture techniques such as integrated multi-trophic aquaculture (IMTA), which utilizes multiple species to mimic natural ecosystems and reduce waste (Johnson et al., 2021; Nissar et al., 2023). Additionally, implementing proper waste management strategies and improving feed efficiency can mitigate pollution and nutrient imbalances in aquaculture operations (Ahmad et al., 2022).

In fishing practices, employing more selective and less destructive methods, such as using sustainable gear and avoiding sensitive habitats, can help minimize environmental damage (Sala et al., 2023). Regulation and oversight play crucial roles in ensuring that aquaculture and fishing operations adhere to environmental standards, protecting ecosystems and biodiversity. Consumers also play a role by choosing sustainably sourced seafood products and supporting certification programs that promote responsible aquaculture and fishing practices. By demanding sustainable products, consumers can encourage the industry to prioritize environmentally friendly methods.

Balancing the growing global demand for seafood with environmental conservation requires a holistic approach that considers both the need for food security and the imperative to protect ecosystems and biodiversity. Innovation, regulation, and consumer awareness are all key elements in achieving sustainable aquaculture and fishing practices that minimize adverse environmental impacts.

5.4 Climate Change and Ocean Acidification

Climate change is exerting profound effects on the world's oceans, significantly impacting fisheries and aquaculture. Rising sea temperatures, ocean acidification, and extreme weather events are among the key manifestations of this global phenomenon, posing significant challenges to the sustainability and productivity of marine environments (Brander et al., 2017).

Elevated sea temperatures alter the behavior, growth rates, and distribution of fish species (Townhill et al., 2019; Alfonso et al., 2021). Many marine organisms have specific temperature ranges within which they thrive, and even small deviations from these optimal conditions can disrupt their life cycles (Ciannelli et al., 2022). Some species may migrate to cooler waters or alter their feeding and breeding behaviors in response to changing temperatures. This redistribution of species can affect the availability of certain fish stocks in traditional fishing areas and impact local economies reliant on those species.

Ocean acidification, caused by the absorption of excess carbon dioxide by seawater, poses a threat to marine life, particularly species that build calcium carbonate shells or skeletons, such as shellfish and corals (Figuerola et al., 2021; Leung et al., 2022). As the acidity of seawater increases, it becomes more difficult for these organisms to form and maintain their calcium-based structures, affecting their growth and survival (Shi & Li, 2023). This not only jeopardizes these species but also disrupts the food web and ecosystems they support.

Extreme weather events, intensified by climate change, can have immediate and devastating impacts on fisheries and aquaculture. Storms, hurricanes, and typhoons can damage aquaculture facilities, disrupt fishing activities, and destroy critical coastal infrastructure (Ramenzoni et al., 2020). Such events not only directly affect the livelihoods of those in the fishing and aquaculture industries but also have broader economic repercussions for communities dependent on seafood. Furthermore, changes in ocean currents, salinity, and nutrient availability due to climate change can affect the productivity of marine ecosystems, impacting the abundance and distribution of fish populations (Talloni-Alvarez et al., 2019; Williamson & Guinder, 2021). Adapting to these challenges requires a multifaceted approach. Developing resilient and adaptive fisheries and aquaculture practices is crucial. This involves diversifying fishing practices, cultivating species that are more resilient to changing conditions, and employing innovative aquaculture techniques that can withstand environmental stressors.

Efforts to mitigate climate change by reducing greenhouse gas emissions are critical. The Paris Agreement and other global initiatives aim to limit temperature increases and reduce the impacts of climate change on marine ecosystems. Additionally, enhancing monitoring and research efforts to better understand the effects of climate change on fisheries and aquaculture can inform adaptive management strategies. Collaboration among governments, scientific institutions, industry stakeholders, and local communities is essential to address the challenges posed by climate change to fisheries and aquaculture. By fostering resilience, implementing adaptive measures, and working toward mitigating climate change impacts, it is possible to safeguard the sustainability and productivity of marine environments for future generations.

5.5 Pollution

Pollution poses a significant challenge to the fisheries and aquaculture industry, threatening marine ecosystems and the sustainability of seafood production. This multifaceted issue encompasses various pollutants, including plastic debris, agricultural and industrial chemical runoff, oil spills, and heavy metals, all of which have detrimental effects on aquatic environments.

Plastic pollution, comprising discarded plastics and microplastics, poses a pervasive threat to marine life. Fish and other marine organisms ingest plastics, leading to physical harm, internal damage, and disruptions in their digestive systems.

Chemical pollutants from industrial and agricultural runoff contaminate water bodies, affecting the health of aquatic organisms and accumulating in the food chain, ultimately reaching consumers. Oil spills, whether from transportation accidents or offshore drilling, have catastrophic effects on marine ecosystems, causing immediate harm to fish, disrupting breeding habitats, and damaging coastal ecosystems. Heavy metals such as mercury and lead, often discharged from industrial sources, accumulate in fish tissues, posing health risks to both marine life and human consumers.

Mitigating pollution demands concerted efforts, including stricter regulations on waste disposal, promoting sustainable practices in industries, and investing in cleaner technologies. Additionally, public awareness campaigns and community engagement are vital in fostering responsible waste management and reducing pollution's impact on fisheries, aquaculture, and the overall health of marine ecosystems.

5.6 Disease and Parasites

Disease and parasites present substantial challenges to the fisheries and aquaculture industry, threatening the health of farmed fish populations and wild species alike. In aquaculture, high-density farming environments can facilitate the spread of diseases and parasites, leading to significant economic losses and environmental impacts.

Bacterial, viral, and parasitic infections (Table 2) can rapidly spread in aquaculture facilities, causing mass mortalities and reducing productivity. Intensive farming practices, often necessary to meet global seafood demands, create ideal conditions for disease outbreaks. Additionally, the introduction of non-native species in aquaculture can lead to the transmission of diseases to wild populations, impacting biodiversity and ecosystem health.

In wild fisheries, disease outbreaks can devastate entire populations. Changing environmental conditions due to climate change can influence the prevalence and spread of diseases, affecting the vulnerability of different species. For instance, warming ocean temperatures may facilitate the expansion of pathogens into new regions, affecting fish health and distribution patterns.

Managing disease and parasites in fisheries and aquaculture involves implementing biosecurity measures, such as proper hygiene, quarantine protocols, and vaccination in aquaculture settings. Enhancing surveillance, monitoring, and research into disease prevention and treatment are crucial for maintaining the health of fish populations, minimizing economic losses, and safeguarding both aquaculture and wild fisheries against the impacts of diseases and parasites.

Table 2 Diseases and parasites that significantly impact aquaculture and wild fish populations

S/N	Disease	Causative organism	Affected organisms	Description
Viral				
1	White Spot Syndrome Virus (WSSV)	White Spot Syndrome Virus	Shrimp (particularly Penaeid shrimp)	WSSV causes a highly contagious disease in shrimp, characterized by whitish spots on the exoskeleton, lethargy, and high mortality rates
2	Infectious Pancreatic Necrosis (IPN)	Infectious Pancreatic Necrosis Virus (IPNV)	Salmonids (such as salmon and trout)	IPN is a viral disease affecting fingerlings and juvenile salmonids, causing pancreatic necrosis, spinal deformities, and high mortality rates
3	Infectious Salmon Anemia (ISA)	Infectious Salmon Anemia Virus (ISAV)	Salmonids, particularly Atlantic salmon	ISA is a highly contagious viral disease affecting salmon, causing anemia, hemorrhaging, and high mortality rates
4	Koi Herpesvirus Disease (KHV)	Cyprinid Herpesvirus 3 (CyHV-3)	Common carp (*Cyprinus carpio*) and koi fish (ornamental varieties of common carp)	KHV is a highly contagious viral disease affecting carp and koi, characterized by lethargy, skin lesions, gill necrosis, and high mortality rates
5	Viral Nervous Necrosis (VNN) or Viral Encephalopathy and Retinopathy (VER)	Betanodavirus	Various marine and freshwater fish species, such as sea bass, grouper, turbot	VNN/VER is a viral disease causing neurological symptoms, including erratic swimming, abnormal behavior, and sometimes blindness
Bacterial				
6	Vibriosis	*Vibrio* spp. (e.g., *Vibrio anguillarum*, *Vibrio harveyi*)	Various marine and freshwater fish species, such as salmon, sea bass, shrimp	Vibriosis is caused by different Vibrio species, leading to systemic infections in fish. Symptoms include skin lesions, hemorrhaging, and mortality
7	Aeromoniasis	*Aeromonas hydrophila*, *Aeromonas salmonicida*	Various freshwater fish species, such as trout, catfish	Aeromoniasis results in skin ulcers, fin rot, and systemic infections in fish
8	Enteric Redmouth Disease (ERM)	*Yersinia ruckeri*	Salmonids, particularly rainbow trout	ERM results in hemorrhaging of the mouth and fins, along with systemic infections in affected fish
9	Francisellosis (Tularemia)	*Francisella* spp. (such as *Francisella noatunensis*)	Various marine and freshwater fish species, such as tilapia, salmon	Francisellosis leads to systemic infections, skin lesions, and high mortality rates in fish

(continued)

Table 2 (continued)

S/N	Disease	Causative organism	Affected organisms	Description
10	Columnaris Disease (Flavobacteriosis)	*Flavobacterium columnare*	Various freshwater fish species, such as catfish, carp	Columnaris disease results in white filamentous growth on the skin, gills, and fins, leading to tissue damage and mortality in affected fish
Parasitic				
11	Sea Lice Infestation	*Lepeophtheirus salmonis*, *Caligus* spp.	Salmonids, such as salmon and trout	Sea lice attach to the skin and mucous membranes of fish, causing irritation, skin lesions, and secondary infections. Infestations can lead to economic losses in salmon aquaculture.
12	Ichthyophthiriasis (Ich)	*Ichthyophthirius multifiliis* (commonly known as Ich or white spot parasite)	Various freshwater fish species, such as trout, goldfish, tilapia	Ich leads to white spots on the skin, gills, and fins of infected fish. It can cause irritation, respiratory distress, and mortality
13	Gyrodactylosis	Monogenean flatworms (*Gyrodactylus* spp.)	Various freshwater fish species, such as salmon, trout	Flatworms attach to the skin and gills, causing irritation, inflammation, and potential mortality in affected fish
14	Monogenean Infections (Dactylogyrosis)	Monogeneans (such as *Dactylogyrus* spp.)	Various freshwater and marine fish species	Monogeneans are ectoparasitic flatworms that attach to fish gills, skin, and fins, leading to irritation, mucus production, and secondary infections
15	Ichthyobodoosis (Costia Disease)	*Ichthyobodo* spp. (e.g., *Ichthyobodo necator*)	Various freshwater and marine fish species	Ichthyobodoosis is caused by protozoan parasites that affect the skin and gills, leading to tissue damage, excess mucus production, and lethargy in infected fish
16	Amoebic Gill Disease (AGD)	*Peramoeba perurans*	Salmonids	The amoeba settles on the surface of the gill, causing irritation, mucus formation and inflammation. In severe cases, it leads to lesions on the gills, reducing oxygen exchange, and occasionally death

5.7 *Invasive Species*

Invasive species present a significant challenge to the fisheries and aquaculture industry, disrupting ecosystems, endangering native species, and posing economic threats. These non-native species, introduced intentionally or accidentally, rapidly proliferate in new environments, outcompeting local species and altering ecological balances.

In aquatic environments, invasive species can wreak havoc by outcompeting native fish for resources, preying upon them, or introducing diseases and parasites. For instance, the introduction of non-native species like the Asian carp in North American waters has led to concerns about their impact on native fish populations and ecosystems due to their rapid reproduction and voracious feeding habits. In aquaculture, invasive species pose risks of escapes or unintentional releases, potentially leading to ecological disruptions when these species establish themselves in local ecosystems. Escaped farmed fish can interbreed with wild populations, diluting genetic diversity and impacting the fitness of native species.

Managing invasive species demands vigilance in monitoring and preventing their introduction, early detection, and rapid response measures to control their spread. Implementing biosecurity protocols, strict regulations on species importation, and public education are vital in preventing the unintentional spread of invasive species in aquaculture and fisheries. Combatting the threats posed by invasive species requires collaborative efforts among governments, industries, and conservation groups to mitigate their impact, protect native biodiversity, and sustain the health of aquatic ecosystems and the fisheries and aquaculture industry.

5.8 Genetics

Genetics presents both a challenge and an opportunity for the fisheries and aquaculture industry. While genetic advancements offer immense potential for improving fish health, growth rates, and disease resistance, they also pose challenges in terms of managing genetic diversity, preventing genetic pollution, and ensuring ethical practices.

Selective breeding and genetic manipulation techniques have been pivotal in enhancing desired traits in farmed fish, leading to increased yields and improved quality. However, the concentration of genetic material within certain species or strains can reduce overall genetic diversity, making these populations more vulnerable to diseases and environmental changes.

Genetic interactions between farmed and wild populations raise concerns about genetic pollution. Escaped farmed fish interbreeding with wild populations can dilute genetic diversity, impacting the fitness and adaptability of native species. Maintaining the integrity of wild gene pools is crucial for preserving natural ecosystems and sustaining biodiversity. Ethical considerations surrounding genetically modified organisms (GMOs) in aquaculture also come into play. Balancing technological advancements with environmental and ethical concerns is essential to ensure responsible and sustainable genetic practices.

Addressing genetic challenges in fisheries and aquaculture demands a balanced approach, incorporating rigorous breeding programs that prioritize genetic diversity, regulatory frameworks that mitigate genetic risks, and ethical guidelines that navigate the responsible use of genetic technologies. Striking this balance will be critical in leveraging genetics to propel the industry forward while safeguarding ecosystems and ethical considerations.

5.9 Social and Economic Issues

Social and economic issues present complex challenges to the fisheries and aquaculture industry, impacting livelihoods, communities, and global food security. In many coastal regions, fishing communities heavily rely on these industries for their sustenance and income. However, overexploitation, mismanagement, and environmental degradation can lead to dwindling fish stocks, posing threats to the livelihoods of those dependent on fishing. In addition, small-scale and artisanal fishers often face challenges related to access to markets, fair prices, lack of infrastructure, and limited access to technology. This situation often exacerbates poverty, especially in developing countries where fishing is a primary source of income.

Additionally, the unequal distribution of resources and access to fishing grounds can spark conflicts among fishing communities or between industrial and small-scale fishers. Disputes over fishing rights, depletion of resources, and competition for markets can strain social cohesion and lead to tensions.

Moreover, fluctuations in market demands, trade regulations, and the globalization of seafood trade impact the economic stability of the fisheries and aquaculture industry. Market price volatility, coupled with increased competition and fluctuating consumer preferences, poses challenges for both large-scale commercial operations and small-scale fishers.

Addressing these social and economic challenges requires a holistic approach that considers the well-being of fishing communities, equitable resource management, and sustainable practices. Investing in alternative livelihoods, empowering local stakeholders through participatory management, and fostering responsible governance frameworks are crucial in ensuring the resilience and sustainability of the fisheries and aquaculture industry for the future.

5.10 Governance and Regulatory Issues

Governance and regulatory issues in fisheries and aquaculture encompass a range of challenges that affect the sustainable management of marine resources and the development of aquaculture.

Many regions lack comprehensive and effective regulatory frameworks for fisheries and aquaculture (Engle & van Senten, 2022). Inadequate laws, regulations, and enforcement mechanisms contribute to overfishing, habitat degradation, and unsustainable aquaculture practices. IUU fishing persists due to weak governance structures and inadequate enforcement of regulations (Lindley & Techera, 2017).

This practice undermines efforts to sustainably manage fisheries, leading to the depletion of fish stocks and impacting marine ecosystems. Inadequate monitoring and surveillance systems in fisheries make it challenging to track fishing activities, monitor compliance with regulations, and prevent IUU fishing. Strengthening monitoring technologies and systems is crucial for effective management. Some fisheries management practices are outdated or ineffective, leading to overfishing and declining fish stocks (Melnychuk et al., 2020). Adopting science-based approaches, such as implementing quotas, size limits, and seasonal closures, is essential for sustainable fisheries management. Issues related to access rights, allocation of fishing quotas, and property rights can lead to conflicts among fishers and communities. Establishing clear and fair access and allocation systems is crucial for equitable and sustainable resource use. Aquaculture governance faces challenges such as inadequate regulations, weak enforcement of environmental standards, and limited spatial planning (Lebel et al., 2018′; Davies et al., 2019). Effective governance frameworks are necessary to ensure sustainable practices and minimize environmental impacts. Many fisheries and aquaculture communities face socioeconomic challenges, including poverty, lack of alternative livelihoods, and unequal access to resources (Ateweberhan et al., 2018). Addressing these issues is essential for the well-being of fishing communities and achieving sustainable development.

Addressing these governance and regulatory challenges requires comprehensive and participatory approaches involving governments, stakeholders, communities, and international organizations. Strengthening regulatory frameworks, enhancing monitoring and enforcement capabilities, promoting community engagement, and adopting science-based management practices are key strategies to ensure the sustainable management of fisheries and aquaculture resources. International cooperation and agreements also play a crucial role in addressing these global challenges.

As a consequence, numerous species face decline or extinction, disrupting food chains and jeopardizing marine biodiversity. Some notable examples of fish species facing decline or extinction due to overfishing include Atlantic bluefin tuna (*Thunnus thynnus*) and Pacific Bluefin Tuna (*Thunnus orientalis*), which are highly sought after for sushi and sashimi; *Orange roughy* (*Hoplostethus atlanticus*); Atlantic cod (Gadus morhua); Grouper species (*Epinephelus* spp. and *Mycteroperca* spp.); Haddock (*Melanogrammus aeglefinus*); Atlantic salmon (*Salmo salar*); and European eel (*Anguilla Anguilla*).

The repercussions of overfishing extend beyond ecological concerns. Coastal communities heavily reliant on fishing for livelihoods face economic hardships as stocks dwindle. Additionally, the industry's sustainability is compromised, jeopardizing its long-term viability and contributing to social and political tensions over diminishing resources.

Addressing overfishing demands a multi-faceted approach involving stricter fishing regulations, adoption of sustainable fishing methods, implementation of marine protected areas, and global cooperation to manage fisheries responsibly. Consumer awareness and support for sustainable seafood choices also play a pivotal role in mitigating overfishing's impacts.

6 Nanochitosan and Its Relevance in Aquaculture

Nanochitosan, a derivative of chitosan obtained from chitin, has gained attention in aquaculture for its versatile applications and potential benefits. Nanochitosan exhibits potent antibacterial and antimicrobial properties, making it useful in controlling and preventing bacterial infections in aquaculture (Abdel-Razek, 2019; Ibrahim et al., 2023). Its ability to inhibit the growth of pathogens can contribute to maintaining healthy aquatic environments and improving the health of cultured fish and shrimp. This helps in preventing secondary infections and diseases, thereby enhancing the overall health of aquaculture species.

Nanochitosan can also be employed in water treatment processes. Its ability to bind with heavy metals, pollutants, and excess nutrients in water can aid in water purification, mitigating environmental stressors and promoting a healthier habitat for aquatic organisms (Ahuekwe et al., 2023). Nanochitosan's properties make it a promising candidate for developing drug delivery systems in aquaculture. It can encapsulate bioactive compounds or medicines, allowing controlled release and targeted delivery to treat specific fish or shrimp diseases more effectively (Yadav et al., 2022). As a derivative of chitin, and a natural biopolymer, nanochitosan is biodegradable and environmentally friendly. Its use aligns with sustainable practices in aquaculture, minimizing environmental impact compared to some synthetic alternatives. Nanochitosan has shown potential in stimulating the immune response in aquatic organisms (Abdel-Tawwab et al., 2019). This can improve their resilience to diseases and stress factors, ultimately contributing to better survival rates in aquaculture settings.

The application of nanochitosan in aquaculture is still evolving, with ongoing research focused on optimizing its formulations, exploring new applications, and understanding its long-term effects on aquatic ecosystems and cultured species. While nanochitosan holds promise, further studies are essential to ensure its safe and effective use, considering factors such as dosages, environmental impacts, and potential interactions with other aquaculture practices.

7 Purpose and Scope of the Book

This book seeks to provide a comprehensive exploration of the multifaceted applications of nanochitosan in revolutionizing and augmenting various facets of fisheries and aquaculture practices. The book aims to elucidate the diverse roles and potential of nanochitosan as a transformative tool, focusing on its contributions to enhancing sustainability, productivity, and the overall health of aquatic ecosystems. The book is aimed at establishing a robust foundation by detailing the fundamental properties, synthesis methods, characterization techniques, and safety aspects of nanochitosan relevant to its application in fisheries and aquaculture. It addressed nanochitosan's pivotal role in combating diseases among fish and other aquatic species. Emphasis will be placed on its antimicrobial properties, wound-healing capabilities, and potential as an immune system enhancer, thereby contributing to improved health and reduced disease prevalence in aquaculture. Exploring nanochitosan's efficacy in water treatment, the book will highlight its capacity to purify aquatic environments by removing contaminants, heavy metals, and pollutants. Furthermore, it is poised to assess nanochitosan's environmental sustainability, considering its biodegradability and eco-friendly attributes. Nanochitosan's applications in improving aquaculture productivity will be thoroughly examined. This will encompass its use in optimizing feed formulations, promoting growth, enhancing nutrient uptake, and contributing to overall efficiency in aquaculture operations. However, the widespread application of nanochitosan raises concerns regarding potential environmental risks associated with its extensive use. These risks primarily revolve around its persistence, mobility, and potential adverse effects on ecosystems. The book thus addressed concerns such as the potential accumulation of nanochitosan in environmental compartments. Nanochitosan may possess some level of resistance to degradation, potentially leading to their accumulation in soil, water bodies, or sediments. This accumulation raises concerns about the long-term environmental impact and the potential disruption of natural ecosystems.

The book explores emerging and innovative applications of nanochitosan in fisheries and aquaculture, envisioning its potential in novel areas such as fish disease prevention, control, and management, water quality enhancement, feed enhancement and nutrition, tagging and barcoding, fish breeding programs, and detoxification and depuration of fish. The book explores the attainment of environmentally friendly aquaculture practices aided by novel nanochitosan. Additionally, it will outline future research directions, addressing challenges and paving the way for further advancements in the field.

By encompassing these diverse aspects, the book aspires to serve as a comprehensive resource for researchers, practitioners, policymakers, and industry stakeholders involved in fisheries and aquaculture. It aims to foster a deeper understanding of nanochitosan's transformative potential and its role in shaping sustainable, efficient, and resilient practices within the fisheries and aquaculture sectors.

References

Abachi, S., Pilon, G., Marette, A., Bazinet, L., & Beaulieu, L. (2023). Beneficial effects of fish and fish peptides on main metabolic syndrome associated risk factors: Diabetes, obesity and lipemia. *Critical Reviews in Food Science and Nutrition, 63*(26), 7896–7944.

Abdel-Razek, N. (2019). Antimicrobial activities of chitosan nanoparticles against pathogenic microorganisms in Nile tilapia, Oreochromis niloticus. *Aquaculture International, 27*(5), 1315–1330.

Abdel-Tawwab, M., Razek, N. A., & Abdel-Rahman, A. M. (2019). Immunostimulatory effect of dietary chitosan nanoparticles on the performance of Nile tilapia, *Oreochromis niloticus* (L.). *Fish and Shellfish Immunology, 88*, 254–258.

Ahmad, A. L., Chin, J. Y., Harun, M. H. Z. M., & Low, S. C. (2022). Environmental impacts and imperative technologies towards sustainable treatment of aquaculture wastewater: A review. *Journal of Water Process Engineering, 46*, 102553.

Ahuekwe, E. F., Isibor, P. O., Akinhanmi, F., Ajiboye, I., Salami, A. O., Adekeye, B. T., Akinpelu, S. O., Kuye, A. D., Bello, A. O., Oyewale, J. O., & George, D. Z. (2023). Utilization of nano-chitosan in the sterilization of ponds and water treatment for aquaculture. In *Next Generation Nanochitosan* (pp. 301–338). Academic Press.

Albert, C. M., Cook, N. R., Pester, J., Vinayagamoorthy, M., Ridge, C., Danik, J. S., Gencer, B., Siddiqi, H. K., Ng, C., Gibson, H., & Mora, S. 2020, December). The vital rhythm trial: omega-3 fatty acid and vitamin D supplementation in the primary prevention of atrial fibrillation. In *Circulation* (Vol. 142, No. 24, pp. E482–E482). Lippincott Williams and Wilkins.

Al-Emran, M., & Griffy-Brown, C. (2023). The role of technology adoption in sustainable development: Overview, opportunities, challenges, and future research agendas. *Technology in Society, 73*, 102240.

Alfonso, S., Gesto, M., & Sadoul, B. (2021). Temperature increase and its effects on fish stress physiology in the context of global warming. *Journal of Fish Biology, 98*(6), 1496–1508.

American Heart Association. (2023). *Fish and Omega-3 Fatty Acids*. Accessed November 24, 2023, from https://www.heart.org/en/healthy-living/healthy-eating/eat-smart/fats/fish-and-omega-3-fatty-acids

Archibong, E. D. (2023). The role of Omega-3 long chain polyunsaturated fatty acids (LCPUFAs) in child development and cognitive function. *Journal of Current Science Research and Review, 1*(05), 158–163.

Astrup, A., Magkos, F., Bier, D. M., Brenna, J. T., de Oliveira Otto, M. C., Hill, J. O., King, J. C., Mente, A., Ordovas, J. M., Volek, J. S., & Yusuf, S. (2020). Saturated fats and health: A reassessment and proposal for food-based recommendations: JACC state-of-the-art review. *Journal of the American College of Cardiology, 76*(7), 844–857.

Ateweberhan, M., Hudson, J., Rougier, A., Jiddawi, N. S., Msuya, F. E., Stead, S. M., & Harris, A. (2018). Community based aquaculture in the western Indian Ocean: Challenges and opportunities for developing sustainable coastal livelihoods. *Ecology and Society, 23*(4).

Auld, K. G. (2021). *Sustainable development of small-scale fisheries and the need for strong measures to protect small-scale fisheries in international trade law.*

Balakrishnan, J., Kannan, S., & Govindasamy, A. (2021). Structured form of DHA prevents neurodegenerative disorders: A better insight into the pathophysiology and the mechanism of DHA transport to the brain. *Nutrition Research, 85*, 119–134.

Balta, I., Stef, L., Pet, I., Iancu, T., Stef, D., & Corcionivoschi, N. (2021). Essential fatty acids as biomedicines in cardiac health. *Biomedicine, 9*(10), 1466.

Batista, V. S., Fabré, N. N., Malhado, A. C., & Ladle, R. J. (2014). Tropical artisanal coastal fisheries: Challenges and future directions. *Reviews in Fisheries Science and Aquaculture, 22*(1), 1–15.

Bjørndal, T., Dey, M., & Tusvik, A. (2024). Economic analysis of the contributions of aquaculture to future food security. *Aquaculture, 578*, 740071.

Bosma, R. H., Debrot, A. O., Rejeki, S., Tonneljck, F., Priyanto, E. B., Susanto, A., Yunlati, W., & Slhombing, W. (2020). *Associated mangrove aquaculture farms; building with nature to restore eroding tropical muddy coasts* (No. 4). Ecoshape.

Brander, K., Cochrane, K., Barange, M., & Soto, D. (2017). Climate change implications for fisheries and aquaculture. *Climate Change Impacts on Fisheries and Aquaculture: A Global Analysis, 1*, 45–62.

Brye, K. (2023). Aquaculture: A sustainable solution for fish cultivation and global food security. *Journal of Agriculture, 6*(4), 91–93.

Carneiro, M., & Martins, R. (2021). *Destructive fishing practices and their impact on the marine ecosystem.*

Charles, A. (2017). The big role of coastal communities and small-scale fishers in ocean conservation. In *Conservation for the Anthropocene Ocean* (pp. 447–461). Academic Press.

Chauhan, A. S., Patel, A. K., Nimker, V., Singhania, R. R., Chen, C. W., Patel, A. K., Raj, T., & Dong, C. D. (2023). Biorefining of essential polyunsaturated fatty acids from microbial sources: Current updates and prospects. *Systems Microbiology and Biomanufacturing*, 1–23.

Chen, X., Xu, Q., & Li, L. (2023). Illegal, unreported, and unregulated fishing governance in disputed maritime areas: Reflections on the international legal obligations of states. *Fishes, 8*(1), 36.

Ciannelli, L., Smith, E., Kearney, K., Hunsicker, M., & McGilliard, C. (2022). Greater exposure of nearshore habitats in the Bering Sea makes fish early life stages vulnerable to climate change. *Marine Ecology Progress Series, 684*, 91–102.

Cohen Kadosh, K., Muhardi, L., Parikh, P., Basso, M., Jan Mohamed, H. J., Prawitasari, T., Samuel, F., Ma, G., & Geurts, J. M. (2021). Nutritional support of neurodevelopment and cognitive function in infants and young children—An update and novel insights. *Nutrients, 13*(1), 199.

Colombo, S. M., & Mazal, X. (2020). Investigation of the nutritional composition of different types of salmon available to Canadian consumers. *Journal of Agriculture and Food Research, 2*, 100056.

Cooke, S. J., Fulton, E. A., Sauer, W. H., Lynch, A. J., Link, J. S., Koning, A. A., Jena, J., Silva, L. G., King, A. J., Kelly, R., & Osborne, M. (2023). Towards vibrant fish populations and sustainable fisheries that benefit all: Learning from the last 30 years to inform the next 30 years. *Reviews in Fish Biology and Fisheries, 33*(2), 317–347.

Dallas, D. C., Sanctuary, M. R., Qu, Y., Khajavi, S. H., Van Zandt, A. E., Dyandra, M., Frese, S. A., Barile, D., & German, J. B. (2017). Personalizing protein nourishment. *Critical Reviews in Food Science and Nutrition, 57*(15), 3313–3331.

Davies, I. P., Carranza, V., Froehlich, H. E., Gentry, R. R., Kareiva, P., & Halpern, B. S. (2019). Governance of marine aquaculture: Pitfalls, potential, and pathways forward. *Marine Policy, 104*, 29–36.

de Boer, J., Schösler, H., & Aiking, H. (2020). Fish as an alternative protein–a consumer-oriented perspective on its role in a transition towards more healthy and sustainable diets. *Appetite, 152*, 104721.

de Seymour, J. V., Beck, K. L., & Conlon, C. A. (2022). Nutrition in pregnancy. *Obstetrics, Gynaecology and Reproductive Medicine, 32*, 253.

DePace, N. L., Colombo, J., DePace, N. L., & Colombo, J. (2019). Omega-3 fatty acids (Prong-1). *Clinical Autonomic and Mitochondrial Disorders: Diagnosis, Prevention, and Treatment for Mind-Body Wellness*, 37–70.

DiNicolantonio, J. J., & O'Keefe, J. H. (2020). The importance of marine omega-3s for brain development and the prevention and treatment of behavior, mood, and other brain disorders. *Nutrients, 12*(8), 2333.

Djuricic, I., & Calder, P. C. (2021). Beneficial outcomes of omega-6 and omega-3 polyunsaturated fatty acids on human health: An update for 2021. *Nutrients, 13*(7), 2421.

Eisenberg, C. (2013). *The wolf's tooth: Keystone predators, trophic cascades, and biodiversity*. Island Press.

Engle, C. R., & van Senten, J. (2022). Resilience of communities and sustainable aquaculture: Governance and regulatory effects. *Fishes, 7*(5), 268.

Figuerola, B., Hancock, A. M., Bax, N., Cummings, V. J., Downey, R., Griffiths, H. J., Smith, J., & Stark, J. S. (2021). A review and meta-analysis of potential impacts of ocean acidification on marine calcifiers from the Southern Ocean. *Frontiers in Marine Science, 8*, 584445.

Ghaly, A. E., Ramakrishnan, V. V., Brooks, M. S., Budge, S. M., & Dave, D. (2013). Fish processing wastes as a potential source of proteins. Amino acids and oils: A critical review. *Journal Microbial and Biochemical Technology, 5*(4), 107–129.

Gioia, C., Lucchino, B., Tarsitano, M. G., Iannuccelli, C., & Di Franco, M. (2020). Dietary habits and nutrition in rheumatoid arthritis: Can diet influence disease development and clinical manifestations? *Nutrients, 12*(5), 1456.

González-Gaya, B., García-Bueno, N., Buelow, E., Marin, A., & Rico, A. (2022). Effects of aquaculture waste feeds and antibiotics on marine benthic ecosystems in the Mediterranean Sea. *Science of the Total Environment, 806*, 151190.

Gray, C. A., & Kennelly, S. J. (2018). Bycatches of endangered, threatened and protected species in marine fisheries. *Reviews in Fish Biology and Fisheries, 28*(3), 521–541.

Greenberg, P. (2014). *Environmental problems of aquaculture*. Accessed November 24, 2023, from https://earthjournalism.net/resources/environmental-problems-of-aquaculture

Haubrock, P. J., Oficialdegui, F. J., Zeng, Y., Patoka, J., Yeo, D. C., & Kouba, A. (2021). The red-claw crayfish: A prominent aquaculture species with invasive potential in tropical and subtropical biodiversity hotspots. *Reviews in Aquaculture, 13*(3), 1488–1530.

Hei, A. (2020, January). Fish as a functional food in human health, diseases and well-being. In *Proceedings of the 107th Indian Science Congress, Banglore, India* (pp. 3–7).

Ibrahim, R. E., Elshopakey, G. E., Abdelwarith, A. A., Younis, E. M., Ismail, S. H., Ahmed, A. I., El-Saber, M. M., Abdelhamid, A. E., Davies, S. J., El-Murr, A., & Rahman, A. N. A. (2023). Chitosan neem nanocapsule enhances immunity and disease resistance in nile tilapia (*Oreochromis niloticus*). *Heliyon, 9*(9), e19354.

Islam, M., & Yasmin, R. (2017). Impact of aquaculture and contemporary environmental issues in Bangladesh. *International Journal of Fisheries and Aquatic Studies, 5*, 100–107.

Johnson, B., Jayakumar, R., Abdul Nazar, A. K., Tamilmani, G., Sakthivel, M., Rameshkumar, P., Anikuttan, K. K., & Sankar, M. (2021). Climate resilient mariculture technologies for food and nutritional security. *Exploring Synergies and Trade-offs between Climate Change and the Sustainable Development Goals*, 83–95.

Kang, B., Vitule, J. R., Li, S., Shuai, F., Huang, L., Huang, X., Fang, J., Shi, X., Zhu, Y., Xu, D., & Yan, Y. (2023). Introduction of non-native fish for aquaculture in China: A systematic review. *Reviews in Aquaculture, 15*(2), 676–703.

Khalid, W., Gill, P., Arshad, M. S., Ali, A., Ranjha, M. M. A. N., Mukhtar, S., Afzal, F., & Maqbool, Z. (2022). Functional behavior of DHA and EPA in the formation of babies brain at different stages of age, and protect from different brain-related diseases. *International Journal of Food Properties, 25*(1), 1021–1044.

Khalili Tilami, S., & Sampels, S. (2018). Nutritional value of fish: Lipids, proteins, vitamins, and minerals. *Reviews in Fisheries Science and Aquaculture, 26*(2), 243–253.

Khan, S. U., Lone, A. N., Khan, M. S., Virani, S. S., Blumenthal, R. S., Nasir, K., Miller, M., Michos, E. D., Ballantyne, C. M., Boden, W. E., & Bhatt, D. L. (2021). Effect of omega-3 fatty acids on cardiovascular outcomes: A systematic review and meta-analysis. *EClinical Medicine, 38*, 100997.

Kuemlangan, B., Amidjogbe, E. R., Nakamura, J., Tomassi, A., Hupperts, R., Bojang, B., & Amador, T. (2023). Enforcement approaches against illegal fishing in national fisheries legislation. *Marine Policy, 149*, 105514.

Kwasek, K., Thorne-Lyman, A. L., & Phillips, M. (2020). Can human nutrition be improved through better fish feeding practices? A review paper. *Critical Reviews in Food Science and Nutrition, 60*(22), 3822–3835.

Lafuente, M., Rodríguez González-Herrero, M. E., Romeo Villadóniga, S., & Domingo, J. C. (2021). Antioxidant activity and neuroprotective role of docosahexaenoic acid (DHA) supplementation in eye diseases that can lead to blindness: A narrative review. *Antioxidants, 10*(3), 386.

Lall, S. P., & Kaushik, S. J. (2021). Nutrition and metabolism of minerals in fish. *Animals, 11*(09), 2711.

Lebel, L., Lebel, P., & Chuah, C. J. (2018). Governance of aquaculture water use. *International Journal of Water Resources Development, 35*, 659.

Leung, J. Y., Zhang, S., & Connell, S. D. (2022). Is ocean acidification really a threat to marine calcifiers? A systematic review and meta-analysis of 980+ studies spanning two decades. *Small, 18*(35), 2107407.

Lewis, S. G., & Boyle, M. (2017). The expanding role of traceability in seafood: Tools and key initiatives. *Journal of Food Science, 82*(S1), A13–A21.

Li, C., & Palmer-Keenan, D. (2016). Assessment of omega-3s knowledge on blood lipids among registered dietitians. *The FASEB Journal, 30*, 675–675.

Li, X., Zheng, S., & Wu, G. (2021). Nutrition and functions of amino acids in fish. *Amino Acids in Nutrition and Health: Amino Acids in the Nutrition of Companion, Zoo and Farm Animals* (pp. 133–168).

Lindley, J., & Techera, E. J. (2017). Overcoming complexity in illegal, unregulated and unreported fishing to achieve effective regulatory pluralism. *Marine Policy, 81*, 71–79.

Little, D. C., Newton, R. W., & Beveridge, M. C. M. (2016). Aquaculture: A rapidly growing and significant source of sustainable food? Status, transitions and potential. *Proceedings of the Nutrition Society, 75*(3), 274–286.

Mallick, R., & Duttaroy, A. K. (2022). Modulation of endothelium function by fatty acids. *Molecular and Cellular Biochemistry, 477*(1), 15–38.

March, A., & Failler, P. (2022). Small-scale fisheries development in Africa: Lessons learned and best practices for enhancing food security and livelihoods. *Marine Policy, 136*, 104925.

Mavraganis, T., Constantina, C., Kolygas, M., Vidalis, K., & Nathanailides, C. (2020). Environmental issues of aquaculture development. *Egyptian Journal Of Aquatic Biology And Fisheries, 24*(2), 441–450.

Melnychuk, M. C., Baker, N., Hively, D., Mistry, K., Pons, M., Ashbrook, C. E., Minto, C., Hilborn, R., & Ye, Y. (2020). *Global trends in status and management of assessed stocks: Achieving sustainable fisheries through effective management* (Vol. 665). Food and Agriculture Organization.

Merdzhanova, A., Stancheva, M., Dobreva, D. A., & Makedonski, L. (2013). Fatty acid and fat soluble vitamins composition of raw and cooked Black Sea horse mackerel. *Ovidius University Annals of Chemistry, 24*(1), 27–34.

Moehl, J., Brummett, R., Boniface, M. K., & Coche, A. (2006). Guiding principles for promoting aquaculture in Africa: Benchmarks for sustainable development. *CIFA Occasional Paper, 28*, I.

Mohanty, B. P., Pati, M. K., Bhattacharjee, S., Hajra, A., & Sharma, A. P. (2013). Small indigenous fishes and their importance in human health. *Advances in Fish Research, 5*, 257–278.

Morell, P., & Fiszman, S. (2017). Revisiting the role of protein-induced satiation and satiety. *Food Hydrocolloids, 68*, 199–210.

Mozaffarian, D., & Wu, J. H. (2011). Omega-3 fatty acids and cardiovascular disease: Effects on risk factors, molecular pathways, and clinical events. *Journal of the American College of Cardiology, 58*(20), 2047–2067.

Msomphora, M. R. (2018). *Interactive governance of small-scale fisheries and aquaculture: Transdisciplinary challenges of community involvement.*

Musalia, L. M., Jonathan, M. M., Liti, D. M., Kyule, D. N., Ogello, E. O., Obiero, K. O., & Kirimi, J. G. (2020). *Aqua-feed wastes: Impact on natural systems and practical mitigations—A review.*

Niner, H. J., Barut, N. C., Baum, T., Diz, D., del Pozo, D. L., Laing, S., Lancaster, A. M. S., McQuaid, K. A., Mendo, T., Morgera, E., & Maharaj, P. N. (2022). Issues of context, capacity and scale: Essential conditions and missing links for a sustainable blue economy. *Environmental Science and Policy, 130*, 25–35.

Nissar, S., Bakhtiyar, Y., Arafat, M. Y., Andrabi, S., Mir, Z. A., Khan, N. A., & Langer, S. (2023). The evolution of integrated multi-trophic aquaculture in context of its design and components paving way to valorization via optimization and diversification. *Aquaculture, 565*, 739074.

Norman, R. A., Crumlish, M., & Stetkiewicz, S. (2019). The importance of fisheries and aquaculture production for nutrition and food security. *Revue Scientifique et Technique (International Office of Epizootics), 38*(2), 395–407.

Okafor-Yarwood, I., Kadagi, N. I., Belhabib, D., & Allison, E. H. (2022). Survival of the richest, not the fittest: How attempts to improve governance impact African small-scale marine fisheries. *Marine Policy, 135*, 104847.

Olaussen, J. O. (2018). Environmental problems and regulation in the aquaculture industry. Insights from Norway. *Marine Policy, 98*, 158–163.

Olopade, O. A., & Dienye, H. E. (2017). Roles of eco-labeling in fisheries conservation and sustainability. *Indonesian Journal of Marine Sciences/Ilmu Kelautan, 22*(2), 99.

Oppedisano, F., Macrì, R., Gliozzi, M., Musolino, V., Carresi, C., Maiuolo, J., Bosco, F., Nucera, S., Caterina Zito, M., Guarnieri, L., & Scarano, F. (2020). The anti-inflammatory and antioxidant properties of n-3 PUFAs: Their role in cardiovascular protection. *Biomedicine, 8*(9), 306.

Pakkiam Muniyasamy, S. (2023). *ATR-MIR and chemometric comparative studies on fish from different environments*.
Pal, J., Shukla, B. N., Maurya, A. K., Verma, H. O., Pandey, G., & Amitha, A. (2018). A review on role of fish in human nutrition with special emphasis to essential fatty acid. *International Journal of Fisheries and Aquatic Studies, 6*(2), 427–430.
Papadopoulou, S. K. (2020). Rehabilitation nutrition for injury recovery of athletes: The role of macronutrient intake. *Nutrients, 12*(8), 2449.
Pradeepkiran, J. A. (2019). Aquaculture role in global food security with nutritional value: A review. *Translational Animal Science, 3*(2), 903–910.
Punia, S., Sandhu, K. S., Siroha, A. K., & Dhull, S. B. (2019). Omega 3-metabolism, absorption, bioavailability and health benefits–a review. *Pharma Nutrition, 10*, 100162.
Ramenzoni, V. C., Borroto Escuela, D., Rangel Rivero, A., González-Díaz, P., Vázquez Sánchez, V., López-Castañeda, L., Falcón Méndez, A., Hernández Ramos, I., Valentín Hernández López, N., Besonen, M. R., & Yoskowitz, D. (2020). Vulnerability of fishery-based livelihoods to extreme events: Local perceptions of damages from hurricane Irma and Tropical storm Alberto in Yaguajay, Central Cuba. *Coastal Management, 48*(5), 354–377.
Roderick, J. (2020). *The FAO and transmission of global aquaculture soft law instruments–A case study on Australian and Norwegian aquaculture perspectives and the role of the FAO* (Master's thesis, University of Stavanger, Norway).
Sala, A., Depestele, J., Gümüş, A., Laffargue, P., Nielsen, J. R., Polet, H., Smith, C. J., Zengin, M., Bastardie, F., Eigaard, O. R., & Hamon, K. G. (2023). Technological innovations to reduce the impact of bottom gears on the seabed. *Marine Policy, 157*, 105861.
Sarojnalini, C., & Hei, A. (2019). Fish as an important functional food for quality life. *U: Functional Foods (Lagouri, V, Ured), 2019*, 77–97.
Schmid, A., & Walther, B. (2013). Natural vitamin D content in animal products. *Advances in Nutrition, 4*(4), 453–462.
Shabbir, H., Shabbir, I., Aslam, M., Sarwar, M. F., Sarwar, M. H., & Sarwar, M. (2020). Fundamental aspects of vitamin B complex in human nourishment and fitness. *American Journal of Food Science and Health, 6*(4), 109–118.
Shahbandeh, M. (2023) .Global fish consumption per capita by region 2019. *Statista*. Accessed November 17, 2023, from https://www.statista.com/statistics/1026312/global-fish-consumption-by-region/
Sharma, C. (2011). Securing economic, social and cultural rights of small-scale and artisanal fishworkers and fishing communities. *Maritime Studies, 10*(2), 41–62.
Shi, Y., & Li, Y. (2023). Impacts of ocean acidification on physiology and ecology of marine invertebrates: A comprehensive review. *Aquatic Ecology*, 1–20.
Simonetto, M., Infante, M., Sacco, R. L., Rundek, T., & Della-Morte, D. (2019). A novel anti-inflammatory role of omega-3 PUFAs in prevention and treatment of atherosclerosis and vascular cognitive impairment and dementia. *Nutrients, 11*(10), 2279.
Singh, J. E. (2020). Dietary sources of omega-3 fatty acids versus omega-3 fatty acid supplementation effects on cognition and inflammation. *Current Nutrition Reports, 9*, 264–277.
Singh, A., Kelkar, N., Natarajan, K., & Selvaraj, S. (2021). Review on the extraction of calcium supplements from eggshells to combat waste generation and chronic calcium deficiency. *Environmental Science and Pollution Research, 28*, 46985–46998.
Siscovick, D. S., Barringer, T. A., Fretts, A. M., Wu, J. H., Lichtenstein, A. H., Costello, R. B., Kris-Etherton, P. M., Jacobson, T. A., Engler, M. B., Alger, H. M., & Appel, L. J. (2017). Omega-3 polyunsaturated fatty acid (fish oil) supplementation and the prevention of clinical cardiovascular disease: A science advisory from the American Heart Association. *Circulation, 135*(15), e867–e884.
Stefaniak, O., Dobrzyńska, M., Drzymała-Czyż, S., & Przysławski, J. (2022). Diet in the prevention of Alzheimer's disease: Current knowledge and future research requirements. *Nutrients, 14*(21), 4564.
Sumaila, U. R., & Tai, T. C. (2020). End overfishing and increase the resilience of the ocean to climate change. *Frontiers in Marine Science, 7*, 523.

Suvarna, V., & Singh, V. (2019). DHA a key supplement for neurologic health—An insight. *Omega Fatty Acids in Brain and Neurological Health*, 409–419.

Talloni-Alvarez, N. E., Sumaila, U. R., Le Billon, P., & Cheung, W. W. (2019). Climate change impact on Canada's Pacific marine ecosystem: The current state of knowledge. *Marine Policy, 104*, 163–176.

Telesetsky, A. (2014). Laundering fish in the global undercurrents: Illegal, unreported, and unregulated fishing and transnational organized crime. *Ecology LQ, 41*, 939.

Thomas, J., Thomas, C. J., Radcliffe, J., & Itsiopoulos, C. (2015). Omega-3 fatty acids in early prevention of inflammatory neurodegenerative disease: A focus on Alzheimer's disease. *BioMed Research International, 2015*.

Townhill, B. L., Radford, Z., Pecl, G., van Putten, I., Pinnegar, J. K., & Hyder, K. (2019). Marine recreational fishing and the implications of climate change. *Fish and Fisheries, 20*(5), 977–992.

Vega, O. M., & Cepeda, C. (2021). Converging evidence in support of omega-3 polyunsaturated fatty acids as a potential therapy for Huntington's disease symptoms. *Reviews in the Neurosciences, 32*(8), 871–886.

Venugopal, V., & Gopakumar, K. (2017). Shellfish: Nutritive value, health benefits, and consumer safety. *Comprehensive Reviews in Food Science and Food Safety, 16*(6), 1219–1242.

von Schacky, C. (2020). Omega 3 index in 2018/19. *Proceedings of the Nutrition Society, 79*(4), 381–387.

Waite, R., Beveridge, M., Brummett, R., Castine, S., Chaiyawannakarn, N., Kaushik, S., Mungkung, R., Nawapakpilai, S. U. P. A. W. A. T., & Phillips, M. I. C. H. A. E. L. (2014). *Improving productivity and environmental performance of aquaculture*. WorldFish.

Watanabe, F. (2007). Vitamin B12 sources and bioavailability. *Experimental Biology and Medicine, 232*(10), 1266–1274.

Weinert, E. J., Miller, S. A., Ikeda, D. M., Chang, K. S., McGinn, J. M., & DuPonte, M. W. (2014). Natural farming: Fish amino acid. *Sustainable Agriculture, SA-12*. University of Hawaiâ€™ i, College of Tropical Agriculture and Human Resources.

Willer, D. F., Brian, J. I., Derrick, C. J., Hicks, M., Pacay, A., McCarthy, A. H., Benbow, S., Brooks, H., Hazin, C., Mukherjee, N., & McOwen, C. J. (2022). 'Destructive fishing'—A ubiquitously used but vague term? Usage and impacts across academic research, media and policy. *Fish and Fisheries, 23*(5), 1039–1054.

Williamson, P., & Guinder, V. A. (2021). Effect of climate change on marine ecosystems. *The Impacts of Climate Change*, 115–176.

Yadav, N., Mudgal, D., Anand, R., Jindal, S., & Mishra, V. (2022). Recent development in nanoencapsulation and delivery of natural bioactives through chitosan scaffolds for various biological applications. *International Journal of Biological Macromolecules, 220*, 537.

Zeng, J., Yu, W., Dong, X., Zhao, S., Wang, Z., Liu, Y., Wong, M. S., & Wang, Y. (2019). A nanoencapsulation suspension biomimetic of milk structure for enhanced maternal and fetal absorptions of DHA to improve early brain development. *Nanomedicine: Nanotechnology, Biology and Medicine, 15*(1), 119–128.

Zhou, L., Xiong, J. Y., Chai, Y. Q., Huang, L., Tang, Z. Y., Zhang, X. F., Liu, B., & Zhang, J. T. (2022). Possible antidepressant mechanisms of omega-3 polyunsaturated fatty acids acting on the central nervous system. *Frontiers in Psychiatry, 13*, 933704.

Zhubi-Bakija, F., Bajraktari, G., Bytyçi, I., Mikhailidis, D. P., Henein, M. Y., Latkovskis, G., Rexhaj, Z., Zhubi, E., Banach, M., Alnouri, F., & Amar, F. (2021). The impact of type of dietary protein, animal versus vegetable, in modifying cardiometabolic risk factors: A position paper from the international lipid expert panel (ILEP). *Clinical Nutrition, 40*(1), 255–276.

Chitosan and Nanotechnology Fundamentals

Oluwadurotimi Samuel Aworunse, Franklyn Nonso Iheagwam, Praise Tomiwa Agbetuyi-Tayo, Ogochukwu Onwaeze, Micheal Bolarinwa Fabiyi, and Samuel Akpoyovware Ejoh

Contents

O. S. Aworunse (✉) · O. Onwaeze · S. A. Ejoh
Department of Biological Sciences, Covenant University, Ota, Nigeria
e-mail: oluwadurotimi.aworunse@covenantuniversity.edu.ng

F. N. Iheagwam
Department of Biochemistry, Covenant University, Ota, Nigeria

Covenant University Public Health and Wellness Research Cluster, Ota, Nigeria

P. T. Agbetuyi-Tayo
Department of Biochemistry, Covenant University, Ota, Nigeria

Covenant Applied Informatics and Communication Africa Centre of Excellence, Ota, Nigeria

M. B. Fabiyi
Universidade Federal do Pará Belem, Belém, Brazil

P. O. Isibor et al. (eds.), *Nanochitosan-Based Enhancement of Fisheries and Aquaculture*, https://doi.org/10.1007/978-3-031-52261-1_2

1 Introduction

Chitosan is a mucopolysaccharide formed from a linear polymer of 1,4-glycosidically linked glucosamine (2-amino-2-deoxy-D-glucopyranose). It is derived from chitin, an aminopolysaccharide polymer that occurs naturally as a building material that gives rigidity and support to insects and shrimps (Aromán-Doval et al., 2023). Chitosan and chitin are important support structures for many organisms and are important renewable macromolecular biomass resources. However, they are different from one another. A substance is categorised as chitosan if the acetyl glucosamines concentration is lower than 50% and chitin if the concentration is 50% or more (Zuma et al., 2015). After cellulose, chitin (1,4-N-acetyl-D- glucosamine) is the most abundant natural biopolymer with at least 1010 tonnes existing in the biosphere. Also, chitin is a component of the exoskeleton of crustaceans, insects and the cell walls of fungi (Jayakumar et al., 2011). The importance of chitin as a source of carbon and nitrogen for marine species and its effects on the marine ecosystem has recently come to light. The primary resources used for the industrial production of chitosan and chitin are marine crustaceans, shrimp, crab and squid bone plates (Rinaudo, 2006). The usefulness of chitosan stems from its inherent qualities of biocompatibility, biodegradability and flexibility; at the nanoscale, chitosan exhibits enhanced functionality when compared to its bulk scale, making it suitable for a wide range of applications as depicted in Fig. 1.

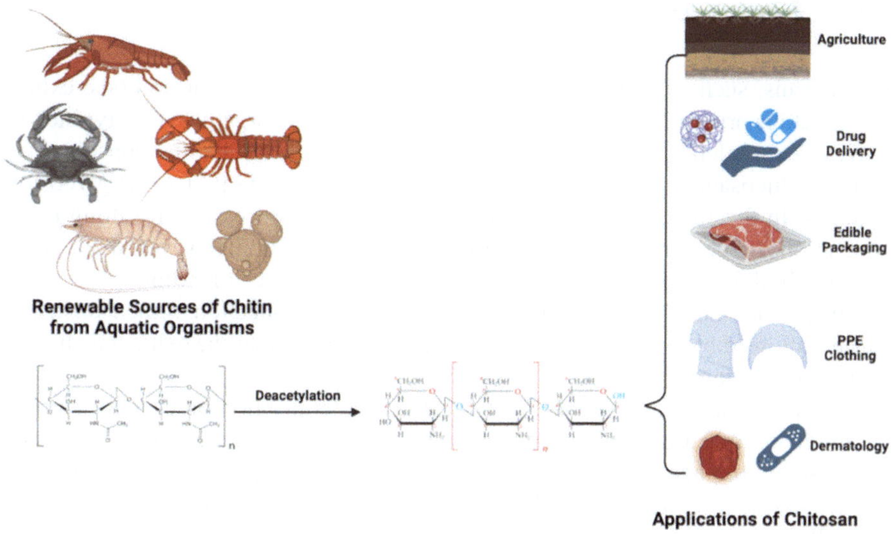

Fig. 1 Overview of chitosan sources and applications

2 Sources of Chitosan

Chitosan is a highly nitrogenous biopolymer with a broad range of applications; it is derived from chitin, a homo polysaccharide consisting of repeated units of N-acetyl-D-glucosamine residues that are held together by β (1–4) linkage. It is predominantly obtained from natural sources, such as the exoskeletons of crustaceans and the fungal cell walls. Chitin can be found in several organisms, alongside other macromolecules. However, it is important to note that chitin or chitosan are not present in the structure of higher animals and higher plants. Shrimp and crab shell wastes have been extensively utilised as key feedstock in industrial settings for the large-scale production of chitin and chitosan. Marine organisms consist of approximately 20–30% chitin, 30–40% proteins, 0–14% lipids and 30–50% minerals (Pellis et al., 2022). In contrast, chitin is predominantly present in the cellular walls and septa of several fungal taxa, including ascomycetes, zygomycetes, basidiomycetes and deuteromycetes. The class of fungus known as zygomycetes is characterised by the presence of significant amounts of chitosan, in addition to chitin, within their cell walls. The composition of the fungal cell wall consists of around 10–20% chitin, 50–60% glucans and 20–30% glycoproteins and smaller quantities of lipids, pigments and inorganic salts (Alemu et al., 2023).

Chitin occurs in three crystalline forms: α-, β- and γ-chitin, with varying physicochemical properties based on hydration, cell size and chain count. These polymorphs differ in how chains of the crystalline regions are reciprocally organised. The α-form is antiparallel, the β-form is parallel, and the γ-form alternates between two parallel and one antiparallel strand. α-Chitin, found in crustaceans and fungi, is the most abundant and easily extracted form of chitin (Namboodiri & Pakshirajan, 2020).

2.1 Chitosan Derived from Crustacean Shells

Crustaceans, such as shrimp, crabs, lobsters and other marine arthropods, exhibit exoskeletons primarily composed of chitin. These aquatic organisms are typified by a rigid exoskeletal structure. Chitin is a linear polysaccharide consisting of N-acetylglucosamine (GlcNAc) units that are connected via β (1 → 4) glycosidic linkages. In the seafood industry, crustaceans undergo processing, resulting in the generation of shells that are commonly discarded as waste. However, these shells possess significant value as they serve as an excellent source of chitin.

The production of chitin from crabs begins with the collection of the exoskeletal waste, followed by a meticulous cleansing step to ensure that the extracted chitin is of the highest possible purity.

Following the cleansing phase, a targeted demineralisation process is deployed to eliminate mineral deposits, particularly calcium carbonate. This step holds immense importance as excessive amounts of minerals can potentially disrupt chitin extraction downstream. Afterwards, any residual protein content within the shells is carefully removed through enzymatic or alkaline methods. This stage is of paramount significance, as the presence of proteins can hinder subsequent deacetylation process (Alemu et al., 2023).

Deacetylation, a fundamental process that removes acetyl groups from chitin is performed by subjecting the chitin to an alkaline treatment using sodium hydroxide (NaOH), under carefully controlled temperature conditions. In this transformative step, acetyl groups are enzymatically cleaved from the GlcNAc moieties within the chitin molecule, effectively converting it into chitosan (Pellis et al., 2022). The degree of deacetylation (which represents the extent of acetyl group elimination), is meticulously regulated, bestowing upon chitosan diverse characteristics.

Following the deacetylation process, the chitosan-containing solution undergoes a series of steps involving neutralisation, filtration and desiccation, resulting in the final chitosan product. It is noteworthy that the choice of drying method profoundly influences the physical attributes of chitosan such as particle size and morphology (Danarto & Distantina, 2016).

2.2 Chitosan Derived from Fungal Sources

The application of fungi in various biotechnology industries such as baking, brewing, antibiotics, organic acid, and enzyme manufacturing, generates enormous fungal biomass wastes. These wastes, along with those from the mushroom industry, can serve as valuable feedstock for the extraction of chitinous polysaccharides. Leveraging fungal cell walls for chitosan extraction has emerged as a viable alternative, presenting an opportunity to advance Sustainable Development Goal 14 (Life Below Water) by alleviating the pressure on marine sources (Aranaz et al., 2021).

Various ascomycetes, zygomycetes, basidiomycetes and deuteromycetes have chitin in their cell walls, making them potential chitosan producers. Due to their high chitosan concentration, zygomycetes are the most promising. Chitosan's physicochemical properties can be standardised by manipulating the growth conditions (Pochanavanich & Suntornsuk, 2002). Mucorales species such as *Cunninghamella*, *Rhizomucor*, *Gongronella*, *Mucor*, *Absidia* and *Rhizopus* have been studied for chitosan production (Ramos Berger et al., 2018). By using the chitin deacetylase they produce, these fungi convert chitin to chitosan. The fungi that produce chitosan are grown in a controlled setting. Chitosan is made from chitin-rich cell walls of fungal biomass recovered during culture. Chitin extraction, sodium hydroxide deacetylation, neutralisation, filtration and drying produce chitosan. This sustainable strategy supports Sustainable Development Goal 14 by minimising marine chitosan extraction and fostering responsible consumption and manufacture.

3 Properties and Characteristics of Chitosan

Chitosan possesses a wide range of properties that makes it a useful and versatile material in several industries, including pharmaceuticals, agriculture, food and biotechnology. The unique attributes of chitosan are derived from its specific chemical composition and interaction with other molecules.

3.1 Structure

Chitosan, the second most abundant biopolymer after cellulose, exhibits a structural similarity to cellulose, but with the substitution of the hydroxyl group at position C-2 by an amino group (Fig. 2). Chitosan possesses a positive ionic charge, a unique characteristic that sets it apart from cellulose. This attribute allows chitosan to effectively interact with molecules that bear a negative charge, including proteins, lipids, fats and ions. The biological identity of chitosan is closely related to its chemical structure. The polysaccharide exhibits a linear configuration consisting of N-acetylglucosamine (GlcNAc) and glucosamine (GlcN) units, which are linked together by $\beta(1 \rightarrow 4)$ glycosidic bonds. The deacetylation process is a crucial stage that distinguishes chitosan from chitin, leading to a higher concentration of glucosamine residues in chitosan. The structural alteration of the material has an impact on its bioactive properties (Alemu et al., 2023).

3.2 Solubility

Chitosan dissolves in acid but is completely insoluble in neutral or alkaline solvents. Chitin is insoluble in solvents, but deacetylation produces a soluble chitosan with primary amino groups and a pKa of 6.5. In acidic solutions, the amine becomes

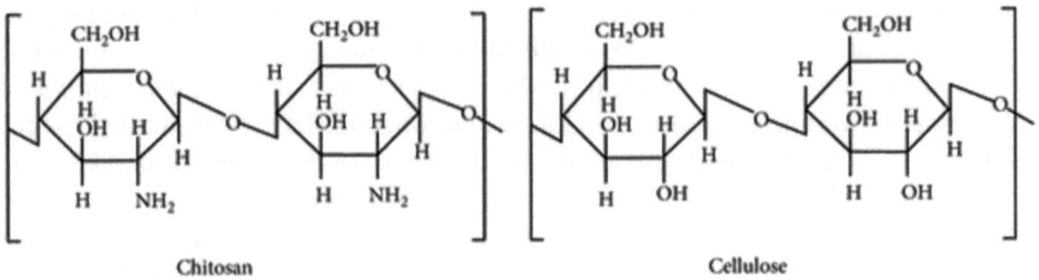

Fig. 2 Schematic diagram of chemical structures of cellulose and chitosan

protonated and positively charged, making chitosan soluble. At pH 6 or above, chitosan loses its charge and becomes insoluble. The solubility of chitosan depends on pH, molecular weight degree of deacetylation, temperature and polymer crystallinity (Vidal et al., 2021).

3.3 Amino Group and Reactivity

During the deacetylation process of chitin to form chitosan, the presence of an increased number of amino groups (-NH2) in chitosan makes it highly reactive. These amino groups enable chitosan functionalisation and derivatisation. Chemical modification can provide chitosan with specialised properties for certain uses. For instance, to vary solubility, stability or charge, amino groups can be acylated, alkylated or connected to functional groups. This chemical structural fine-tuning confers on chitosan versatility for applications in many fields. Their reactivity allows strong bonding, forming hydrogels, beads and membranes for drug delivery and tissue engineering. Chitosan's plasticity and adaptability arise from the amino group reactivity (Piekarska et al., 2023; Khan & Alamry, 2021).

3.4 Antibacterial Property

Chitosan exhibits antibacterial properties, making it a good alternative to commonly used antibiotics in aquaculture. Chitosan is polycationic molecule, with positively charged amino groups at physiological pH, that enables the formation of electrostatic interactions with phosphate and carboxylate groups in bacterial cell membranes, which are negatively charged (Yan et al., 2021). This interaction leads to the damage of the membrane. Antibiotics can cause antibiotic resistance and environmental damage, whereas chitosan is more sustainable (Yilmaz Atay, 2019).

3.5 Decomposition

Enzymatic decomposition of chitosan by bacteria in natural environments produces harmless degradation products, making it ideal for ecologically safe operations. This fits the sustainability pattern, finding use in bioplastics, agriculture and wastewater treatment (Gohil et al., 2021). Chitosan's low cytotoxicity and biocompatibility makes it an ideal bioengineered component. Drug delivery devices, tissue scaffolds and wound dressings benefit from their biocompatibility with living systems, including humans (Desai et al., 2023). Controlled deacetylation confers variability in molecular weight, thereby contributing to the biochemical plasticity of chitosan. This adaptability allows biochemical properties to be customised, with greater molecular weight variants for drug administration and lower molecular weight ones for water purification (Piekarska et al., 2023).

4 Nanotechnology Basics: Nanoparticles, Nanomaterials and Nanostructured Materials

4.1 Nanoparticles

Nanoparticles, as defined by ISO (2010), encompass units or entities with all nanoscale external dimensions, i.e., 1–100 nm. They exhibit distinct physical and chemical characteristics as a result of their considerable surface area and diminutive dimensions at the nanoscale. Various classifications can be assigned to them, depending on their distinct qualities, forms or sizes. The optical properties of nanoparticles are said to be size-dependent, resulting in various colours due to absorption within the visible spectrum. The reactivity, hardness and other qualities of nanoparticles are contingent upon their distinct size, shape and structure (Khan et al., 2019).

5 Nanoparticle Classification Based on the Nature of Particles

5.1 Organic Nanoparticles

Organic nanoparticles (ONPs) are made from 100 nm and smaller organic molecules. Ferritin, nanochitosan, micelles, dendrimers and liposomes are familiar organic nanoparticles or polymers. Biodegradable and non-toxic micelles and liposomes are nanocapsules with a hollow interior that are sensitive to heat and electromagnetic radiation. Nanochitosan; an organic nanoparticle derived from chitosan, shows remarkable biodegradability and biocompatibility (Ijaz et al., 2020). Similar to micelles and liposomes, nanochitosan offers advantages for pharmaceutical delivery as it possesses a unique hollow core/shell structure for the encapsulation of drugs (Mikušová & Mikuš, 2021).

5.2 Inorganic Nanoparticles

Inorganic nanoparticles are particles that lack carbon. They are typically composed of metals or metal oxides. The inorganic nanoparticles include metallic nanoparticles and metal oxide nanoparticles.

5.2.1 Metallic Nanoparticles

Nanoparticles of most metals can be synthesised using destructive or constructive methods. Most metal-based nanoparticles are synthesised using aluminium (Al), cadmium (Cd), cobalt (Co), copper (Cu), gold (Au), iron (Fe), lead (Pb), silver (Ag) and zinc (Zn) (Mekuye & Abera, 2023). Quantum effects and a high surface-to-volume ratio give metal nanoparticles excellent ultraviolet-visible sensitivity, electrical, catalytic, thermal and antibacterial properties (Mekuye & Abera, 2023).

5.2.2 Metal Oxide Nanoparticles

Researchers have become interested in metal oxides in recent decades. Metal oxides are formed from positive metallic and negative oxygen ions. Strong and persistent ionic connections result from electrostatic interactions between positive metal and negative oxygen ions. When exposed to oxygen at normal temperature, iron nanoparticles (Fe) transform to iron oxide (Fe_2O_3), which is far more reactive than iron nanoparticles. Synthesised oxide-based nanoparticles change metal-based characteristics. Metal oxide nanoparticles are manufactured for their increased reactivity and efficiency. Silicon dioxide, titanium oxide, zinc oxide and aluminium oxide are often synthesised oxides (Fontana et al., 2022).

5.3 Carbon Nanoparticles

When carbon is mixed with other materials, it forms bonds that are unrivalled in strength. Due to their odd shape and different characteristics, they are used in many industries. Carbon nanomaterials can store and produce energy, cleanse water and wastewater, and be used biologically. The most common carbon-based nanoparticles are fullerenes and carbon nanotubes (CNTs). Fullerenes are hollow spherical cages. Their high electrical conductivity, structural strength, electron affinity and adaptability makes them commercially attractive (Altammar, 2023).

6 Nanoparticle Classification Based on the Dimensionality

Based on dimensionality, nanoparticles are mostly one-dimensional particles. Thin films or surface coatings are commonly considered materials with nanoscale dimensions in 1D. Thin films have been extensively researched and applied in many disciplines such as electronics, information storage systems, chemical and biological sensors, fibre-optic systems, as well as magneto-optic and optical devices for several decades. Thin films can be deposited using a variety of techniques and can be developed with precise control at the atomic scale (monolayer level) (Jeevanandam et al., 2018).

6.1 Two-Dimensional Nanoparticles

Two-dimension nanoparticles have two dimensions that are measured on the nanometre scale. Examples of such materials encompass nanotubes, dendrimers, nanowires, fibres and fibrils. Particles that possess a significant aspect ratio and have diameters within the nanoscale range are also classified as 2D nanomaterials. The understanding of the properties of two-dimensional (2D) systems is comparatively limited, and their manufacturing capabilities are not as advanced (Afolalu et al., 2019).

6.2 Three-Dimensional Nanoparticles

Three dimensional nanomaterials encompass materials that possess nanoscale dimensions in all three spatial directions. These include quantum dots or nanocrystals, fullerenes, particles, precipitates and colloids. Certain three-dimensional (3D) systems, such as natural nanomaterials and combustion products, metallic oxides, carbon black, titanium oxide (TiO_2) and zinc oxide (ZnO), have been extensively studied and are widely recognised. However, there are other 3D systems, including fullerenes, dendrimers and quantum dots, which present significant difficulties in both production and comprehension of properties (Afolalu et al., 2019).

7 Nanomaterials

Nanomaterials are defined as materials with at least one of their dimensions is in the nanoscale, i.e., smaller than 100 nm (Baig et al., 2021). Based on their dimensionalities, nanomaterials are placed into four different classes, summarised in Fig. 3.

7.1 Zero-Dimensional Nanomaterials (0-D)

The nanomaterials in this class have all three dimensions in the nanoscale range. Examples are quantum dots, fullerenes and nanoparticles.

7.2 One-Dimensional Nanomaterials (1-D)

This group of nanomaterials have one dimension outside the nanoscale. Examples are nanotubes, nanofibers, nanorods, nanowires and nanohorns.

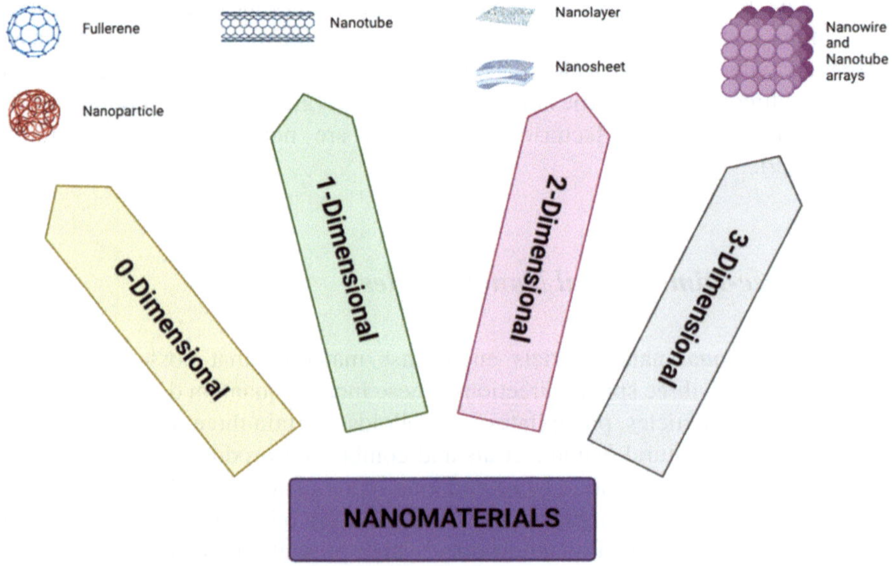

Fig. 3 Classification of nanomaterials based on dimensionality

7.3 Two-Dimensional Nanomaterials (2-D)

The nanomaterials in this category have two dimensions outside the nanoscale. Examples are nanosheets, nanofilms and nanolayers.

7.4 Three-Dimensional Nanomaterials (3-D) or Bulk Nanomaterials

In this class, the materials are not confined to the nanoscale in any dimension. This class contains bulk powders, dispersions of nanoparticles, arrays of nanowires and nanotubes, and others (Baig et al., 2021).

8 Typical Synthesis Method of Nanomaterials

Traditional physical, chemical and biological syntheses have been modified for the production of nanoparticles and nanomaterials (Katti & Sharon, 2019). These methods have been empirically tested extensively, and their several benefits and drawbacks may vary depending on the context of nanomaterial production being undertaken (Salem et al., 2022). Regardless of the type, they may all assume one of two unique approaches (Fig. 4): a top-down or bottom-up orientation to nanoparticle formation (Barhoum et al., 2022). The former involves reducing macroscale materials into their constituent nanoparticle clusters, while the latter involves building up into nanoparticle clusters from respective atoms (Sharon, 2019). Typically, the top-down approach is poorly suited for making evenly structured products, and even when using substantial power, it is especially difficult to generate extremely tiny particles (Khan et al., 2022). The flaws of the bulk materials are likely to have an enormous effect on the physical and chemical composition of the products (Singh et al., 2020).

8.1 Top-Down Syntheses

In this approach, a destructive method is utilised. The process begins with a larger molecule that undergoes decomposition into smaller units, which are subsequently transformed into appropriate nanoparticles. Various procedures, including grinding/milling, chemical vapour deposition (CVD), physical vapour deposition (PVD) and other decomposition methods, have been documented as examples of this particular technology (Khan et al., 2019). Top-down nanoparticle synthesis uses milling, lithography and repeated quenching. These methods use controlled

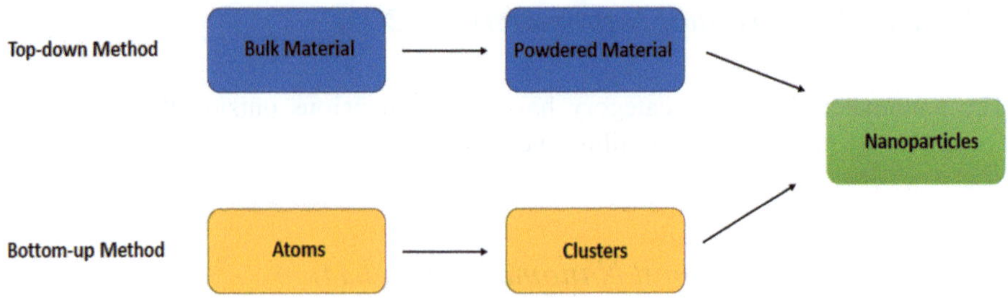

Fig. 4 Top-down and bottom-up synthesis method of nanomaterials

mechanical or thermal processes to reduce larger structures or materials to nanoparticles. The milling involves grinding or crushing bulk material into smaller particles. However, lithography precisely modifies or removes material from a larger substrate to create a nanopattern or structure. Repeated quenching causes phase changes or structural modifications and nanoparticle formation by rapidly cooling a material (Cele, 2020).

A major drawback with the top-down method is the difficulty of controlling particle size and structure. Variations in milling conditions and material properties can make it difficult to obtain uniform nanoparticles with specific dimensions, and even when using substantial power, it is especially difficult to generate extremely tiny particles (Khan et al., 2022). The flaws of the bulk materials are likely to have an enormous effect on the physical and chemical composition of the products (Singh et al., 2020). Moreover, replicating intricate patterns using nano-scale lithography can be problematic. Repeated quenching can also cause particle size and structure variations, making homogeneity difficult (Bello et al., 2015).

8.2 Bottom-Up Approach

The bottom-up approach is widely employed in nanoparticle synthesis because it meticulously builds materials atom-by-atom, molecule-by-molecule and cluster-by-cluster. This differs from the top-down method of breaking down complexes to obtain nanoparticles. Starting at the molecular or atomic level, the bottom-up approach gradually assembles these tiny units into the desired nanoparticle structure (Cele, 2020). This method allows precise nanoparticle creation, enabling the design of materials with specific sizes, shapes, compositions and surface properties (Cele, 2020). The bottom-up approach is ideal for generating uniformly structured nanoparticles allowing adequate freedom for controlling physical and chemical properties (Barhoum et al., 2022). Although useful, it is resource inefficient and therefore unideal for mass production (Salem et al., 2022). The top-down nanoparticle formation involves processes like thermal evaporation, ball milling, sputtering and laser ablation, whereas the bottom-up approaches include hydrothermal production, combustion, co-precipitation and So-Gel techniques (Table 1) (Singh et al., 2020).

Table 1 Nanoparticle synthesis approaches and processes

Process	Type	Approach	Benefits	Drawbacks
Ball milling	Physical	Top-down	Resource efficient and scalable (El-Eskandarany et al., 2021) Simple, systematic process (Elkhatib et al., 2015) Depending on the size and velocity of the mill, it can produce about 2–200 nm sized products (Baig et al., 2021)	Nanostructured materials produced are typically irregularly shaped (Baig et al., 2021) Products are easily contaminated in the process (Abid et al., 2022)
Melt mixing	Physical	Top-down	Cost efficient and scalable (El-Eskandarany et al., 2021). Simple process with wide applicability (Elkhatib et al., 2015) Creates highly homogeneous nanoproducts (Abid et al., 2022)	Products are highly reactive and prone to agglomeration (Taki and Sharon, 2019) Most types are energy demanding (Andrade-Guel et al., 2022) Due to the required temperature, only nanoproducts that are resistant to thermal degradation may be used (Abid et al., 2022)
Colloidal technique	Chemical	Bottom-up	Allows for the control of nanoproduct properties (Natsuki et al., 2015) Large output size (Quinson et al., 2021) Creates highly homogeneous nanoproducts (Jamkhande et al., 2019) Wide applicability (Quinson et al., 2021)	Resource inefficient and limited in scalability (Natsuki et al., 2015) Nanoproducts are highly reactive and prone to agglomeration (Natsuki et al., 2015) Potential to generate environmentally unfriendly bye-products (Quinson et al., 2021)
Hydrothermal synthesis	Physical	Bottom-up	Allows for the control of nanoproduct properties in the chemical reactions (Gan et al., 2020) Environmentally friendly and sustainable process (Darr et al., 2017) Nanoproducts are often pure, enhanced and homogeneous (Gan et al., 2020)	Nanoproducts are unstable at high temperatures, which limits its applicability (Darr et al., 2017) Complex process with lowered applicability (Jamkhande et al., 2019) Resource inefficient and limited in scalability (Darr et al., 2017)

(continued)

Table 1 (continued)

Process	Type	Approach	Benefits	Drawbacks
Lithography	Physical	Top-down	Allows for the control of nanoproduct properties (Sekhri et al., 2023) Nanoproducts are often pure, enhanced and homogeneous (Mukasyan & Manukyan, 2015) High consistency, which makes it ideal for academic and commercial uses requiring uniformity in nanoparticle qualities (Sekhri et al., 2023)	Complex process with lowered applicability (Jamkhande et al., 2019) Resource inefficient and limited in scalability (Sekhri et al., 2023) Raw materials are difficult and the technology is not readily accessible (Sekhri et al., 2023) Limited in the size of Nanoproducts it can produce (Sekhri et al., 2023)
Combustion	Physical	Top-down	Fast and efficient, with low energy requirements which makes it ideal for academic uses (Mukasyan et al., 2015) Simple process with wide applicability (Jamkhande et al., 2019). Cost Efficient and optimisable (Jamkhande et al., 2019)	Nanoproducts are highly reactive and prone to agglomeration (Rahinov et al., 2020) Limited allowance for the control of nanoproduct properties (Mukasyan et al., 2015). Technique requires extreme environmental conditions that are potentially hazardous and limits the labour force (Rahinov et al., 2020)
Sputtering	Physical	Top-down	Environmental conditions are tightly controlled; hence, it allows for precise development of Nanoproducts (Sekhri et al., 2023) Wide applicability (Jamkhande et al., 2019) Particularly useful for thin film coatings, microfabrication and etching (Zhao et al., 2021)	Limited allowance for control of nanoproduct size and properties (Abid et al., 2022) The process is slow and energy-intensive (Rane et al., 2018) Output is considerably less than other Synthesis techniques (Zhao et al., 2021) Cost inefficient and poor scalability (Jamkhande et al., 2019)

(continued)

Table 1 (continued)

Process	Type	Approach	Benefits	Drawbacks
Laser ablation	Physical	Top-down	Has a minimal environmental impact (Rashid et al., 2021) Widely applicable in producing many types of Nanoproducts (Elkhatib et al., 2015) Allows for the control of nanoproduct size (Abid et al., 2022)	Process is slow (Abid et al., 2022) with low replicability (Jamkhande et al., 2019) Resource inefficient and limited in scalability (Sekhri et al., 2023) Nanoproducts are not typically contamination-free (Rane et al., 2018)
Sol-gel technique	Chemical	Bottom-up	Products are protected against deoxygenation and corrosion (Sekhri et al., 2023) Widely applicable in producing many types of Nanoproducts (Jamkhande et al., 2019) Nanoproducts are often pure, enhanced and homogeneous (Rahinov et al., 2020)	Multiple factors contribute to nanoproduct properties; hence, allowance for controlling the outcome is limited (Zhao et al., 2021). Nanoproducts are often unstable, impure, highly reactive and prone to agglomeration (Abid et al., 2022)
Biosynthesis	Biological	Top-down	A green synthesis technique; hence, it presents the most environmentally friendly nanomaterial production process (Koul et al., 2021) Products are highly biocompatible and widely applicable (Nguyen et al., 2022) Process is cost efficient (Koul et al., 2021) Owing to biological diversity of producing organisms, nanoproducts are multi-functional, typically with novel applications (Nguyen et al., 2022)	Limited allowance for the control of nanoproduct properties (Mukasyan et al., 2015) Processes may be complex, slow and pose significant challenges for commercial upscaling (Nguyen et al., 2022) Products are prone to contamination (Koul et al., 2021)

9 Nanostructured Systems

Nanostructured systems encompass materials or technologies that exhibit a distinct structural arrangement at the nanoscale. The term 'nanostructured' denotes the deliberate arrangement or configuration of materials at the nanoscale to attain specific qualities and capabilities. These structures can take diverse forms, including nanoparticles, nanocomposites, nanofibers, nanowires, quantum dots and other variations. Nanostructures may occur naturally or be deliberately designed and constructed (Jeevanandam et al., 2018). The behaviour and characteristics of materials at the nanoscale frequently exhibit notable disparities compared to their larger scale counterparts. These disparities arise from various factors, including quantum effects, augmented surface area, modified surface reactivity, enhanced mechanical capabilities, and better electrical, optical and magnetic attributes. The process of nanostructuring confers distinct properties upon materials, which can be effectively utilised in a wide range of applications spanning multiple disciplines (Mekuye & Abera, 2023).

Several examples of nanostructured materials have been deployed in fisheries and aquaculture to improve various aspects of the industry and promote sustainable aquaculture practices. Nanoencapsulation of nutrients and drugs is a candid example, and it involves incorporating nutrients, vitamins or drugs into nanoscale capsules or particles. In aquaculture, essential nutrients or drugs can be nanoencapsulated and mixed with fish feed, allowing for better absorption, targeted delivery and improved feed efficiency, ultimately enhancing the growth and health of aquatic organisms (Muhammad Mudassar Shahzad, 2022). Examples of such are nanostructured carriers, like lipid nanoparticles or polymeric nanoparticles. These nanostructures made from biodegradable and biocompatible polymers, such as chitosan, poly (lactic-co-glycolic acid) (PLGA), polyethylene glycol (PEG) and polylactic acid (PLA), are commonly used for polymeric nanoparticle synthesis (Perinelli et al., 2019). Nanochitosan is a highly promising nanomaterial due to its inherent biocompatibility, non-toxicity and biodegradability. This suggests that nanochitosan has potential biomedical applications, such as drug administration and the reconstruction of biological tissues (Kravanja et al., 2019). The production of nanochitosan specifically designed for aquaculture applications entails the precise generation of chitosan particles at the nanometre scale, offering a flexible solution for various purposes within the aquaculture industry.

10 Nanotechnology in Aquaculture

In recent years, aquaculture has received significant attention owing to its potential to increase access to affordable sources of protein, healthy fat and essential micronutrients in developing countries and to support food security amidst a burgeoning global population (Igwegbe et al., 2021). According to Fajardo et al. (2022), the aquaculture sector employs about 20.5 million people globally. In addition to bolstering rural employment and livelihood (Sarkar et al., 2022), the sector contributes to the GDP of economies around the world through the generation of revenues (Ogunfowora et al., 2021; Sarkar et al., 2022). For instance, China since 2002, has maintained the top spot as the world's largest producer of fish and fishery products, generating a total revenue of USD 21.7 billion from exports alone (Guggisberg,

2022). Similarly, Brazil with an annual output of 563,000 tonnes is the 14th largest exporter of fish products, earning USD 1.6 billion in foreign revenue in 2018 (Pauly & Zeller, 2017; Coldebella et al., 2017). Direct farm sales of aquaculture products were valued at an estimated USD 281.5 billion in 2020, an increase of USD 18.5 billion from 2018 and USD 6.7 billion from 2019. In 2020, the aquaculture production encompasses 35.1 million tonnes of algae for both non-food and food uses, 87.5 million tonnes of aquatic animals largely used as food by humans, 700 tonnes of pearls and shells for ornamental use, amounting to an overall 122.6 million tonnes in live weight (FAO, 2022). While aquaculture represents a major food production system with notable economic impacts, its sustainability is threatened by the problems of disease outbreak, environmental contamination, ineffective diagnostic and therapeutic tools, and inefficient feed utilisation (Shah & Mraz, 2020; Nasr-Eldahan et al., 2021; Sarkar et al., 2022).

The exploitation of nanotechnology to transform the aquaculture and seafood industry is attracting huge interest, with well over a thousand products comprising nanomaterials presently in the market (Fajardo et al., 2022). Various categories of these nanotechnology-based systems have been developed by leveraging the properties of nanoparticles such as small size, antimicrobial activity, high adsorption and bioavailability, large surface area, better solubility and dispersion, high target activity, controlled release dynamics and improved stability (Guo et al., 2013; Fajardo et al., 2022; Khan & Hossain, 2022; Su et al., 2022). Current applications of nanotechnology in aquaculture to enhance sustainability, efficiency and production (Fajardo et al., 2022) include but not limited to fish packaging, drug delivery, pathogen detection, water treatment and purification, dietary supplements and nutraceuticals delivery, fish breeding and fish vaccination.

10.1 Fish Packaging

The perishable nature of fresh fish is a major challenge. Therefore, any packaging solution that can extend shelf life, while maintaining the nutritional integrity of fish products, is desirable (Selvaraj et al., 2014). Nanopackaging made from natural nanoscale polymers like starch, cellulose and chitosan particles is used to strengthen packaging to reduce the incidence of bruising or mechanical damage to packed fish fillets (De Azeredo, 2009; Handy, 2012; Selvaraj et al., 2014).

Chitosan nanocomposites have been employed for the fabrication and strengthening of edible films of packaging to extend shelf life, reduce the deterioration of fish meat and retain fish flavour by reducing the formation of oxidation products and volatile bases during cold storage (De Moura et al., 2008; Yu et al., 2018; Ahmed et al., 2019). Bionanocomposites comprising a biopolymer matrix such as nanochitosan reinforced with low fractions of nanoparticles and fish gelatin have been demonstrated to improve the barrier properties, mechanical strength and heat resistance of fish packaging materials compared to regular micro- or macroscale composites or pristine biopolymers due to their high surface area and aspect ratio (Rhim & Kim,

2014). In addition to preserving the fish, the bionanocomposites reduce packaging weight, as less material is required to achieve superior barrier properties, thus reducing packaging cost and generating minimal waste (Hosseini & Gómez-Guillén, 2018).

Active packing consisting of materials capable of releasing substances to or absorbing and scavenging substances from food to retain quality or delay degradation has been developed. This packaging technology involves the integration of active compounds with antimicrobial or antioxidant activities into a polymeric coating, matrix or in pads, sachets or labels. Active packaging usually contains antioxidants (e.g. vitamin C, vitamin E, butylated hydroxytoluene), antimicrobials (e.g. chitosan, phenolic compounds, peptides, essential oils), absorbers or scavengers (e.g. carbon dioxide emitters or absorbers, oxygen scavengers, ethylene adsorbers or absorbers and moisture control agents) (Rodrigues et al., 2021). Nanochitosan-based active packaging films with enhanced antioxidants and antimicrobial activities have been well-documented (Homayounpour et al., 2021; Kumar et al., 2020; Sadadekar et al., 2023; Jiang et al., 2023). In addition to inhibiting the growth of food-borne pathogens, active nanopackages can also act as thermal insulators, thus maintaining the quality and prolonging the shelf life of the stored fish (De Azeredo, 2009; Anvar et al., 2021).

The past decade has witnessed the development of biosensors for detecting different harmful substances in foods. This sensitivity has been significantly enhanced through the application of nanomaterials that exhibit high mechanical flexibility, conductivity, surface functionalisation, surface area and biocompatibility (Mohammadpour & Naghib, 2021). Nanosensors transformed into films and embedded within flexible packaging (Sundramoorthy et al., 2018) have been used to develop intelligent packaging (a smart packaging technology that exploits internal molecules or external conditions of the packed fish for real-time monitoring of quality and sensing of microorganisms that can cause spoilage or disease at different stages along the supply chain) (Vanderroost et al., 2014; Drago et al., 2020). Nanosensors offer selective and sensitive platforms for detecting deteriorative markers (Mohammadpour & Naghib, 2021) in packed aquaculture products. For instance, chitosan nanoparticles used for the detection of volatile nitrogen compounds through the sensing of pH change in Salmon (Rodrigues et al., 2021). Intelligent packing makes use of nanosensors categorised as time-temperature indicators, freshness indicators, optical oxygen sensors, moisture indicators, toxins indicators, pH contaminants indicators, optochemical CO_2 indicators, and spoilage and pathogens indicators (Alfei et al., 2020; Anvar et al., 2021).

10.2 Drug Delivery

Disease outbreaks occasioned by the impact of climate change and deteriorating environmental quality are major concerns to the development and sustainability of aquaculture (Fajardo et al., 2022; Sarkar et al., 2022). Since 1990, the global shrimp industry has incurred losses to the tune of USD 10 billion owing to the outbreak of infectious myonecrotic virus and white spot disease (Sarkar et al., 2022). In addition, the global ornamental fish business worth USD 15 billion is burdened with the problem of antibiotic resistance. Currently, available traditional drug delivery methods are ineffective as a result of low bioavailability in aquatic medium. More so, the conventional mode of disease detection is in the long term not feasible. As such, it becomes imperative to adopt innovative technological solutions to address this issue (Sarkar et al., 2022). Nanotechnology-based delivery media facilitates precise cell and tissue target by enhancing bioavailability, solubility and sustained release of hydrophobic drugs while conferring protection from degradation (Ahmed et al., 2019). Polymeric nanoparticles such as nanochitosan have been considerably studied as carriers for drug delivery (Fan et al., 2012) due to their biodegradability and biocompatibility (Okeke et al., 2022). For example, nanochitosan emulsion-based edible coatings have been employed for targeted drug delivery against bacteria and virus-induced fish diseases (Shah & Mraz, 2020; Nasr-Eldahan et al., 2021). Additionally, chitosan-based nanoencapsulation has been reported to be effective in delivering drugs to control epizootic ulcerative syndrome and vibriosis in fish and white-spot syndrome in shrimps, as well as for the production of pathogen-free fish fingerlings, prawn and shrimp post-larvae (Muruganandam et al., 2019). Solid core drug delivery system incorporating solid nanoparticles with a fatty acid shell enclosing the drug of interest can function at relatively low temperature and pressure, making it particularly useful for heat sensitive or labile fish medicines (Mitchell & Trivedi, 2010).

10.3 Fish Vaccination

Vaccination is implemented in modern aquaculture facilities to prevent the spread of infectious diseases and their attendant economic impacts (Tattiyapong et al., 2022). Nanoparticles surface-engineered with proteins, polymers, cell-penetrating peptides and other targeting ligands are gaining significant attention as versatile delivery systems for fish vaccine formulations (Biswas, 2020). Nanotechnology-based vaccines offer several advantages including ease of administration in young fishes, less labour intensity, enhanced protection against degradation in the gut of fishes, ease of absorption and delivery to target cells, and mass vaccination in commercial aquaculture systems (Selvaraj et al., 2014; Tattiyapong et al., 2022). Synthetic polylactide-co-glycolide acid (PGLA) and chitosan are the most explored nanocarriers for the delivery of adjuvants and antigens to immune cells (Biswas, 2020; Okeke et al., 2022). Nanochitosan immersion-based vaccine elicited better antibody response against tilapia lake virus (the causative agent for tilapia lake virus disease), with increased survivability in farmed tilapia under laboratory and field trials. In addition, the nano-delivery system enhanced mucoadhesive properties through the gills of the fish (Tattiyapong et al., 2022). Chitosan nanoparticles have been deployed for the delivery of inactivated viruses against infectious salmon anaemia virus (ISAV). The vaccine, which incorporates DNA encoding the ISAV replicase as an adjuvant conferred > 77% protection against ISAV (Rivas-Aravena et al., 2015). Similarly, oral DNA vaccine developed from nanochitosan and chitosan/tripolyphosphate nanoparticles exhibited moderate protection against against *Vibrio anguillarum* in Asian Lates calcarifer (Vimal et al., 2012). Oral DNA vaccine comprising nanochitosan loaded with *Vibrio parahemolyticus* gene encoding outer membrane protein K (ompK) was capable of eliciting a protective immune response against *Vibrio parahemolyticus* in black seabream (*Acanthopagrus schlegelii*) (Li et al., 2013). In the same vein, the recombinant DNA-nanochitosan vaccine boosted shrimp immunity against white spot syndrome virus (WSSV) when orally administered (Sekhon, 2014; Okeke et al., 2022).

10.4 Pathogen Detection and Control

Disease outbreak is considered a prime threat to intensive aquaculture systems (Toranzo et al., 2005; Shah & Mraz, 2020). Nanotechnology-based biosensors can be employed in the aquaculture industry for microbe detection and control (Sekhon, 2014; Kamalii et al., 2018). Nanochitosan has been employed for the development of an extremely sensitive electrochemical genosensor for the detection of pathogenic *Aeromonas* sp. in spiked tap water. Constructed from multi-layered carbon nanotubes–chitosan–bismuth complex and lead sulphides nanoparticles, the genosensor holds significant potential for the diagnosis of fish diseases (Fernandes et al., 2015). The detection of etiological agents is important, particularly for combating disease outbreaks early on and minimising the economic impacts of disease in commercial aquaculture facilities, as it can take a long time for the devastation caused by pathogens to manifest before their presence is detected, thus delaying control response (Fajardo et al., 2022).

10.5 Water Treatment and Purification

Concerns about impaired water quality due to the disposal of agricultural, industrial and municipal waste and abuse of antibiotics and other synthetic compounds are growing globally. Therefore, ensuring good water quality is a critical task required for maintaining fish health and sustainable aquaculture management (Toranzo et al., 2005; Shah & Mraz, 2020). Nanotechnology-based adsorption and photocatalysis are two affordable and efficient strategies deployed for the elimination of environmental pollutants in aquaculture facilities to provide safe and favourable conditions for fish farming (Shah & Mraz, 2020; Sarkar et al., 2022; Fajardo et al., 2022). Chitosan nanoparticles, magnetite-chitosan and chitosan-clay nanocomposites have emerged as adsorbents for the removal of heavy metals from water (Futalan et al., 2011; Namdeo & Bajpai, 2008; Fang et al., 2017).

10.6 Fish Quality Testing

Post-harvest fish losses resulting from quality deterioration are a major problem along the supply chain of fish and fishery products. About 10 to 12 million tonnes of fish are lost annually from aquaculture and capture fisheries (Maulu et al., 2020). Improved post-harvest technology could considerably close the gap between the demand and supply of fish by minimising or completely eliminating post-harvest fish losses (Getu et al., 2015; Otuya et al., 2017). Bionanosensors incorporating formaldehyde hydrogenase and nanomaterials such as nanochitosan and carbon nanotubes have emerged for the precise detection of formalin (a harmful preservative that is applied to maintain the freshness of fish in transit), with high sensitivity, quick response time and high reproducibility. A similar technology that integrates deposits of ionic liquid, gold nanoparticles and chitosan on a glassy carbon electrode for sensing formalin in fish tissue has been developed (Noor Aini et al., 2016). More so, nanochitosan has been integrated into biosensors for xanthine detection in fish meat (Devi et al., 2013; Ahmed et al., 2019). Biosensors based on graphene oxide-chitosan nanocomposites catalytic film have also been developed for the detection of phenylalanine in the fluid samples of Tuna fish. The nanobiosensor is rapid and can be used for non-destructive fish freshness assessment in a large number of samples in a short period of time (Fazial & Tan, 2021).

10.7 Supplements and Nutraceuticals Delivery

Dietary supplements and nutraceuticals are recognised to perform a crucial role in boosting growth and immunological functions (Shah & Mraz, 2020; Sarkar et al., 2022). Nanotechnology can aid the efficient delivery of these materials by enhancing nutrient stability, solubility, bioavailability and bioaccessibility across the digestive tract of fishes (Shah & Mraz, 2020; Fajardo et al., 2022). Quercetin, trace minerals and water-insoluble vitamins can be solubilised by encapsulating with nanoparticles for use as a nutritional supplement with improved bioavailability across the gut of fishes (Handy, 2012; Singha et al., 2017; Armobin et al., 2023). A diet enriched with chitosan nanoparticles markedly enhanced the growth, survival and meat quality of African catfish (*Clarius gariepinus*) fingerlings (Udo et al., 2018) and Nile tilapia (*Oreochromis nilotica*) (Wang & Li, 2011; Abdel-Tawwab et al., 2019). Nanochitosan in combination with dietary thymol significantly improved health and feed utilisation in Nile tilapia. In addition to increasing intestinal villus length, the co-supplemented diet promoted catalase, protease and lipase activities in the fish species (Abd El-Naby et al., 2020). Similarly, dietary enrichment with chitosan nanoparticles promoted feed utilisation and growth in Nile tilapia (*O. niloticus*) by enhancing the activities of digestive enzymes. The nanochitosan diet also inhibited the growth of intestinal microbial populations and improved innate immunity in *O. niloticus* (Abd El-Naby et al., 2019). A mixture of nanochitosan with vitamin C attenuated pesticide-induced toxicological stress, while also improving growth in *O. niloticus* (Okeke et al., 2022). Again, chitosan-vitamin C nanocomposite dietary supplementation promoted growth, antioxidant profile, immune response, disease resistance and intestinal histology in the fingerlings of Nile tilapia (Ibrahim et al., 2021). A similar result was also reported with a blend of nanocurcumin and nanochitosan in Nile tilapia (Elabd et al., 2023). Folic acid-coated chitosan nanoparticles improved feed utilisation, growth, immune response and antioxidant profile of rainbow trout (Farahnak Roudsari et al., 2021). Furthermore, water-soluble N, N, N-trimethyl nanochitosan can used as a stable carrier system for the delivery of vitamins B9, B12 and C (Katata-Seru et al., 2019). Nanochitosan loaded with Selenium (Se) has been demonstrated to be efficient for the delivery of dietary Se in Nile tilapia, for improved feed efficiency and antioxidant activity (Araujo et al., 2021). Chitosan nanocarriers are efficient media for the slow delivery of proteins and proteolytic enzymes in fish (Kumari et al., 2013).

10.8 Fish Breeding

Broodstock management is a key step in fish breeding and reproduction. To support a broodstock, maturation of the gonads is achieved by employing augmented feed or multiphase hormone delivery through injections. The latter, which is administered during the pre-spawning phase, can be challenging. On the other hand, supplementation of diet with maturation hormones like testosterone or progesterone can cause additional issues, as they can be leached into the surrounding water during delivery. To circumvent these problems, hormonal pellets are implanted subcutaneously (Kailasam et al., 1998). Subcutaneous hormonal delivery through nanoencapsulation has emerged as a more efficient strategy (Kumari et al., 2013). It is now possible to subcutaneously implant a nanocarrier loaded with hormone during the pre-spawning stage to stimulate maturation ascribable to gradual release and extended retention time (Sarkar et al., 2022). This approach has been successfully used to overcome the need for multiple administrations of luteinising hormone-releasing hormone (LHRH) in fish. Compared to bare LHRH, half the dose of nanochitosan and chitosan-gold conjugated with salmon LHRH can elicit higher reproductive efficiency in female *Cyprinus carpio* with minimal accumulation in the tissues (Rather et al., 2013; Khosravi-Katuli et al., 2017). Furthermore, the delivery of vitamins, micro- and macronutrients using nanocarriers can enhance breeding performance and reproduction in fishes. Fish feed enriched with nanoscale selenium is emerging as an efficient solution to address male sterility in fishes, producing broodstocks with superior fecundity.

11 Conclusion

Chitosan, abundantly sourced from nature, exhibits remarkable properties that renders it a versatile biopolymer. Deacetylation produces chitosan with enhanced attributes such as biocompatibility, biodegradability and flexibility, forming the basis for further exploitation at the nanoscale. Nanonisation of chitosan significantly amplifies its potential with expanded applications in drug delivery, vaccination and water treatment in aquaculture practice. Nanochitosan integration addresses critical challenges in the fishing sector while aligning seamlessly with Sustainable Development Goal 14, fostering sustainable aquaculture practices. With recent advances in the strategies deployed in nanotechnology for modifying chitosan, the practice can become more sustainable, reducing waste and promoting efficient utilisation of resources. This holds great promise for establishing environmentally conscious processes within the realm of nanochitosan applications.

References

Abd El-Naby, A. S., Al-Sagheer, A. A., Negm, S. S., & Naiel, M. A. (2020). Dietary combination of chitosan nanoparticle and thymol affects feed utilization, digestive enzymes, antioxidant status, and intestinal morphology of *Oreochromis niloticus*. *Aquaculture, 515*, 734577.
Abd El-Naby, F. S., Naiel, M. A., Al-Sagheer, A. A., & Negm, S. S. (2019). Dietary chitosan nanoparticles enhance the growth, production performance, and immunity in Oreochromis niloticus. *Aquaculture, 501*, 82–89.
Abdel-Tawwab, M., Razek, N. A., & Abdel-Rahman, A. M. (2019). Immunostimulatory effect of dietary chitosan nanoparticles on the performance of Nile tilapia, *Oreochromis niloticus* (L.). *Fish & Shellfish Immunology, 88*, 254–258.
Abid, N., Khan, A. M., Shujait, S., Chaudhary, K., Ikram, M., Imran, M., et al. (2022). Synthesis of nanomaterials using various top-down and bottom-up approaches, influencing factors, advantages, and disadvantages: A review. *Advances in Colloid and Interface Science, 300*, 102597.
Afolalu, S. A., Soetan, S. B., Ongbali, S. O., Abioye, A. A., & Oni, A. S. (2019). Morphological characterization and physio-chemical properties of nanoparticle–review. *IOP Conference Series: Materials Science and Engineering, 640*(1), 012065.
Ahmed, F., Soliman, F. M., Adly, M. A., Soliman, H. A., El-Matbouli, M., & Saleh, M. (2019). Recent progress in biomedical applications of chitosan and its nanocomposites in aquaculture: A review. *Research in Veterinary Science, 126*, 68–82.
Alemu, D., Getachew, E., & Mondal, A. K. (2023). Study on the physicochemical properties of chitosan and their applications in the biomedical sector Y. Yang (ed.). *International Journal of Polymer Science., 2023*, 1–13.
Alfei, S., Marengo, B., & Zuccari, G. (2020). Nanotechnology application in food packaging: A plethora of opportunities versus pending risks assessment and public concerns. *Food Research International, 137*, 109664.
Altammar, K. A. (2023). A review on nanoparticles: Characteristics, synthesis, applications, and challenges. *Frontiers in Microbiology, 14*, 1155622.
Andrade-Guel, M., Cabello-Alvarado, C., Avila-Orta, C. A., Pérez-Alvarez, M., Cadenas-Pliego, G., Reyes-Rodríguez, P. Y., & Rios-González, L. (2022). Green flame-retardant composites based on PP/TiO2/lignin obtained by melt-mixing extrusion. *Polymers, 14*(7), 1300.

Anvar, A. A., Ahari, H., & Ataee, M. (2021). Antimicrobial properties of food nanopackaging: A new focus on foodborne pathogens. *Frontiers in Microbiology, 12*, 690706.
Aranaz, I., Alcántara, A. R., Civera, M. C., Arias, C., Elorza, B., Heras Caballero, A., & Acosta, N. (2021). Chitosan: An overview of its properties and applications. *Polymers, 13*(19), 3256.
Araujo, J. M., Fortes-Silva, R., Pola, C. C., Yamamoto, F. Y., Gatlin, D. M., III, & Gomes, C. L. (2021). Delivery of selenium using chitosan nanoparticles: Synthesis, characterization, and antioxidant and growth effects in Nile tilapia (Orechromis niloticus). *PLoS One, 16*(5), e0251786.
Armobin, K., Ahmadifar, E., Adineh, H., Samani, M. N., Kalhor, N., Yilmaz, S., et al. (2023). Quercetin application for common carp (*Cyprinus carpio*): I. effects on growth performance, humoral immunity, antioxidant status, immune-related genes, and resistance against heat stress. *Aquaculture Nutrition, 2023*, 1168262.
Aromán-Doval, R., Torres-Arellanes, S. P., Tenorio-Barajas, A. Y., Gómez-Sánchez, A., & Valencia-Lazcano, A. A. (2023). Chitosan: Properties and its application in agriculture in context of molecular weight. *Polymers, 15*(13), 2867.
Baig, N., Kammakakam, I., & Falath, W. (2021). Nanomaterials: A review of synthesis methods, properties, recent progress, and challenges. *Materials Advances, 2*(6), 1821–1871.
Barhoum, A., García-Betancourt, M. L., Jeevanandam, J., Hussien, E. A., Mekkawy, S. A., Mostafa, M., et al. (2022). Review on natural, incidental, bioinspired, and engineered nanomaterials: History, definitions, classifications, synthesis, properties, market, toxicities, risks, and regulations. *Nanomaterials, 12*(2), 177.
Bello, S. A., Agunsoye, J. O., & Hassan, S. B. (2015). Synthesis of coconut shell nanoparticles via a top-down approach: Assessment of milling duration on the particle sizes and morphologies of coconut shell nanoparticles. *Materials Letters, 159*, 514–519.
Biswas, P. P. (2020). Overview of fish vaccines with focus on nanovaccines. *International Research Journal of Basic and Applied Sciences, 5*, 33–42.
Cele, T. (2020). Preparation of nanoparticles. In S. Marius Avramescu, K. Akhtar, I. Fierascu, S. Bahadar Khan, F. Ali, & A. M. Asiri (Eds.), *Engineered nanomaterials–health and safety*. IntechOpen.
Coldebella, A., Gentelini, A. L., Piana, P. A., Coldebella, P. F., Boscolo, W. R., & Feiden, A. (2017). Effluents from fish farming ponds: A view from the perspective of its main components. *Sustainability, 10*(1), 3.
Danarto, Y. C., & Distantina, S. (2016). *Optimizing deacetylation process for chitosan production from green mussel (perna viridis) shell* (p. 030028).
Darr, J. A., Zhang, J., Makwana, N. M., & Weng, X. (2017). Continuous hydrothermal synthesis of inorganic nanoparticles: Applications and future directions. *Chemical Reviews, 117*(17), 11125–11238.
De Azeredo, H. M. (2009). Nanocomposites for food packaging applications. *Food Research International, 42*(9), 1240–1253.
De Moura, M. R., Avena-Bustillos, R. J., McHugh, T. H., Krochta, J. M., & Mattoso, L. H. C. (2008). Properties of novel hydroxypropyl methylcellulose films containing chitosan nanoparticles. *Journal of Food Science, 73*(7), N31–N37.
Desai, N., Rana, D., Salave, S., Gupta, R., Patel, P., Karunakaran, B., Sharma, A., Giri, J., Benival, D., & Kommineni, N. (2023). Chitosan: A potential biopolymer in drug delivery and biomedical applications. *Pharmaceutics, 15*(4), 1313.
Devi, R., Yadav, S., Nehra, R., Yadav, S., & Pundir, C. S. (2013). Electrochemical biosensor based on gold coated iron nanoparticles/chitosan composite bound xanthine oxidase for detection of xanthine in fish meat. *Journal of Food Engineering, 115*(2), 207–214.
Drago, E., Campardelli, R., Pettinato, M., & Perego, P. (2020). Innovations in smart packaging concepts for food: An extensive review. *Food, 9*(11), 1628.
Elabd, H., Mahboub, H. H., Salem, S. M., Abdelwahab, A. M., Alwutayd, K. M., Shaalan, M., et al. (2023). Nano-curcumin/chitosan modulates growth, biochemical, immune, and Antioxidative profiles, and the expression of related genes in Nile tilapia, Oreochromis niloticus. *Fishes, 8*(7), 333.

El-Eskandarany, M. S., Al-Hazza, A., Al-Hajji, L. A., Ali, N., Al-Duweesh, A. A., Banyan, M., & Al-Ajmi, F. (2021). Mechanical milling: A superior nanotechnological tool for fabrication of nanocrystalline and nanocomposite materials. *Nanomaterials, 11*(10), 2484.
Elkhatib, E. A., Mahdy, A. M., & Salama, K. A. (2015). Green synthesis of nanoparticles by milling residues of water treatment. *Environmental Chemistry Letters, 13*, 333–339.
FAO. (2022). *The state of the world fisheries and aquaculture*. https://www.fao.org/3/cc0461en/online/sofia/2022/aquaculture-production.html#:~:text=SOURCE%3A%20FAO.%20World%20aquaculture%20production%20of%20animal%20species,comparable%20to%20some%20years%20in%20the%20last%20decade
Fajardo, C., Martinez-Rodriguez, G., Blasco, J., Mancera, J. M., Thomas, B., & De Donato, M. (2022). Nanotechnology in aquaculture: Applications, perspectives and regulatory challenges. *Aquaculture and Fisheries, 7*(2), 185–200.
Fan, W., Yan, W., Xu, Z., & Ni, H. (2012). Erythrocytes load of low molecular weight chitosan nanoparticles as a potential vascular drug delivery system. *Colloids and Surfaces B: Biointerfaces, 95*, 258–265.
Fang, Z., Zhao, Y., Warner, R. D., & Johnson, S. K. (2017). Active and intelligent packaging in meat industry. *Trends in Food Science & Technology, 61*, 60–71.
Farahnak Roudsari, S., Rajabi Islami, H., Mousavi, S. A., & Shamsaie Mehrgan, M. (2021). Folic acid-coated nanochitosan ameliorated the growth performance, hematological parameters, antioxidant status, and immune responses of rainbow trout (Oncorhynchus mykiss). *Frontiers in Veterinary Science, 8*, 647722.
Fazial, F. F., & Tan, L. L. (2021). Phenylalanine-responsive fluorescent biosensor based on graphene oxide-chitosan nanocomposites catalytic film for non-destructive fish freshness grading. *Food Control, 125*, 107995.
Fernandes, A. M., Abdalhai, M. H., Ji, J., Xi, B. W., Xie, J., Sun, J., et al. (2015). Development of highly sensitive electrochemical genosensor based on multiwalled carbon nanotubes–chitosan–bismuth and lead sulfide nanoparticles for the detection of pathogenic Aeromonas. *Biosensors and Bioelectronics, 63*, 399–406.
Fontana, F., Anselmi, M., & Limonta, P. (2022). Molecular mechanisms and genetic alterations in prostate cancer: From diagnosis to targeted therapy. *Cancer Letters, 534*, 215619.
Futalan, C. M., Kan, C. C., Dalida, M. L., Pascua, C., & Wan, M. W. (2011). Fixed-bed column studies on the removal of copper using chitosan immobilized on bentonite. *Carbohydrate Polymers, 83*(2), 697–704.
Gan, Y. X., Jayatissa, A. H., Yu, Z., Chen, X., & Li, M. (2020). Hydrothermal synthesis of nanomaterials. *Journal of Nanomaterials, 2020*, 1–3.
Getu, A., Misganaw, K., & Bazezew, M. (2015). Post-harvesting and major related problems of fish production. *Fish and Aquatic Journal, 6*(04), 10–4172.
Gohil, S. V., Padmanabhan, A., Kan, H.-M., Khanal, M., & Nair, L. S. (2021). Degradation-dependent protein release from enzyme sensitive injectable glycol chitosan hydrogel. *Tissue Engineering Part A, 27*(13–14), 867–880.
Guggisberg, S. (2022). Transparency in the activities of the food and agriculture organization for sustainable fisheries. *Marine Policy, 136*, 104498.
Guo, D., Xie, G., & Luo, J. (2013). Mechanical properties of nanoparticles: Basics and applications. *Journal of Physics D: Applied Physics, 47*(1), 013001.
Handy, R. D. (2012). FSBI briefing paper: Nanotechnology in fisheries and aquaculture. *Fisheries Society of the British Isles, 1*, 29.
Homayounpour, P., Shariatifar, N., & Alizadeh-Sani, M. (2021). Development of nanochitosan-based active packaging films containing free and nanoliposome caraway (Carum carvi. L) seed extract. *Food Science & Nutrition, 9*(1), 553–563.
Hosseini, S. F., & Gómez-Guillén, M. C. (2018). A state-of-the-art review on the elaboration of fish gelatin as bioactive packaging: Special emphasis on nanotechnology-based approaches. *Trends in Food Science & Technology, 79*, 125–135.
Ibrahim, R. E., Amer, S. A., Farroh, K. Y., Al-Gabri, N. A., Ahmed, A. I., El-Araby, D. A., & Ahmed, S. A. (2021). The effects of chitosan-vitamin C nanocomposite supplementation on

the growth performance, antioxidant status, immune response, and disease resistance of Nile tilapia (Oreochromis niloticus) fingerlings. *Aquaculture, 534*, 736269.
Ijaz, I., Gilani, E., Nazir, A., & Bukhari, A. (2020). Detail review on chemical, physical and green synthesis, classification, characterizations and applications of nanoparticles. *Green Chemistry Letters and Reviews, 13*(3), 223–245.
Igwegbe, C. A., Onukwuli, O. D., Ighalo, J. O., & Umembamalu, C. J. (2021). Electrocoagulation-flocculation of aquaculture effluent using hybrid iron and aluminiumelectrodes: A comparative study. *Chemical Engineering Journal Advances, 6*, 100107
ISO 10801. (2010). *Nanotechnologies—Generation of metal nanoparticles for inhalation toxicity testing using the evaporation/condensation method.*
Jamkhande, P. G., Ghule, N. W., Bamer, A. H., & Kalaskar, M. G. (2019). Metal nanoparticles synthesis: An overview on methods of preparation, advantages and disadvantages, and applications. *Journal of Drug Delivery Science and Technology, 53*, 101174.
Jayakumar, R., Prabaharan, M., Sudheesh Kumar, P. T., Nair, S. V., & Tamura, H. (2011). Biomaterials based on chitin and chitosan in wound dressing applications. *Biotechnology Advances, 29*(3), 322–337.
Jeevanandam, J., Barhoum, A., Chan, Y. S., Dufresne, A., & Danquah, M. K. (2018). Review on nanoparticles and nanostructured materials: History, sources, toxicity and regulations. *Beilstein Journal of Nanotechnology, 9*, 1050–1074.
Jiang, A., Patel, R., Padhan, B., Palimkar, S., Galgali, P., Adhikari, A., et al. (2023). Chitosan based biodegradable composite for antibacterial food packaging application. *Polymers, 15*(10), 2235.
Kailasam, M., Thirunavukkarasu, A. R., Chandra, P. K., Pereira, S., & Rajendran, K. V. (1998). *Induction of maturity and spontaneous spawning of captive broodstock of Bhetki, Lates calcarifer (Bloch) through hormonal manipulation.*
Kamalii, A., Ahilan, B., Felix, N., Kannan, B., & Prabu, E. (2018). Applications of nanotechnology in fisheries and aquaculture. *Journal of Aquaculture in the Tropics, 33*(3/4), 111–117.
Katata-Seru, L., Ramalapa, B., & Tshweu, L. (2019). Nanoformulated delivery systems of essential nutraceuticals and their applications. In *Nanoemulsions-properties, fabrications and applications*. IntechOpen.
Katti, A. K. S., & Sharon, M. (2019). European Nano knowledge that led to Faraday's understanding of gold nanoparticles. In *History of nanotechnology: From pre-historic to modern times* (pp. 141–212).
Khan, A., & Alamry, K. A. (2021). Recent advances of emerging green chitosan-based biomaterials with potential biomedical applications: A review. *Carbohydrate Research, 506*, 108368.
Khan, I., Saeed, K., & Khan, I. (2019). Nanoparticles: Properties, applications and toxicities. *Arabian Journal of Chemistry, 12*(7), 908–931.
Khan, Y., Sadia, H., Ali Shah, S. Z., Khan, M. N., Shah, A. A., Ullah, N., Ullah, M. F., Bibi, H., Bafakeeh, O. T., Khedher, N. B., & Eldin, S. M. (2022). Classification, synthetic, and characterization approaches to nanoparticles, and their applications in various fields of nanotechnology: A review. *Catalysts, 12*(11), 1386.
Khan, S., & Hossain, M. K. (2022). Classification and properties of nanoparticles. In *Nanoparticle-based polymer composites* (pp. 15–54). Woodhead Publishing.
Khosravi-Katuli, K., Prato, E., Lofrano, G., Guida, M., Vale, G., & Libralato, G. (2017). Effects of nanoparticles in species of aquaculture interest. *Environmental Science and Pollution Research, 24*, 17326–17346.
Koul, B., Poonia, A. K., Yadav, D., & Jin, J. O. (2021). Microbe-mediated biosynthesis of nanoparticles: Applications and future prospects. *Biomolecules, 11*(6), 886.
Kravanja, G., Primožič, M., Knez, Ž., & Leitgeb, M. (2019). Chitosan-based (nano) materials for novel biomedical applications. *Molecules, 24*(10), 1960.
Kumar, S., Mukherjee, A., & Dutta, J. (2020). Chitosan based nanocomposite films and coatings: Emerging antimicrobial food packaging alternatives. *Trends in Food Science & Technology, 97*, 196–209.

Kumari, R., Gupta, S., Singh, A. R., Ferosekhan, S., Kothari, D. C., Pal, A. K., & Jadhao, S. B. (2013). Chitosan nanoencapsulated exogenous trypsin biomimics zymogen-like enzyme in fish gastrointestinal tract. *PLoS One, 8*(9), e74743.

Li, L., Lin, S. L., Deng, L., & Liu, Z. G. (2013). Potential use of chitosan nanoparticles for oral delivery of DNA vaccine in black seabream Acanthopagrus schlegelii Bleeker to protect from Vibrio parahaemolyticus. *Journal of Fish Diseases, 36*(12), 987–995.

Maulu, S., Hasimuna, O. J., Monde, C., & Mweemba, M. (2020). An assessment of post-harvest fish losses and preservation practices in Siavonga district, southern Zambia. *Fisheries and aquatic sciences, 23*, 1–9.

Mekuye, B., & Abera, B. (2023). Nanomaterials: An overview of synthesis, classification, characterization, and applications. *Nano Select, 4*(8), 486–501.

Mikušová, V., & Mikuš, P. (2021). Advances in chitosan-based nanoparticles for drug delivery. *International Journal of Molecular Sciences, 22*(17), 9652.

Mitchell, J., & Trivedi, V. (2010). Pharmaceutical nanomaterials: The preparation of solid core drug delivery systems (SCDDS). *Journal of Pharmacy and Pharmacology, 62*(10), 1457–1458.

Mohammadpour, Z., & Naghib, S. M. (2021). Smart nanosensors for intelligent packaging. In *Nanosensors for smart manufacturing* (pp. 323–346). Elsevier.

Muhammad Mudassar Shahzad, F. L. (2022). Applications of nanotechnology in fish health. *International Journal of Veterinary Science, 2*, 244–253.

Mukasyan, A. S., & Manukyan, K. V. (2015). Combustion/micropyretic synthesis of atomically thin two-dimensional materials for energy applications. *Current Opinion in ChemicalEngineering, 7*, 16–22.

Mukasyan, A. S., Rogachev, A. S., & Aruna, S. T. (2015). Combustion synthesis in nanostructured reactive systems. *Advanced Powder Technology, 26*(3), 954–976.

Muruganandam, M., Chipps, S. R., & Ojasvi, P. R. (2019). On the advanced technologies to enhance fisheries production and management. *Acta Scientific Agriculture, 3*(8), 216–222.

Namboodiri, M. T., & Pakshirajan, K. (2020). Valorization of waste biomass for chitin and chitosan production. In *Waste biorefinery* (pp. 241–266). Elsevier.

Namdeo, M., & Bajpai, S. K. (2008). Chitosan–magnetite nanocomposites (CMNs) as magnetic carrier particles for removal of Fe (III) from aqueous solutions. *Colloids and Surfaces A: Physicochemical and Engineering Aspects, 320*(1–3), 161–168.

Nasr-Eldahan, S., Nabil-Adam, A., Shreadah, M. A., Maher, A. M., & El-Sayed Ali, T. (2021). A review article on nanotechnology in aquaculture sustainability as a novel tool in fish disease control. *Aquaculture International, 29*, 1459–1480.

Natsuki, J., Natsuki, T., & Hashimoto, Y. (2015). A review of silver nanoparticles: Synthesis methods, properties and applications. *International Journal of Materials Science and Applications, 4*(5), 325–332.

Nguyen, N. T. T., Nguyen, L. M., Nguyen, T. T. T., Liew, R. K., Nguyen, D. T. C., & Van Tran, T. (2022). Recent advances on botanical biosynthesis of nanoparticles for catalytic, water treatment and agricultural applications: A review. *Science of the Total Environment, 827*, 154160.

Noor Aini, B., Siddiquee, S., & Ampon, K. (2016). Development of formaldehyde biosensor for determination of formalin in fish samples; Malabar red snapper (*Lutjanus malabaricus*) and longtail tuna (Thunnus tonggol). *Biosensors, 6*(3), 32.

Ogunfowora, L. A., Iwuozor, K. O., Ighalo, J. O., & Igwegbe, C. A. (2021). Trends in the treatment of aquaculture effluents using nanotechnology. *Cleaner Materials, 2*, 100024.

Okeke, E. S., Chukwudozie, K. I., Nyaruaba, R., Ita, R. E., Oladipo, A., Ejeromedoghene, O., et al. (2022). Antibiotic resistance in aquaculture and aquatic organisms: A review of current nanotechnology applications for sustainable management. *Environmental Science and Pollution Research, 29*(46), 69241–69274.

Otuya, P., Sifuna, A. W., Onyango, D. M., Owigar, R., Kowenje, C., Lung'ayia, H. B. O., & Oduor, A. O. (2017). Evaluation of fish processing and preservation systems along the shores of Lake Victoria towards enhancement of sun drying technology. *International Journal of Food Science and Nutrition Engineering, 7*(5), 111–118.

Pauly, D., & Zeller, D. (2017). Comments on FAOs state of world fisheries and aquaculture (SOFIA 2016). *Marine Policy, 77*, 176–181.

Pellis, A., Guebitz, G. M., & Nyanhongo, G. S. (2022). Chitosan: Sources, processing and modification techniques. *Gels, 8*(7), 393.

Perinelli, D. R., Cespi, M., Bonacucina, G., & Palmieri, G. F. (2019). PEGylated polylactide (PLA) and poly (lactic-co-glycolic acid) (PLGA) copolymers for the design of drug delivery systems. *Journal of Pharmaceutical Investigation, 49*(4), 443–458.

Piekarska, K., Sikora, M., Owczarek, M., Jóźwik-Pruska, J., & Wiśniewska-Wrona, M. (2023). Chitin and chitosan as polymers of the future—Obtaining, modification, life cycle assessment and Main directions of application. *Polymers, 15*(4), 793.

Pochanavanich, P., & Suntornsuk, W. (2002). Fungal chitosan production and its characterization. *Letters in Applied Microbiology, 35*(1), 17–21.

Quinson, J., Simonsen, S. B., Theil Kuhn, L., & Arenz, M. (2021). Commercial spirits for surfactant-free syntheses of electro-active platinum nanoparticles. *Sustainable chemistry, 2*(1), 1–7.

Rahinov, I., Sellmann, J., Lalanne, M. R., Nanjaiah, M., Dreier, T., Cheskis, S., & Wlokas, I. (2020). Insights into the mechanism of combustion synthesis of iron oxide nanoparticles gained by laser diagnostics, mass spectrometry, and numerical simulations: A mini-review. *Energy & Fuels, 35*(1), 137–160.

Ramos Berger, L. R., Montenegro Stamford, T. C., De Oliveira, K. Á. R., De Miranda Pereira Pessoa, A., De Lima, M. A. B., Estevez Pintado, M. M., Saraiva Câmara, M. P., De Oliveira Franco, L., Magnani, M., & De Souza, E. L. (2018). Chitosan produced from *Mucorales* fungi using agroindustrial by-products and its efficacy to inhibit Colletotrichum species. *International Journal of Biological Macromolecules, 108*, 635–641.

Rane, A. V., Kanny, K., Abitha, V. K., & Thomas, S. (2018). Methods for synthesis of nanoparticles and fabrication of nanocomposites. In *Synthesis of inorganic nanomaterials* (pp. 121–139). Woodhead Publishing.

Rashid, T. M., Nayef, U. M., Jabir, M. S., & Mutlak, F. A. H. (2021). Synthesis and characterization of Au: ZnO (core: shell) nanoparticles via laser ablation. *Optik, 244*,167569.

Rather, M. A., Sharma, R., Gupta, S., Ferosekhan, S., Ramya, V. L., & Jadhao, S. B. (2013). Chitosan-nanoconjugated hormone nanoparticles for sustained surge of gonadotropins and enhanced reproductive output in female fish. *PLoS One, 8*(2), e57094.

Rhim, J. W., & Kim, Y. T. (2014). Biopolymer-based composite packaging materials with nanoparticles. In *Innovations in food packaging* (pp. 413–442). Academic Press.

Rinaudo, M. (2006). *Chitin and chitosan: Properties and applications* (Vol. 38). Elsevier Ltd.

Rivas-Aravena, A., Fuentes, Y., Cartagena, J., Brito, T., Poggio, V., La Torre, J., et al. (2015). Development of a nanoparticle-based oral vaccine for Atlantic salmon against ISAV using an alphavirus replicon as adjuvant. *Fish & Shellfish Immunology, 45*(1), 157–166.

Rodrigues, C., Souza, V. G. L., Coelhoso, I., & Fernando, A. L. (2021). Bio-based sensors for smart food packaging—Current applications and future trends. *Sensors, 21*(6), 2148.

Sadadekar, A. S., Shruthy, R., Preetha, R., Kumar, N., Pande, K. R., & Nagamaniammai, G. (2023). Enhanced antimicrobial and antioxidant properties of Nano chitosan and pectin based biodegradable active packaging films incorporated with fennel (*Foeniculum vulgare*) essential oil and potato (*Solanum tuberosum*) peel extracts. *Journal of Food Science and Technology, 60*(3), 938–946.

Salem, S. S., Hammad, E. N., Mohamed, A. A., & El-Dougdoug, W. (2022). A comprehensive review of nanomaterials: Types, synthesis, characterization, and applications. *Biointerface Research in Applied Chemistry, 13*(1), 41.

Sarkar, B., Mahanty, A., Gupta, S. K., Choudhury, A. R., Daware, A., & Bhattacharjee, S. (2022). Nanotechnology: A next-generation tool for sustainable aquaculture. *Aquaculture, 546*, 737330.

Sekhon, B. S. (2014). Nanotechnology in Agri-food production: An overview. *Nanotechnology, Science and Applications, 7*, 31–53.

Sekhri, A., Sharma, R., & Bharti, R. (2023). Nanoparticles: Synthetic techniques, surface deposition and applications. In *AIP Conference Proceedings (Vol. 2535, No. 1)*. AIP Publishing.

Selvaraj, B., Subramanian, K., Gopal, S., & Renuga, P. S. (2014). Nanotechnology as a novel tool for aquaculture industry: A review. *World Journal of Pharmaceutical Sciences*, 1089–1096.

Shah, B. R., & Mraz, J. (2020). Advances in nanotechnology for sustainable aquaculture and fisheries. *Reviews in Aquaculture, 12*(2), 925–942.

Sharon, M. (Ed.). (2019). *History of nanotechnology: From prehistoric to modern times*. John Wiley & Sons.

Singh, V., Yadav, P., & Mishra, V. (2020). Recent advances on classification, properties, synthesis, and characterization of nanomaterials. In *Green synthesis of nanomaterials for bioenergy applications* (pp. 83–97).

Singha, S., Das, K., & Jha, N. (2017). Nano-systems for micro-nutrient delivery in aquaculture: A critical analysis. *Ann Aquaculture Research, 4*(4), 1046.

Su, Q., Zhao, X., Zhang, X., Wang, Y., Zeng, Z., Cui, H., & Wang, C. (2022). Nano functional food: Opportunities, development, and future perspectives. *International Journal of Molecular Sciences, 24*(1), 234.

Sundramoorthy, A. K., Kumar, T. H. V., & Gunasekaran, S. (2018). Graphene-based nanosensors and smart food packaging systems for food safety and quality monitoring. In *Graphene bioelectronics* (pp. 267–306). Elsevier.

Tattiyapong, P., Kitiyodom, S., Yata, T., Jantharadej, K., Adamek, M., & Surachetpong, W. (2022). Chitosan nanoparticle immersion vaccine offers protection against tilapia lake virus in laboratory and field studies. *Fish & Shellfish Immunology, 131*, 972–979.

Toranzo, A. E., Magariños, B., & Romalde, J. L. (2005). A review of the main bacterial fish diseases in mariculture systems. *Aquaculture, 246*(1–4), 37–61.

Udo, I. U., Etukudo, U., & Anwana, U. I. U. (2018). Effects of chitosan and chitosan nanoparticles on water quality, growth performance, survival rate and meat quality of the African catfish, Clarias gariepinus. *Nanoscience, 1*(1), 12–25.

Vanderroost, M., Ragaert, P., Devlieghere, F., & De Meulenaer, B. (2014). Intelligent food packaging: The next generation. *Trends in Food Science & Technology, 39*(1), 47–62.

Vidal, I., Zheng, Q., Hicks, J. L., Chen, J., Platz, E. A., Trock, B. J., Kulac, I., Baena-Del Valle, J. A., Sfanos, K. S., Ernst, S., Jones, T., Maynard, J. P., Glavaris, S. A., Nelson, W. G., Yegnasubramanian, S., & De Marzo, A. M. (2021). GSTP1 positive prostatic adenocarcinomas are more common in black than white men in the United States C.N. Robson (ed.). *Plos One, 16*(6), e0241934.

Vimal, S., Taju, G., Nambi, K. N., Majeed, S. A., Babu, V. S., Ravi, M., & Hameed, A. S. (2012). Synthesis and characterization of CS/TPP nanoparticles for oral delivery of gene in fish. *Aquaculture, 358*, 14–22.

Wang, Y., & Li, J. (2011). Effects of chitosan nanoparticles on survival, growth and meat quality of tilapia, Oreochromis nilotica. *Nanotoxicology, 5*(3), 425–431.

Yan, D., Li, Y., Liu, Y., Li, N., Zhang, X., & Yan, C. (2021). Antimicrobial properties of chitosan and chitosan derivatives in the treatment of enteric infections. *Molecules, 26*(23), 7136.

Yilmaz Atay, H. (2019). Antibacterial activity of chitosan-based systems. In S. Jana & S. Jana (Eds.), *Functional chitosan* (pp. 457–489).

Yu, D., Xu, Y., Regenstein, J. M., Xia, W., Yang, F., Jiang, Q., & Wang, B. (2018). The effects of edible chitosan-based coatings on flavor quality of raw grass carp (Ctenopharyngodon idellus) fillets during refrigerated storage. *Food Chemistry, 242*, 412–420.

Zhao, L., Liu, Z., Chen, D., Liu, F., Yang, Z., Li, X., et al. (2021). Laser synthesis and microfabrication of micro/nanostructured materials toward energy conversion and storage. *Nano-Micro Letters, 13*, 1–48.

Zuma, K., Izumi, R., Osaki, T., Ifuku, S., Morimoto, M., Saimoto, H., et al. (2015). Chitin, chitosan, and its derivatives for wound healing: Old and new materials. *Journal of Functional Biomaterials, 6*(1), 104–142.

Nanochitosan Synthesis, Optimization, and Characterization

Patrick Omoregie Isibor

Contents

P. O. Isibor (✉)
Department of Biological Sciences, College of Science and Technology,
Covenant University, Ota, Ogun State, Nigeria
e-mail: patrick.isibor@covenantuniversity.edu.ng

P. O. Isibor et al. (eds.), *Nanochitosan-Based Enhancement of Fisheries and Aquaculture*, https://doi.org/10.1007/978-3-031-52261-1_3

1 Introduction

Nanochitosan optimization plays a crucial role in the field of aquaculture, providing several benefits that contribute to the sustainability and efficiency of aquaculture operations (El-Naggar et al., 2019).

Optimization allows for fine-tuning the synthesis parameters of nanochitosan, leading to enhanced and tailored properties. This includes improved particle size, surface area, and other physicochemical characteristics, which are crucial for specific applications (Cho et al., 2012). Nanochitosan, when optimized, can exhibit superior biocompatibility, making it more suitable for biomedical applications. This is particularly important for drug delivery systems, tissue engineering, and other medical applications where the material interacts with biological systems.

Different applications demand nanochitosan with specific properties. Optimization enables the customization of nanochitosan for targeted uses, such as controlled drug release, wound healing, or environmental remediation (El-Naggar et al., 2020). Tailoring the material to specific requirements enhances its effectiveness in diverse applications. Nanochitosan has diverse applications across industries, including medicine, agriculture, water treatment, and food technology. Optimization allows for versatility, enabling the material to be applied across a wide range of fields with improved efficacy.

Optimized nanochitosan serves as a valuable tool for researchers and developers. Consistent and well-characterized nanochitosan samples enable more reliable and reproducible experiments, facilitating advancements in research and development. Nanochitosan can be optimized for its ability to absorb and remove pollutants from water, such as heavy metals and organic contaminants. This is essential for maintaining high water quality in aquaculture systems, creating a healthier environment for aquatic organisms (El-Naggar et al., 2022).

Optimized nanochitosan formulations can be employed for disease prevention and treatment in aquaculture. Nanochitosan's antimicrobial properties help control the growth of pathogenic microorganisms, reducing the risk of diseases that can adversely affect fish and shrimp populations (El-Naggar et al., 2020, 2021, 2022). Nanochitosan can be incorporated into aquaculture feeds to enhance nutrient absorption and utilization by aquatic organisms. Optimized nanochitosan formulations can improve feed efficiency, leading to better growth rates and overall health of the cultured species (El-Naggar et al., 2021). Optimization of nanochitosan can result in materials with immunostimulant properties. These formulations can boost the immune response of aquatic organisms, making them more resilient to diseases and environmental stressors.

Nanochitosan, when optimized, can effectively control the formation of biofilms in aquaculture facilities. Biofilms can harbor pathogens and negatively impact water quality (Shard et al., 2014). Nanochitosan's anti-biofilm properties contribute to a cleaner and healthier aquaculture environment. Optimized nanochitosan can be used for pond and water management in aquaculture systems (El-Naggar et al., 2019). Its natural origin and biodegradable properties make it an eco-friendly alternative for pond treatment, helping to maintain a balanced and sustainable aquatic ecosystem.

Nanochitosan's chelating properties can be optimized for the removal of heavy metals from water. This is particularly important in areas where water sources may be contaminated with metals that can be harmful to aquatic organisms, and subsequently, consumers of aquaculture products. Nanochitosan can contribute to the efficient management of waste generated in aquaculture operations. Optimized nanochitosan formulations can be used to treat and solidify organic waste, facilitating its removal and reducing the environmental impact of aquaculture activities. The optimization of nanochitosan aligns with the broader goals of sustainable aquaculture (El-Naggar et al., 2019; Shard et al., 2014). By improving water quality, disease resistance, and overall efficiency, nanochitosan contributes to environmentally responsible and economically viable aquaculture practices. Ongoing optimization efforts in nanochitosan open up opportunities for further research and development. This includes exploring new applications, refining existing formulations, and addressing specific challenges faced by the aquaculture industry.

The optimization of nanochitosan in aquaculture offers a range of benefits, including improved water quality, disease management, feed efficiency, and overall sustainability. It represents a valuable tool for addressing key challenges and promoting the responsible and efficient cultivation of aquatic organisms.

2 Methods of Nanochitosan Synthesis

Several methods are employed for the synthesis of nanochitosan, each with its own advantages and limitations. The choice of method depends on the desired properties and intended applications of the nanochitosan. The methods for efficient nanochitosan synthesis include:

2.1 Acid Hydrolysis

Acid hydrolysis is a pivotal process in the synthesis of nanochitosan, a nanoscale derivative of chitosan, which is renowned for its diverse applications across various fields. This method involves subjecting chitosan to acidic conditions to induce the breakdown of its larger molecular chains, resulting in the formation of smaller nanoparticles (Levitin et al., 2014; Wijesena et al., 2015). Two commonly employed acids for this purpose are hydrochloric acid (HCl) and acetic acid (CH_3COOH). The utilization of these acids facilitates the cleavage of glycosidic linkages within the chitosan structure, leading to the production of nanoscale particles with unique properties (Figs. 1 and 2).

The choice of acid plays a crucial role in determining the characteristics of the resulting nanochitosan. Hydrochloric acid is a strong mineral acid that efficiently catalyzes the hydrolysis process, leading to the formation of smaller and more uniform nanoparticles. On the other hand, acetic acid, a milder organic acid, allows for a more controlled hydrolysis, which is often desirable for preserving certain features of the chitosan structure (Yanat & Schroën, 2021).

The success of the acid hydrolysis process is contingent upon carefully manipulating key reaction parameters. Acid concentration, temperature, and reaction time are critical factors that can be fine-tuned to optimize the size and properties of the resulting nanochitosan. Higher acid concentrations generally lead to more rapid hydrolysis but may also pose challenges such as increased degradation. The temperature of the reaction influences the reaction rate, with elevated temperatures accelerating the hydrolysis process. However, extreme temperatures can potentially lead to undesired side reactions or degradation of the nanochitosan. The duration of the reaction, expressed as reaction time, is another parameter that directly impacts

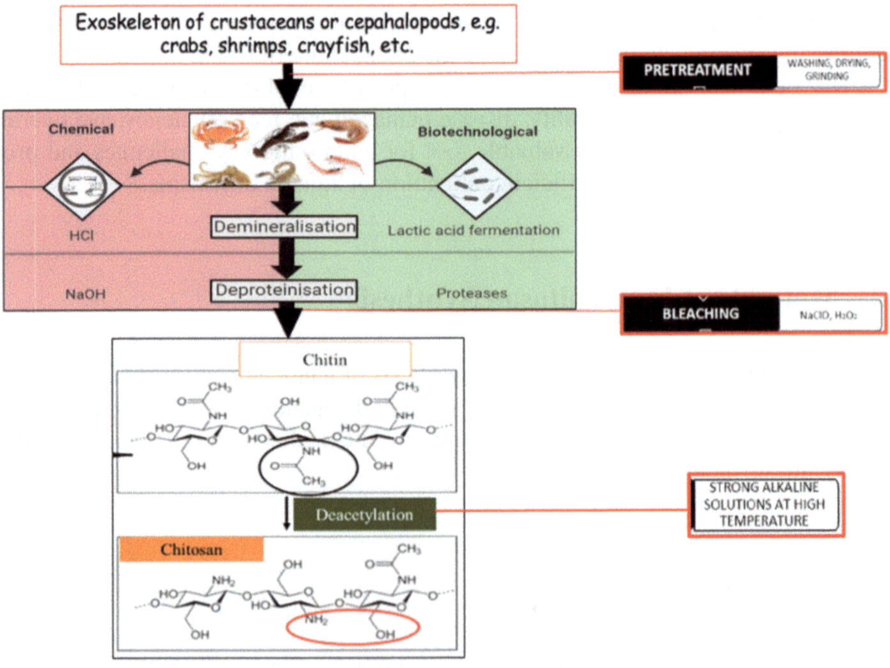

Fig. 1 Comparative analysis of chemical and biotechnological chitin extraction

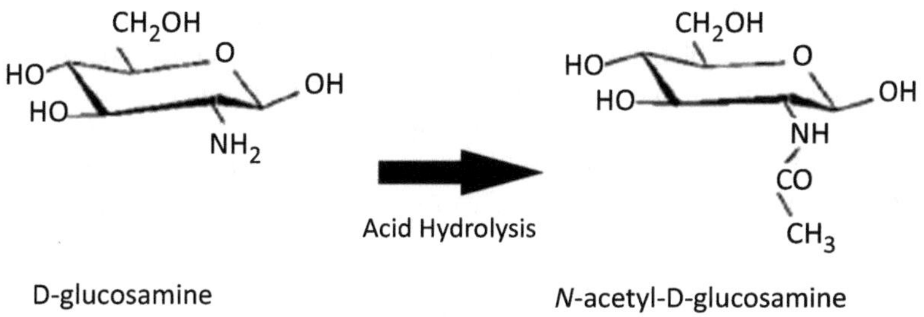

Fig. 2 Chemical reaction of deacetylation

the extent of hydrolysis, with longer reaction times generally resulting in smaller nanoparticles (Zhang et al., 2010).

In essence, the interplay of these parameters allows for the precise control and customization of nanochitosan characteristics. The resulting nanoparticles exhibit unique features such as increased surface area, improved solubility, and altered chemical reactivity compared to their macro-scale counterpart, chitosan. These tailored properties make nanochitosan particularly desirable for a wide array of applications, ranging from biomedicine to environmental remediation and nanotechnology.

Acid hydrolysis is a strategic method for the production of nanochitosan, involving the judicious selection of acid type and careful optimization of reaction parameters. This process not only facilitates the transformation of chitosan into nanoscale particles but also allows for the fine-tuning of critical properties, opening avenues for the tailored application of nanochitosan in various scientific and industrial domains.

2.2 *Ionic Gelation*

Ionic gelation stands as a prominent method in the synthesis of chitosan nanoparticles, characterized by the cross-linking of chitosan molecules through ionic interactions (Shard et al., 2014). This process is frequently accomplished using tripolyphosphate (TPP) as a cross-linking agent, which facilitates the formation of a stable gel network. The method is particularly valued for its simplicity, efficiency, and versatility in producing nanoparticles with controlled properties. At the core of ionic gelation is the interaction between the positively charged amino groups on chitosan and the negatively charged groups on the TPP molecules (Hejjaji et al., 2018). The electrostatic attraction between these oppositely charged species leads to the formation of a gel structure, resulting in the entrapment of chitosan within the network. This cross-linking imparts stability and structural integrity to the nanoparticles, making them suitable for various applications.

The choice of TPP as a cross-linking agent in ionic gelation is driven by its ability to form stable complexes with chitosan. TPP molecules act as bridges, connecting individual chitosan chains and creating a three-dimensional network. This not only enhances the stability of the resulting nanoparticles but also contributes to their uniformity and size control. One of the distinctive features of the ionic gelation method is the ability to modulate the size of the chitosan nanoparticles by adjusting key parameters. The chitosan-to-TPP ratio plays a crucial role in determining the extent of cross-linking, and consequently, the size of the nanoparticles. A higher chitosan-to-TPP ratio tends to result in larger particles due to increased cross-linking density, while a lower ratio leads to smaller particles (Hejjaji et al., 2018). Furthermore, reaction conditions, such as pH and temperature, can be fine-tuned to further influence the size and characteristics of the nanoparticles. Optimal conditions for ionic gelation ensure efficient cross-linking without compromising the structural integrity of the chitosan or the stability of the resulting nanoparticles. The versatility of the ionic gelation method extends beyond size control, as it allows for the encapsulation of various bioactive compounds within the chitosan nanoparticles. This feature makes it a valuable technique in drug delivery systems, where controlled release and targeted delivery are essential.

Ionic gelation represents a robust and flexible approach for the synthesis of chitosan nanoparticles. Through the manipulation of the chitosan-to-TPP ratio and reaction conditions, this method enables precise control over the size and properties of the nanoparticles. The resulting chitosan nanoparticles find applications in drug delivery, biomaterials, and other fields where controlled and tailored characteristics are paramount.

2.3 Nanoprecipitation

Nanoprecipitation is a versatile method employed for the synthesis of chitosan nanoparticles, relying on the rapid mixing of a chitosan solution with a non-solvent to induce precipitation and subsequent nanoparticle formation. This technique has gained prominence due to its simplicity, scalability, and the ability to finely control the size and characteristics of the resulting nanoparticles (Kaya et al., 2013). The key to nanoprecipitation lies in the judicious selection of a non-solvent, with common choices being ethanol or acetone.

The process initiates with the creation of a chitosan solution, typically in a solvent where chitosan is soluble, such as acetic acid. This chitosan solution is then rapidly mixed with a non-solvent, which is miscible with the solvent but not with chitosan. The abrupt introduction of the non-solvent induces a sudden decrease in the solubility of chitosan, leading to its precipitation in the form of nanoparticles. The non-solvent serves a dual purpose of facilitating the rapid dissolution of chitosan and acts as a stabilizing agent during nanoparticle formation (Kaya et al., 2014).

Ethanol and acetone are commonly chosen as non-solvents in nanoprecipitation due to their compatibility with chitosan and their ability to promote the formation of stable nanoparticles. The choice of non-solvent can influence the rate of precipitation, the size distribution, and the overall stability of the chitosan nanoparticles.

One of the notable advantages of the nanoprecipitation method is the precise control it affords over the size of the chitosan nanoparticles. This control is achieved by manipulating parameters such as the concentration of the chitosan solution and the speed of mixing. Higher concentrations of the chitosan solution tend to result in larger nanoparticles due to increased chitosan availability during precipitation. The mixing conditions, including the speed of mixing and the duration of the process, also play a pivotal role in determining the final nanoparticle size. Rapid mixing is often associated with smaller particle sizes, as it minimizes the time available for chitosan particles to aggregate. The nanoprecipitation method's flexibility and ease of implementation make it suitable for various applications, especially in the pharmaceutical and biomedical fields. The controlled size and uniformity of the chitosan nanoparticles make them ideal candidates for drug delivery systems, where precise dosage control and targeted release are critical (Elsawy et al., 2016).

Nanoprecipitation is a powerful method for the synthesis of chitosan nanoparticles, offering a straightforward yet highly controllable approach to nanoparticle formation. By adjusting parameters such as solution concentration and mixing conditions, researchers can tailor the size and properties of the chitosan nanoparticles to meet the specific requirements of diverse applications, particularly in the realm of drug delivery and nanomedicine.

2.4 Coacervation

Coacervation is a distinctive method employed in the synthesis of chitosan nanoparticles, distinguished by the induction of phase separation in a chitosan solution through alterations in environmental conditions, such as pH or temperature (Thomasin et al., 1997). This process leads to the formation of a coacervate, a condensed, liquid-rich phase that emerges from the original solution. The coacervate, rich in chitosan, serves as the precursor for the subsequent fabrication of chitosan nanoparticles (Dubey et al., 2016).

The mechanism of coacervation relies on the manipulation of solution conditions to surpass the solubility threshold of chitosan, prompting its transition from a homogeneous solution to a biphasic system comprising a polymer-rich coacervate and a polymer-poor supernatant. Commonly, changes in pH or temperature induce this phase separation. For instance, adjusting the pH of a chitosan solution to a level where it becomes less soluble can initiate coacervation. Similarly, alterations in temperature can trigger phase separation by influencing the polymer's solubility characteristics (Bhatia et al., 2011).

The coacervate obtained during this process encapsulates chitosan in a concentrated form, providing a basis for the subsequent fabrication of nanoparticles. The coacervate phase can be further processed to solidify and stabilize the chitosan nanoparticles, leading to the creation of a well-defined and controlled nanostructure. A significant advantage of the coacervation method lies in its versatility. This technique allows for the encapsulation of various substances within the chitosan nanoparticles during their formation. This property is particularly valuable in applications such as drug delivery, where active pharmaceutical ingredients or other bioactive compounds can be incorporated into the nanoparticles during their fabrication (Hijazi et al., 2019). The coacervation process thus serves as a platform for creating chitosan nanoparticles with tailored functionalities.

Moreover, the size and properties of the resulting nanoparticles can be finely tuned by adjusting the parameters of the coacervation process. This includes manipulating factors like the initial chitosan concentration, the rate of pH or temperature change, and the overall process duration. These adjustments offer researchers a high degree of control over the characteristics of the chitosan nanoparticles, making coacervation a versatile and customizable method for various applications (Ngan et al., 2014).

Coacervation represents a powerful approach for the synthesis of chitosan nanoparticles, leveraging the phase separation of chitosan from its solution under controlled environmental conditions. This method's adaptability, coupled with its capability to encapsulate diverse substances within the nanoparticles, positions coacervation as a valuable tool in nanotechnology, drug delivery, and other fields requiring precise control over nanoparticle properties.

2.5 *Emulsion Cross-Linking*

Emulsion cross-linking is a specialized method utilized in the synthesis of chitosan nanoparticles, distinguished by the creation of an emulsion, followed by subsequent cross-linking to produce stable nanoparticles. This technique involves the dispersion of a chitosan solution in an oil phase, resulting in the formation of tiny droplets or globules within the oil. The cross-linking step further solidifies these dispersed chitosan entities into nanoparticles, imparting them with enhanced stability and specific properties (Riegger et al., 2018). The emulsion is typically formed by vigorously mixing a chitosan solution with an oil phase, creating a stable dispersion of chitosan droplets within the oil. Commonly used oils include mineral oil or vegetable oils. This emulsion serves as the template for the subsequent fabrication of chitosan nanoparticles.

Cross-linking is a crucial step in the emulsion cross-linking method, and it involves the introduction of cross-linking agents that foster the formation of stable connections between chitosan molecules (Liu & Gao, 2009). Glutaraldehyde and genipin are commonly employed as cross-linking agents in this process. These agents interact with the amino groups on chitosan, creating bridges or cross-links that solidify the nanoparticle structure. The choice of cross-linking agent can influence the stability, biocompatibility, and properties of the resulting nanoparticles.

One notable advantage of emulsion cross-linking is its utility in encapsulating hydrophobic substances within the chitosan nanoparticles (Ribeiro et al., 2020).

The oil phase of the emulsion provides a conducive environment for incorporating hydrophobic compounds, ensuring their effective encapsulation during the nanoparticle formation. This property is particularly valuable in applications such as drug delivery, where hydrophobic drugs or bioactive compounds can be efficiently encapsulated within the chitosan nanoparticles for targeted and controlled release. The size and properties of the chitosan nanoparticles obtained through emulsion cross-linking can be tailored by adjusting various parameters. The ratio of chitosan to the oil phase, the type of oil used, the concentration of the chitosan solution, and the cross-linking conditions are all factors that can be fine-tuned to achieve desired nanoparticle characteristics (Riegger et al., 2018).

Emulsion cross-linking is a valuable method for the synthesis of chitosan nanoparticles, offering a unique approach through the creation of an emulsion followed by cross-linking for nanoparticle formation. Its versatility, especially in encapsulating hydrophobic substances, makes it particularly useful in pharmaceutical and biomedical applications, where controlled release and targeted delivery are critical considerations. The method's ability to be tailored for specific applications underscores its importance in the diverse field of nanotechnology. However, the challenges involved in the technique require some attention. Achieving uniform cross-linking throughout the nanochitosan particles can be challenging. Variations in cross-linking density within the material may impact its mechanical and functional properties. Achieving precise control over the release of encapsulated substances in drug delivery applications can be challenging. The release kinetics may be influenced by factors such as the cross-linking agent, particle size, and cross-linking conditions. Some cross-linking agents used in emulsion cross-linking may introduce cytotoxicity or compromise the biocompatibility of nanochitosan. Addressing these concerns is crucial, especially for biomedical applications. Scaling up the emulsion cross-linking process for industrial production can be challenging. Maintaining consistency in particle properties and achieving cost-effective large-scale production are ongoing challenges. Cross-linked nanochitosan may face stability challenges in harsh environments, such as extreme pH conditions or high temperatures. Developing cross-linking strategies that enhance stability in diverse conditions is important (Riegger et al., 2018). Ensuring reproducibility and standardization of the emulsion cross-linking process is crucial for consistent product quality. Variations in raw materials, emulsification techniques, and cross-linking conditions may impact the final product. Future Prospects in Emulsion Cross-Exploration of novel and advanced cross-linking agents that offer improved biocompatibility, controlled release, and enhanced stability will likely play a significant role in the future of emulsion cross-linking. Developing techniques to precisely control the release kinetics of substances encapsulated within cross-linked nanochitosan opens up opportunities for personalized and targeted drug delivery systems. Integration of smart and responsive materials that can undergo changes in response to specific stimuli (such as pH, temperature, or biological signals) may lead to the development of advanced functional materials with controlled properties. Expanding the use of cross-linked nanochitosan in biomedical applications, such as tissue engineering, wound healing, and regenerative medicine, holds significant promise.

Further research into biocompatibility and in vivo behavior will be crucial. The development of environmentally friendly and sustainable emulsion cross-linking processes, including the use of green cross-linking agents and eco-friendly conditions, aligns with the growing focus on green technologies. Exploring synergies by combining cross-linked nanochitosan with other nanomaterials can lead to the development of multifunctional materials with enhanced properties for various applications (Ribeiro et al., 2020). Emulsion cross-linking has the potential to contribute to the field of precision medicine by enabling the design of nanocarriers with tailored properties for individualized therapeutic approaches. Continued advancements in characterization techniques will aid in better understanding the structure and properties of cross-linked nanochitosan, facilitating more precise optimization and quality control. Addressing the current challenges and exploring these future prospects will contribute to the continued advancement and broader adoption of emulsion cross-linking techniques for nanochitosan, unlocking its potential in diverse industrial and biomedical applications. There are several alternatives to emulsion cross-linking for modifying the properties of nanochitosan. The choice of method depends on the desired application and specific requirements. Chemical cross-linking involves the use of chemical agents to form covalent bonds between chitosan molecules. Common cross-linking agents include glutaraldehyde, genipin, and epichlorohydrin. Chemical cross-linking is versatile and can be tailored for different applications. Physical cross-linking methods involve inducing cross-linking through physical interactions such as hydrogen bonding, van der Waals forces, or electrostatic interactions. Techniques include freeze-drying, irradiation, and ionotropic gelation. Physical cross-linking methods are often milder compared to chemical methods. Enzymatic cross-linking uses enzymes to catalyze the formation of bonds between chitosan molecules. Transglutaminase and tyrosinase are examples of enzymes used for this purpose. Enzymatic methods are often more specific and environmentally friendly. Covalent Bond Formation entails formation of covalent bonds between chitosan molecules can be achieved through various methods such as carbodiimide chemistry or click chemistry. These methods provide a high degree of control over the cross-linking process. Ultraviolet (UV) Cross-Linking: UV cross-linking involves exposing chitosan to ultraviolet light in the presence of a photoinitiator. This method is often used in combination with photopolymerizable cross-linkers and is suitable for applications where controlled spatial and temporal cross-linking is required. Supercritical fluid methods, such as supercritical carbon dioxide or supercritical ethanol treatment, can induce cross-linking by modifying the chitosan structure. These methods offer advantages in terms of mild processing conditions and environmentally friendly outcomes. Layer-by-layer assembly involves depositing alternating layers of oppositely charged materials onto a substrate, including chitosan. By incorporating cross-linking agents or conditions between layers, controlled cross-linking can be achieved. Physical adsorption involves the attachment of molecules onto the surface of chitosan through non-covalent interactions. This method is simple and suitable for certain applications such as drug delivery and sensing. In situ gelation methods involve inducing gelation or cross-linking within the application environment. This is particularly useful for applications such as in vivo drug delivery or wound healing where the gelation occurs at the target site. Electrostatic assembly utilizes electrostatic interactions to induce cross-linking between charged species. This method is often employed in the layer-by-layer assembly technique and can be used to create films and coatings. Each cross-linking method has its advantages and limitations, and the choice depends on the specific requirements of the intended application, the desired properties of the modified nanochitosan, and considerations such as biocompatibility and environmental impact. Researchers often explore multiple methods to optimize the properties of nanochitosan for different applications.

2.6 Supercritical Fluid Technology

Supercritical fluid technology, specifically the use of supercritical carbon dioxide ($scCO_2$), represents an advanced and innovative method for the production of chitosan nanoparticles. In this technique, chitosan is dissolved in a supercritical fluid, typically supercritical carbon dioxide due to its unique properties, and the subsequent rapid depressurization of the system induces the formation of chitosan nanoparticles. This method capitalizes on the distinctive characteristics of supercritical fluids, where they exist in a state that combines the properties of both liquids and gases under specific temperature and pressure conditions (Cardoso et al., 2022).

The process begins with the dissolution of chitosan in supercritical carbon dioxide, creating a homogeneous solution. Supercritical carbon dioxide serves as an excellent solvent due to its tunable properties—adjusting temperature and pressure allows precise control over its density and solvation capabilities. This feature facilitates the efficient dissolution of chitosan in a manner that preserves its structure and properties. The nanoparticle formation is triggered by rapidly depressurizing the system, leading to the expansion of the supercritical fluid and causing the chitosan to undergo precipitation and aggregation, resulting in the formation of nanoparticles. The sudden change in pressure induces the nucleation and growth of chitosan nanoparticles within the supercritical fluid, allowing for the creation of a controlled and uniform nanostructure.

One of the key advantages of supercritical fluid technology is its ability to offer precise control over the size and morphology of the chitosan nanoparticles. By adjusting parameters such as pressure, temperature, and chitosan concentration, researchers can tailor the characteristics of the nanoparticles according to specific application requirements. The controlled and tunable nature of this method makes it particularly advantageous for applications where uniformity in particle size and morphology is crucial.

Furthermore, the use of supercritical carbon dioxide as a solvent has environmental benefits, as it is a non-toxic, non-flammable, and readily available substance. The absence of organic solvents makes this method more sustainable and appealing for applications in pharmaceuticals, food, and other industries where the potential residues of solvents are of concern (Cardoso et al., 2022).

Supercritical fluid technology, particularly utilizing supercritical carbon dioxide, emerges as a sophisticated and versatile method for producing chitosan nanoparticles. Its ability to dissolve chitosan efficiently and induce controlled nanoparticle formation through rapid depressurization offers unique advantages, including precise size and morphology control, as well as environmental sustainability. The method's versatility positions it as a promising avenue for the production of chitosan nanoparticles with tailored properties for various applications in nanotechnology, medicine, and beyond.

2.7 *Enzymatic Hydrolysis*

Enzymatic hydrolysis is a method employed in the production of nanochitosan, involving the use of enzymes to cleave chitosan into smaller particles. This enzymatic process is particularly advantageous for its controlled and mild nature, and it offers a unique approach to obtaining nanochitosan with specific properties (Fonseca et al., 2020).

In this method, enzymes, such as lysozyme, are utilized to catalyze the hydrolysis of chitosan. Lysozyme is a naturally occurring enzyme found in various organisms, including humans, and it possesses the ability to selectively break the glycosidic linkages present in chitosan molecules. This controlled enzymatic degradation leads to the production of nanoscale chitosan particles. The enzymatic hydrolysis method is characterized by its mild reaction conditions compared to other hydrolysis approaches. The enzymatic process occurs under relatively gentle temperatures and pH levels, minimizing the risk of degradation or undesirable modifications to the chitosan structure. This mildness is especially beneficial when dealing with sensitive biomolecules or when aiming to preserve specific properties of chitosan, such as its bioactivity or molecular weight.

The controlled nature of enzymatic hydrolysis allows for precise manipulation of reaction parameters to tailor the properties of the resulting nanochitosan. By adjusting factors such as enzyme concentration, reaction time, and temperature, researchers can modulate the extent of hydrolysis, leading to the production of nanochitosan particles with desired sizes, molecular weights, and functional characteristics. The specificity of enzymes like lysozyme in targeting chitosan's glycosidic linkages ensures a controlled breakdown of the polymer into smaller fragments without compromising its essential properties. This makes enzymatic hydrolysis particularly useful in applications where the preservation of chitosan's inherent characteristics is crucial, such as in biomedical and pharmaceutical applications. Furthermore, the environmentally friendly nature of enzymatic hydrolysis adds to its appeal (Fonseca et al., 2020). Enzymes are biodegradable and often sourced from renewable materials, contributing to the sustainability of the nanochitosan production process. The absence of harsh chemicals or extreme reaction conditions aligns with the principles of green chemistry, making this method an eco-friendly choice (Roncal et al., 2007).

Enzymatic hydrolysis offers a controlled, mild, and environmentally friendly approach to producing nanochitosan. The use of enzymes like lysozyme enables the targeted cleavage of chitosan, leading to the formation of nanoscale particles with specific properties. This method finds application in various fields, including biomedicine, due to its ability to generate nanochitosan while preserving the desirable features of the polymer.

2.8 *Electrostatic Assembly*

Electrostatic assembly is a sophisticated method utilized in the fabrication of chitosan-based structures, involving the layer-by-layer deposition of chitosan and oppositely charged molecules or nanoparticles. This technique capitalizes on the attractive forces between charged entities to create multilayered structures with controlled thickness, composition, and functionalities. Electrostatic assembly offers a versatile and precise approach to engineering complex materials, making it particularly valuable for applications such as coatings and controlled-release systems (dos Santos et al., 2022).

The electrostatic assembly process begins with a substrate, which can be a solid surface or a pre-existing layer of material. Chitosan, with its positively charged amino groups, serves as a cationic component in this assembly. The substrate is first coated with a layer of chitosan. Subsequently, an anionic counterpart, often another molecule or nanoparticle with a negatively charged surface, is introduced onto the chitosan layer. This creates a new layer through the electrostatic attraction between the positively charged chitosan and the negatively charged species. The process can be repeated, alternating between positively and negatively charged components, to build up multiple layers.

One of the significant advantages of electrostatic assembly lies in the ability to precisely control the thickness and composition of the resulting multilayered structures. By adjusting parameters such as the charge density, concentration, and molecular weight of the components, researchers can tailor the properties of the layers. This tunability enables the creation of coatings with specific functionalities or the development of intricate architectures for controlled-release systems. The multilayered structures obtained through electrostatic assembly exhibit a high degree of organization and uniformity. This orderliness arises from the sequential deposition of charged components, allowing for the precise control of layer thickness and arrangement. The resulting films or coatings often display excellent adhesion to the substrate and enhanced mechanical properties. Electrostatic assembly finds extensive use in coatings for various applications, including biomedical devices, sensors, and packaging materials. The controlled-release systems, enabled by this method, are particularly valuable in drug delivery, where the release rate of therapeutic agents can be finely tuned by modifying the composition and thickness of the layers (Ferreira et al., 2020). In addition to controlled release, electrostatically assembled chitosan structures are employed in diverse fields such as tissue engineering and sensing devices.

The versatility of this method makes it suitable for creating functional coatings with tailored properties to meet specific requirements in different industries.

Electrostatic assembly stands out as a powerful and versatile method for fabricating chitosan-based multilayered structures. Its ability to create organized, controlled architectures makes it valuable for applications demanding precise coatings, such as in controlled-release systems and advanced materials in various technological and biomedical fields. Electrostatic assembly is a versatile technique for the production of nanochitosan-based materials. However, it comes with its set of challenges, particularly when applied to nanochitosan production. Achieving a uniform charge distribution on chitosan nanoparticles can be challenging. Variations in charge density may result in non-uniform assembly and affect the overall properties of the nanochitosan product. The stability of chitosan nanoparticles during the electrostatic assembly process is crucial. Aggregation or precipitation of nanoparticles can occur, leading to difficulties in obtaining a well-defined and stable assembly. Electrostatic assembly may introduce impurities from the charged species or stabilizers used in the process. Maintaining the purity of nanochitosan is essential, especially for applications in biomedicine and pharmaceuticals. Transitioning from laboratory-scale experiments to large-scale production can be challenging. Maintaining the same assembly efficiency and product quality at larger scales requires careful consideration of the electrostatic assembly process. Electrostatic assembly is sensitive to environmental conditions such as humidity and temperature. Variations in these conditions may impact the reproducibility and consistency of the assembly process. Achieving precise control over the assembly dynamics, such as the rate of nanoparticle deposition, can be challenging. Fine-tuning these parameters is crucial for tailoring the properties of the nanochitosan product. The choice of stabilizers or surfactants used in the electrostatic assembly process may introduce concerns related to biocompatibility and cytotoxicity. Ensuring the final nanochitosan product is safe for biomedical applications is essential. Electrostatic assembly often involves surface modification of nanoparticles for specific applications. Achieving controlled and well-defined surface modifications can be challenging, impacting the functionality of the nanochitosan. The robustness and stability of films or coatings produced through electrostatic assembly can be a challenge. Ensuring the mechanical strength and durability of the assembled nanochitosan structures is important for practical applications. Integration of nanochitosan assemblies with other materials may present compatibility challenges. Achieving strong adhesion or interactions with diverse substrates can be crucial for certain applications. Regulatory Compliance: Meeting regulatory standards for products developed through electrostatic assembly is essential, particularly in industries like healthcare. Compliance with quality and safety standards is crucial for the acceptance of nanochitosan products in the market. Addressing these challenges requires a thorough understanding of the electrostatic assembly process, careful selection of materials and conditions, and ongoing research to optimize and overcome these limitations. Advances in nanotechnology and materials science continue to contribute to the refinement and broader applicability of electrostatic assembly techniques for nanochitosan production.

2.9 High-Pressure Homogenization

High-pressure homogenization is a technique employed in the production of nanochitosan particles by subjecting chitosan dispersions to elevated pressures. This method is known for its effectiveness in generating small and uniform nanoparticles, making it a valuable approach for various applications in nanotechnology and materials science (Ding & Kan, 2017).

The high-pressure homogenization process involves forcing a chitosan dispersion through a narrow gap or a series of nozzles under high pressure. This intense pressure disrupts the chitosan particles, leading to their breakdown into smaller entities in the nanometer range. The mechanical forces generated during high-pressure homogenization induce particle size reduction and dispersion, resulting in the formation of nanosized chitosan particles. One of the primary advantages of high-pressure homogenization is its capability to produce nanochitosan particles with a high degree of uniformity. The intense forces applied during the process contribute to breaking down larger chitosan aggregates and promoting a more consistent size distribution among the resulting nanoparticles. This uniformity is crucial in applications such as drug delivery and biomaterials, where particle size plays a significant role in determining their behavior and interactions with biological systems.

The size of the nanochitosan particles can be controlled by adjusting key parameters during high-pressure homogenization, including the pressure applied, the number of passes through the homogenizer, and the concentration of the chitosan dispersion. Higher pressures and multiple passes generally lead to further size reduction, while careful optimization allows for the production of nanochitosan particles with specific size ranges tailored to the desired application. The small size and uniformity achieved through high-pressure homogenization contribute to the increased surface area of the nanochitosan particles. This enhanced surface area can be advantageous in applications such as drug delivery, where a higher surface area allows for better interaction with drugs or other bioactive compounds. Additionally, the small particle size improves the dispersibility and stability of the nanochitosan particles in various formulations. The versatility of high-pressure homogenization extends its applicability to a range of chitosan dispersions, including those with different concentrations and viscosities. This adaptability makes the method suitable for a variety of industrial and research applications (Shi et al., 2011).

High-pressure homogenization is an effective and versatile method for producing nanochitosan particles. The process's ability to generate small and uniform nanoparticles with controlled size distribution makes it valuable for applications demanding precision in particle characteristics, such as drug delivery, biomaterials, and other nanotechnology-related fields.

2.10 *Hydrothermal Synthesis*

Hydrothermal synthesis is a method for producing nanochitosan by subjecting chitosan to specific conditions of high temperature and high pressure in an aqueous solution. This process induces chemical reactions and structural transformations within chitosan, resulting in the formation of nanoscale particles. Hydrothermal synthesis is known for its ability to produce nanochitosan with unique properties, such as enhanced crystallinity and stability, making it a valuable approach in materials science and nanotechnology (Huang et al., 2019).

The hydrothermal synthesis process typically involves placing chitosan in an aqueous solution and subjecting it to elevated temperatures and pressures in a sealed reaction vessel. The high-pressure and high-temperature conditions alter the chemical and physical properties of chitosan, leading to the formation of nanoscale structures. The water acts as both a reaction medium and a solvent, facilitating the transformation of chitosan into nanosized particles.

One significant advantage of hydrothermal synthesis is its ability to enhance the crystallinity of nanochitosan. Crystallinity refers to the degree of order in the arrangement of atoms or molecules within a material. The controlled high-pressure and high-temperature conditions of hydrothermal synthesis promote the formation of well-organized structures in the nanochitosan particles, resulting in improved crystallinity. This enhanced crystallinity can influence the mechanical, thermal, and functional properties of the nanochitosan, making it suitable for specific applications where crystalline structures are advantageous. Another notable property of nanochitosan produced through hydrothermal synthesis is increased stability. The controlled conditions during the synthesis process lead to the formation of nanoparticles with enhanced stability in aqueous environments. This stability is crucial for applications such as drug delivery, where nanochitosan particles need to maintain their integrity and functionality in physiological conditions.

The size and morphology of the nanochitosan particles obtained through hydrothermal synthesis can be influenced by adjusting key parameters, including temperature, pressure, and reaction time. Controlling these parameters allows researchers to tailor the properties of the nanochitosan particles for specific applications, providing versatility in the design of nanomaterials. Hydrothermal synthesis is employed in various fields, including biomaterials, catalysis, and drug delivery, due to its ability to produce nanochitosan with unique and desirable properties. The controlled and reproducible nature of this method makes it suitable for large-scale production of nanochitosan with consistent characteristics.

Hydrothermal synthesis is a powerful method for the production of nanochitosan, utilizing high-temperature and high-pressure conditions to induce structural transformations and enhance the crystallinity and stability of the resulting nanoparticles. This method's versatility and the ability to tailor nanochitosan properties make it a valuable tool in materials science and nanotechnology for applications that benefit from unique and well-defined nanomaterials (Hao et al., 2010).

The selection of the appropriate method depends on the specific requirements of the intended application, including the desired particle size, stability, and the incorporation of additional functionalities. Researchers often choose and optimize these methods based on the unique characteristics needed for their nanochitosan applications.

3 Factors Affecting Nanochitosan Optimization

Several factors play a crucial role in the optimization of nanochitosan, influencing its properties and suitability for specific applications. The optimization process involves adjusting these factors to achieve the desired characteristics. Key factors affecting nanochitosan optimization include:

3.1 Chitosan Source and Characteristics

The quality of nanochitosan is influenced by the source of chitosan, and different sources may result in nanochitosan with varying characteristics. The primary sources of chitosan are crustacean shells, fungal biomass, and microbial fermentation, each with its own associated quality considerations.

(i) *Crustacean Shells*:

Chitosan obtained from crustacean shells, such as shrimp and crab shells, is a traditional and widely used source (Fadlaoui et al., 2019).

Quality Considerations:

- High Purity: Crustacean-derived chitosan is generally of high purity, especially if the starting material is sourced from reputable seafood processing industries.
- Variable Quality: The quality of crustacean-derived chitosan can be affected by factors such as the species, age, and processing methods of the crustacean shells.
- Allergenic Concerns: Potential allergenic reactions in individuals with shellfish allergies may limit its use in certain applications.

(ii) *Fungal Sources*:

Chitosan derived from fungal biomass, particularly through fermentation processes using fungi like *Aspergillus niger*.

Quality Considerations:

- Controlled Production: Fungal sources offer a more controlled environment for chitosan production, allowing for more predictable and consistent quality.
- Customizable: The fermentation process can be tailored to produce chitosan with specific characteristics, providing a degree of customization.

- Purity: Fungal-derived chitosan may have fewer impurities compared to crustacean-derived chitosan.

(iii) *Microbial Sources*:

Chitosan-like polymers produced by certain bacteria, such as Vibrio and Bacillus strains.

Quality Considerations:

- Lower Yield: Microbial sources generally yield lower amounts of chitosan-like polymers compared to crustacean shells or fungal biomass.
- Controlled Environment: Similar to fungal sources, microbial fermentation allows for controlled production.

(iv) *Synthetic Methods*:

Chitosan synthesized through chemical depolymerization of chitin.

Quality Considerations:

- Controlled Production: Offers control over the production process but involves the use of chemicals, which may impact environmental considerations.
- Consistency: Provides consistent chitosan quality but may lack some of the natural characteristics found in biologically sourced chitosan.

Quality Parameters for Nanochitosan:

Degree of Deacetylation (DD): The extent to which chitin has been deacetylated to form chitosan. Higher DD often results in enhanced solubility and reactivity.

Molecular Weight: The size of chitosan molecules, influencing properties such as viscosity, film-forming ability, and bioactivity.

Purity: The presence of impurities, such as proteins, minerals, and residual chemicals, can affect the performance of nanochitosan in various applications.

Particle Size and Distribution: For nanochitosan, the size and uniformity of particles are critical parameters that impact properties like surface area, dispersion, and bioavailability.

The quality of nanochitosan is intricately linked to the source of chitosan. While crustacean shells provide a traditional and high-purity source, fungal and microbial sources offer more controlled production processes. Researchers and industries must carefully consider their specific application requirements when choosing the source of chitosan for nanochitosan production (Fadlaoui et al., 2019).

Understanding and controlling these parameters during the production of nanochitosan allow researchers to tailor the material for specific applications. For instance, nanochitosan from fungal sources with precise DD and molecular weight can be designed for targeted drug delivery or wound-healing applications. The choice of chitosan source, along with careful control of DD and molecular weight, provides a versatile platform for designing nanochitosan with desired properties for the aspect of aquaculture to be addressed.

3.2 Degree of Deacetylation (DD)

The degree of deacetylation (DD) is a fundamental parameter that characterizes the extent to which chitin, a linear polysaccharide composed of N-acetylglucosamine units, undergoes deacetylation to form chitosan. This process involves the removal of acetyl groups from chitin, resulting in the exposure of amino groups. The degree of deacetylation has profound effects on the properties of chitosan, and consequently, on its nanoscale counterpart, nanochitosan (Ghannam et al., 2016).

The solubility of chitosan in aqueous solutions is significantly influenced by its degree of deacetylation. As chitosan becomes more deacetylated, it tends to be more soluble in acidic conditions. Chitosan with a higher degree of deacetylation is generally more soluble over a broader pH range, including mildly acidic to neutral conditions. This solubility characteristic is crucial when considering the applications of nanochitosan in various formulations and processes.

The charge density of chitosan is determined by the presence of amino groups on its molecular structure. A higher degree of deacetylation corresponds to an increased density of amino groups, contributing to a more positively charged polymer. This positive charge is particularly valuable in applications where electrostatic interactions are important, such as in the binding of negatively charged substances. For nanochitosan, the charge density can impact its interactions with other particles, surfaces, or biomolecules (Hu et al., 2002).

The reactivity of chitosan is closely related to its amino group content, which is influenced by the degree of deacetylation. Higher DD results in more amino groups being available for chemical reactions. These amino groups can participate in reactions with various compounds, allowing for the functionalization and modification of nanochitosan to tailor its properties for specific applications. The reactivity of nanochitosan is vital in the synthesis of advanced materials, drug delivery systems, and other nanotechnology applications.

The physicochemical properties of nanochitosan, including its size, charge, and surface characteristics, are intricately connected to the degree of deacetylation. Nanochitosan derived from chitosan with a higher degree of deacetylation tends to exhibit different surface properties, zeta potential, and dispersion behavior compared to less deacetylated counterparts. These properties are critical in determining the performance of nanochitosan in various applications, such as drug delivery, nanocomposites, and biomedical materials (Jang et al., 2004; Lavall et al., 2007; Ostolska & Wisniewska, 2014).

Understanding the impact of the degree of deacetylation on nanochitosan properties is essential for tailoring its behavior for specific applications. For example, nanochitosan with a higher DD might be preferred for applications requiring enhanced solubility, reactivity, and positively charged surfaces, whereas lower DD might be suitable for applications where controlled release or specific surface interactions are desired. The degree of deacetylation plays a pivotal role in shaping the properties of chitosan, and consequently, nanochitosan. It influences solubility, charge density, and reactivity, impacting the behavior of nanochitosan in various applications (Kumar et al., 2016). Understanding and controlling the degree of deacetylation allow researchers and industries to customize nanochitosan for specific uses in nanotechnology, materials science, and biomedical fields.

3.3 Molecular Weight

The molecular weight of chitosan is a critical parameter that significantly influences its properties, and by extension it plays a crucial role in shaping the characteristics of nanochitosan. The molecular weight refers to the size of chitosan molecules, which is determined by the degree of polymerization—essentially, the number of monomeric units (glucosamine and N-acetylglucosamine) in the polymer chain. This molecular weight has profound effects on the mechanical, rheological, and biological properties of chitosan and becomes a key consideration in optimizing nanochitosan for specific applications (Paulino et al., 2006).

The mechanical strength and integrity of chitosan-based materials are directly influenced by the molecular weight. High molecular weight chitosan tends to form stronger and more resilient structures, making it suitable for applications where mechanical strength is crucial, such as in the development of nanocomposites, films, and scaffolds for tissue engineering.

Rheology refers to the study of the flow and deformation of materials. The molecular weight of chitosan affects its viscosity and flow behavior. Higher molecular weight chitosan solutions typically exhibit higher viscosity. This rheological property is important in various applications, including the formulation of nanochitosan dispersions, where viscosity can impact processing and application methods. In biological applications, the molecular weight of chitosan can influence its interactions with biological systems. For example, lower molecular weight chitosan may be more easily biodegradable, which can be advantageous for certain drug delivery systems or implantable devices. On the other hand, higher molecular weight chitosan may exhibit prolonged biological activity and slower degradation rates (Sakuma et al., 2011).

The optimization of nanochitosan involves carefully controlling its molecular weight to achieve specific functionalities. This optimization is critical for tailoring nanochitosan for diverse applications, including drug delivery, wound healing, and nanocomposite materials (Sivashankari & Prabaharan, 2017). For drug delivery, the molecular weight can influence the release kinetics of therapeutic agents from nanochitosan carriers. Lower molecular weight chitosan may facilitate faster drug release, while higher molecular weight chitosan may result in sustained release over an extended period. In nanocomposites, the molecular weight can impact the dispersion of nanochitosan within the composite matrix, affecting the overall mechanical and barrier properties of the material (Shukla et al., 2013).

The molecular weight of chitosan can be modulated during the production of nanochitosan to achieve specific functionalities. Techniques such as enzymatic hydrolysis, chemical degradation, or controlled synthesis methods allow researchers to obtain nanochitosan with tailored molecular weights. By carefully selecting the appropriate molecular weight range, researchers can design nanochitosan with desired properties for targeted applications, ensuring optimal performance in various environments. The molecular weight of chitosan is a key determinant in shaping its properties, and this influence extends to nanochitosan. The optimization of nanochitosan involves a nuanced control of its molecular weight to achieve specific functionalities and behaviors, making it a versatile material for diverse applications in nanotechnology, biomedicine, and materials science (Wang et al., 2013).

3.4 Particle Size and Morphology

The size of nanochitosan particles is a crucial factor that significantly influences their performance and applicability in various fields. Nanoparticle size is a key parameter that impacts several important properties, including surface area, bioavailability, and interactions with other substances. The optimization of nanochitosan involves careful control of particle size to achieve specific functionalities, and achieving a uniform distribution of particle sizes is essential for consistent and reliable performance (Wang & Li, 2011).

The interplay of reaction time, temperature, and pressure directly influences the size and morphology of nanochitosan particles. These factors determine the physical structure of the nanoparticles, impacting their surface area, reactivity, and behavior in specific applications. Researchers optimize these parameters to achieve a controlled and reproducible size and morphology of nanochitosan particles. This optimization is critical for tailoring nanochitosan to meet the specific requirements of diverse applications, such as drug delivery or materials synthesis. The surface area-to-volume ratio increases as the size of nanochitosan particles decreases. This elevated surface area is particularly advantageous in applications where a high surface area is desirable, such as in catalysis, adsorption, and drug delivery. Increased surface area enhances the potential for interactions with other substances, making nanochitosan more effective in various applications (Wong et al., 2020).

In biomedical applications, the size of nanochitosan particles is critical for their bioavailability. Nanoparticles with smaller sizes can exhibit improved bioavailability as they may be more readily taken up by cells or tissues. This is particularly relevant in drug delivery systems, where nanochitosan is employed to encapsulate and deliver therapeutic agents to specific targets within the body. The size of nanochitosan particles influences their interactions with other substances, such as drugs, proteins, or cells. For example, smaller particles may have enhanced permeability through biological barriers, making them more effective in drug delivery. Additionally, the size of nanochitosan can impact its ability to interact with pathogens, pollutants, or other materials in applications like water purification or antimicrobial coatings. Achieving optimal performance often involves controlling and optimizing the particle size of nanochitosan (El-Sayed, 2019). Various methods, such as chemical synthesis, mechanical milling, or precipitation techniques, can be employed to control particle size during the production of nanochitosan. Uniformity in particle size distribution is critical for ensuring consistent performance. Narrow size distributions contribute to reproducibility and predictability in the behavior of nanochitosan in different applications (Yin et al., 2017).

The size of nanochitosan particles can be tailored based on specific application requirements. For instance, in drug delivery, the particle size can be optimized to ensure controlled release kinetics, targeted delivery, and enhanced cellular uptake. In nanocomposites and coatings, optimizing particle size ensures uniform dispersion within the matrix, influencing mechanical and barrier properties (El-Sayed, 2019). In environmental applications, such as water treatment, the particle size of nanochitosan is crucial for its adsorption capacity. Smaller particles with increased surface area are more effective in adsorbing pollutants and contaminants from water.

The size of nanochitosan particles is a pivotal factor in determining their performance across various applications. Optimization involves precise control over particle size to achieve desired functionalities, and a uniform distribution of particle sizes is essential for consistent and reliable results. Understanding the impact of particle size on properties such as surface area, bioavailability, and interactions allows researchers and industries to tailor nanochitosan for specific applications in areas ranging from medicine to environmental remediation (Zhao et al., 2018).

3.5 Preparation Method

The method employed to prepare nanochitosan is a critical determinant that strongly influences the properties of the nanoparticles. Different preparation techniques, such as high-pressure homogenization, ionic gelation, or hydrothermal synthesis, can impart distinct characteristics to nanochitosan, impacting its size, morphology, surface properties, and ultimately its performance in various applications. Optimization involves selecting the most suitable preparation method based on the desired properties and intended application of nanochitosan.

High-Pressure Homogenization:
High-pressure homogenization involves subjecting chitosan dispersions to elevated pressures, resulting in nanosized particles. This method is known for producing small and uniform nanochitosan particles. The choice of pressure, number of passes through the homogenizer, and concentration of the chitosan dispersion can be optimized to control the size and distribution of nanochitosan particles. High-pressure homogenization is suitable for applications requiring uniform nanoparticles with consistent properties.

Ionic Gelation:
Ionic gelation involves the cross-linking of chitosan molecules using ionic interactions, often with agents like tripolyphosphate. This method is commonly employed

for synthesizing chitosan nanoparticles with controlled sizes. The ratio of chitosan to cross-linking agent, reaction conditions, and the choice of cross-linking agent can be optimized to control the size and characteristics of nanochitosan particles (Lavall et al., 2007; Riegger et al., 2018). Ionic gelation is often chosen for applications where precise control over particle size and stability is critical.

Hydrothermal Synthesis:
Hydrothermal synthesis involves the reaction of chitosan in an aqueous solution under high-temperature and high-pressure conditions, leading to nanoscale particle formation. This method can enhance the crystallinity and stability of nanochitosan. The temperature, pressure, and reaction time during hydrothermal synthesis can be optimized to control the size and crystallinity of nanochitosan particles. This method is suitable for applications requiring nanochitosan with unique properties, such as improved stability.

Selection of the Most Suitable Method:
The choice of the preparation method is often application-specific. For example, high-pressure homogenization might be preferred for drug delivery applications where particle size uniformity is crucial. Ionic gelation may be chosen for controlled-release systems and hydrothermal synthesis for applications where enhanced stability and crystallinity are required. By selecting the most suitable preparation method, researchers can tailor the properties of nanochitosan to meet the specific requirements of different applications. This includes fine-tuning particle size, surface characteristics, and other relevant attributes.

Nanochitosan, prepared through various methods, finds applications in aquaculture and environmental remediation. The versatility of nanochitosan allows for its optimization to suit the unique demands of different industries and research domains. The method used to prepare nanochitosan is a crucial factor in determining its properties, and optimization involves carefully selecting the most suitable preparation method based on the targeted application. Each method has its advantages and considerations, and the choice depends on the desired characteristics of nanochitosan for a specific use case. Through proper optimization, researchers can harness the unique properties of nanochitosan to address challenges and advance applications across various domains (Hejjaji et al., 2018).

3.6 Reaction Parameters

The synthesis of nanochitosan involves several key parameters, including reaction time, temperature, and pressure, which collectively play a crucial role in determining the properties of the resulting nanoparticles. Optimization in nanochitosan synthesis refers to the systematic adjustment and fine-tuning of these parameters to achieve the desired characteristics in terms of size, morphology, stability, and other relevant properties (Shard et al., 2014).

Reaction Time:
The duration of the reaction time influences the extent of the chemical and physical transformations occurring during nanochitosan synthesis. Longer reaction times may lead to increased particle growth or aggregation. Optimization involves determining the optimal reaction time to achieve the desired nanoparticle size and distribution. Controlling reaction time is crucial for maintaining reproducibility and consistency in nanochitosan synthesis. The reaction time can influence the particle size and morphology of nanochitosan. Longer reaction times may allow for more extensive nucleation and growth, potentially resulting in larger particles or aggregates. The degree of deacetylation, which represents the extent of acetyl group removal from chitosan, is affected by the reaction time. Prolonged reaction times may lead to increased deacetylation, altering the chemical structure of nanochitosan. The reaction time can impact the zeta potential and surface charge of nanochitosan particles. Extended reaction times may affect the protonation of amino groups on the surface, influencing the colloidal stability and interactions with other substances. Reaction time plays a role in the crystallinity of nanochitosan. Longer reaction times may allow for more extensive rearrangement of polymer chains, potentially influencing the crystalline structure and thermal stability of nanochitosan. Prolonged reaction times can lead to chemical instability or undesired reactions. It is crucial to optimize the reaction time to ensure the stability of nanochitosan and minimize the risk of degradation. The biocompatibility of nanochitosan can be influenced by the reaction time, particularly in biomedical applications. Optimizing the reaction time is essential to maintain the desired biological properties of nanochitosan. In drug delivery applications, reaction time can affect the drug loading capacity and release kinetics of nanochitosan carriers. Controlling reaction time is important for achieving the desired drug delivery profiles. Reaction time can impact the rheological properties of nanochitosan solutions. Changes in reaction time may affect the viscosity, gelation behavior, and flow properties of nanochitosan solutions, which are crucial in applications such as coatings and films. Prolonged reaction times may have implications for the environmental sustainability of nanochitosan production. Optimizing reaction times is essential to minimize energy consumption and environmental impact. Reaction time influences the overall cost of nanochitosan production. Balancing the need for specific properties with reaction efficiency is crucial for achieving cost-effective processes. Researchers and manufacturers should carefully optimize the reaction time based on the desired properties for specific applications, considering factors such as biocompatibility, stability, and cost-effectiveness. It is important to conduct systematic studies to understand the impact of varying reaction times on nanochitosan properties and performance.

Temperature:
Temperature is a critical parameter that affects the rate of chemical reactions and the stability of the nanochitosan particles. Higher temperatures may lead to faster reaction kinetics but could also influence the size and morphology of the nanoparticles. Fine-tuning the temperature allows researchers to control the kinetics of nanochitosan synthesis. The choice of temperature is often guided by the desired

properties of the nanoparticles and the specific application requirements (Hijazi et al., 2019). Temperature influences the kinetics of the chemical reactions involved in nanochitosan synthesis. Higher temperatures may accelerate the reaction rates, potentially leading to smaller particle sizes and altered morphologies due to changes in nucleation and growth processes. The temperature can affect the degree of deacetylation of chitosan during the synthesis process. Higher temperatures may promote more efficient deacetylation, resulting in increased conversion of chitosan to chitin, which is an important parameter in nanochitosan optimization. Temperature can impact the zeta potential and surface charge of nanochitosan particles. Variations in temperature may alter the protonation of amino groups on the chitosan surface, influencing the colloidal stability and interactions with other materials. The crystallinity of nanochitosan can be affected by temperature. Higher temperatures may lead to changes in the arrangement of polymer chains, potentially influencing the crystalline structure of nanochitosan and its thermal stability. Elevated temperatures can affect the chemical stability of nanochitosan, leading to degradation or undesired chemical reactions. It is essential to optimize the temperature to ensure the stability of nanochitosan during synthesis and subsequent processing. Temperature plays a role in determining the biocompatibility of nanochitosan, especially in biomedical applications. Careful control of temperature is necessary to avoid adverse effects on the biological properties of nanochitosan. In drug delivery applications, temperature can influence the drug loading capacity and release kinetics of nanochitosan carriers. Controlling temperature during the synthesis process is important for achieving the desired drug delivery profiles. Temperature affects the rheological properties of nanochitosan solutions. Changes in temperature can impact the viscosity, gelation behavior, and flow properties of nanochitosan solutions, which are important considerations in applications such as coatings and films. The temperature used in the synthesis process may have implications for the environmental sustainability of nanochitosan production. Optimizing temperature conditions is essential to minimize energy consumption and environmental impact. Temperature control during the synthesis process can influence the overall cost of nanochitosan production. Balancing the need for specific properties with energy efficiency is crucial for achieving cost-effective processes. Researchers and manufacturers need to carefully optimize and control the temperature during nanochitosan synthesis to achieve the desired properties for specific applications while considering factors such as biocompatibility, stability, and cost-effectiveness.

Pressure:

Pressure is particularly relevant in synthesis methods involving high-pressure conditions, such as hydrothermal synthesis. It influences the solubility of reactants, reaction rates, and the formation of nanoscale structures. Adjusting pressure parameters allows for the optimization of nanochitosan synthesis, especially in methods where pressure plays a significant role. Optimization ensures that the pressure conditions are conducive to the formation of nanoparticles with the desired characteristics (Shard et al., 2014).

pH and Other Reaction Parameters:
Additional parameters, including pH, concentration of reactants, and choice of solvents, can influence the properties of nanochitosan. pH, for instance, affects the degree of deacetylation during synthesis. The acid concentration used in the synthesis of nanochitosan can have significant effects on its properties and applications. Acid concentration influences the size and morphology of nanochitosan particles. Higher acid concentrations may result in smaller particle sizes due to increased protonation of amino groups in chitosan, affecting nucleation and growth during synthesis. Acid concentration can impact the degree of deacetylation of chitosan, which is the extent to which acetyl groups are removed from the chitosan structure. Higher acid concentrations may lead to increased deacetylation, affecting the overall chemical structure of nanochitosan. Acid concentration affects the surface charge of nanochitosan particles. Higher concentrations may lead to increased positive charge on the particles due to the protonation of amino groups, influencing their colloidal stability, dispersion, and interaction with other substances. Acid concentration can influence the crystallinity of nanochitosan. Higher concentrations may promote a more crystalline structure, affecting its mechanical properties, thermal stability, and susceptibility to enzymatic degradation. The choice of acid concentration can impact the chemical and physical stability of nanochitosan. Optimizing the acid concentration is crucial to ensure that the nanochitosan product remains stable over time, especially in applications where long-term stability is required. The acid concentration used in the synthesis process can affect the biocompatibility of nanochitosan, which is crucial for its use in biomedical applications. Balancing the degree of deacetylation and maintaining a suitable surface charge is essential to ensure compatibility with biological systems. In drug delivery applications, the acid concentration can influence the ability of nanochitosan to encapsulate and release drugs. Optimizing the acid concentration helps achieve the desired drug loading capacity and controlled release kinetics. The acid concentration can impact the rheological properties of nanochitosan solutions, affecting their viscosity and flow behavior. This is important for applications such as coatings, where the ability to form a uniform film is influenced by the rheological characteristics of the nanochitosan solution. The acid concentration used in the synthesis process may have environmental implications. Sustainable and eco-friendly approaches need to be considered to minimize the environmental impact of nanochitosan production. The choice of acid concentration can affect the overall cost of nanochitosan synthesis. Optimizing the acid concentration for efficiency and cost-effectiveness is crucial for large-scale production and commercial viability. The acid concentration is a critical parameter in nanochitosan optimization, and careful control of this parameter can tailor the properties of nanochitosan for specific applications. Researchers and manufacturers need to consider these effects to achieve the desired performance and characteristics in nanochitosan-based materials. Optimization involves carefully selecting and adjusting these additional parameters to achieve the desired nanoparticle characteristics. Maintaining optimal pH conditions is crucial for controlling the deacetylation process and other chemical reactions.

3.7 Stabilizers and Surfactants

The incorporation of stabilizers and surfactants during the synthesis of nanochitosan plays a pivotal role in influencing the stability and dispersibility of the resulting nanoparticles. Optimization in this context involves selecting and utilizing appropriate stabilizers to prevent undesirable phenomena such as particle aggregation and to enhance the overall performance of nanochitosan in various applications. Hence, stabilizers and surfactants are added to nanochitosan synthesis processes to mitigate challenges related to particle aggregation, agglomeration, and poor dispersibility. These agents help maintain the stability of the nanochitosan particles by preventing them from clumping together, which can affect their uniformity and performance.

Influence on Stability:
Stabilizers work by creating a protective layer around the nanochitosan particles, preventing them from coming into direct contact with each other. This prevents aggregation and ensures that the nanoparticles remain well-dispersed in the solution or matrix. Surfactants, in particular, can improve the colloidal stability of nanochitosan by reducing the surface tension between the particles and the surrounding medium. This reduction in surface tension helps disperse the nanoparticles more effectively.

Optimization for Dispersibility:
Optimization involves carefully selecting the most suitable stabilizers and surfactants for the specific synthesis method and application of nanochitosan. The choice depends on factors such as the chemistry of the stabilizer, compatibility with the chitosan matrix, and the intended use of nanochitosan. The concentration of stabilizers is a critical parameter in achieving optimal stability and dispersibility. Too little stabilizer may not provide adequate protection against aggregation, while excessive amounts may adversely affect the properties of nanochitosan.

Compatibility with Chitosan:
The selected stabilizers and surfactants should be chemically compatible with chitosan to ensure effective stabilization without causing unwanted side reactions or altering the properties of the nanoparticles. Compatibility considerations are crucial for maintaining the integrity and functionality of nanochitosan. The choice of stabilizers can be tailored based on the specific requirements of different applications. For instance, in biomedical applications, stabilizers that are biocompatible and do not interfere with biological systems may be preferred, while in materials science, stabilizers optimizing nanoparticle dispersion in a specific matrix could be crucial.

Characterization of Nanoparticles:
Optimization also involves the characterization of nanochitosan particles to assess their stability and dispersibility. Techniques such as dynamic light scattering (DLS), zeta potential measurements, and electron microscopy can provide insights into the size distribution, surface charge, and morphology of the nanoparticles. Stabilizers play a role in ensuring the long-term stability of nanochitosan, especially during

storage. Optimizing the stabilizer formulation takes into account the conditions under which nanochitosan will be stored and used, preventing issues such as sedimentation or irreversible aggregation over time (El-Naggar et al., 2019).

The incorporation of stabilizers and surfactants is a critical aspect of optimizing nanochitosan synthesis. The careful selection and concentration optimization of these agents contribute to enhanced stability, colloidal dispersibility, and overall performance of nanochitosan in diverse applications. The choice of stabilizers is application-specific, and their effective utilization is crucial for harnessing the unique properties of nanochitosan in areas such as drug delivery, materials science, and environmental remediation.

3.8 Cross-Linking Agents

In methodologies like ionic gelation, the selection and concentration of cross-linking agents play a critical role in shaping the size and stability of nanochitosan particles. Optimization in this context involves careful consideration of suitable cross-linking agents and the precise control of their concentration during the synthesis process. Ionic gelation involves the cross-linking of chitosan molecules using ionic interactions, often with the assistance of cross-linking agents. These agents facilitate the formation of a network structure, leading to the creation of nanochitosan particles.

The choice of cross-linking agent and its concentration can significantly impact the size of nanochitosan particles. Cross-linking agents contribute to the formation of a stable structure, and the degree of cross-linking influences the final size of the nanoparticles. Cross-linking agents also contribute to the stability of nanochitosan particles by preventing their aggregation. The right choice and concentration of cross-linking agents help maintain the structural integrity of the nanoparticles. The chemical nature of the cross-linking agent is crucial in determining its compatibility with chitosan and its influence on the properties of nanochitosan. Common cross-linking agents in ionic gelation include tripolyphosphate (TPP) and sodium sulfate. The strength of the ionic interaction between the cross-linking agent and chitosan impacts the effectiveness of the cross-linking process. This interaction strength contributes to the stability and structure of the nanochitosan particles.

Optimization involves precisely controlling the concentration of the cross-linking agent. The concentration directly affects the extent of cross-linking and consequently, the size and stability of the resulting nanochitosan particles. Optimization requires a balance; too little cross-linking may lead to insufficient stabilization and larger particle sizes, while excessive cross-linking can result in overly rigid structures and reduced stability (Riegger et al., 2018).

To ensure the success of optimization, researchers use various analytical techniques, including spectroscopy, microscopy, and chromatography, to characterize the effects of post-synthesis treatments. These techniques provide insights into changes in size distribution, surface chemistry, and overall morphology. The optimization process includes the use of analytical techniques such as dynamic light scattering (DLS), transmission electron microscopy (TEM), and scanning electron microscopy (SEM) to characterize the size, distribution, and morphology of nanochitosan particles. These tools provide insights into the effects of varying cross-linking agent concentrations.

The choice and concentration of cross-linking agents can be tailored based on the specific requirements of different applications. For instance, in drug delivery systems, optimizing the size and stability of nanochitosan particles is critical for controlled release and targeting. In the context of environmental considerations, optimizing the use of cross-linking agents may involve exploring green chemistry approaches, such as using eco-friendly cross-linkers that do not introduce harmful byproducts or residues.

The optimization of nanochitosan synthesis through methods like ionic gelation requires a strategic approach to the selection and concentration of cross-linking agents. The careful consideration of these factors ensures the desired particle size and stability, contributing to the overall effectiveness of nanochitosan in various applications, from drug delivery to materials science.

3.9 Post-Treatment Processes

Post-synthesis treatments, including drying methods and surface modifications, are crucial steps in shaping the final properties of nanochitosan. Optimization in this context involves the thoughtful selection and application of appropriate post-treatment processes to enhance specific characteristics, ensuring that the nanochitosan is tailored for intended applications.

The choice of drying method following synthesis can significantly influence the properties of nanochitosan. Different drying techniques, such as freeze-drying or spray-drying, can impact the particle size, morphology, and overall stability of the nanochitosan. The optimization process involves selecting the most suitable drying method based on the desired characteristics of nanochitosan. Freeze-drying, for example, can be advantageous for preserving the structure of nanoparticles, while spray-drying may be preferred for large-scale production and enhanced stability (Shard et al., 2014).

Surface modifications are performed to tailor the surface properties of nanochitosan for specific applications. These modifications can include the introduction of functional groups, coatings, or the attachment of bioactive molecules to enhance the interactions of nanochitosan with other substances. The optimization of surface modifications involves choosing the most effective modification techniques and ensuring that the introduced changes align with the targeted application. For instance, functionalizing the surface of nanochitosan can enhance its compatibility with targeted cells in biomedical applications. Optimization also considers environmental sustainability. Green synthesis approaches, such as using eco-friendly surface modification agents or minimizing energy-intensive drying processes, contribute to the eco-friendliness of nanochitosan production.

The optimization of nanochitosan involves careful consideration of post-synthesis treatments, including drying methods and surface modifications. The selection of these treatments is driven by the specific requirements of the intended applications, and optimization ensures that the final nanochitosan product possesses the desired properties for enhanced performance across diverse fields such as medicine, materials science, and environmental applications.

3.10 Application-Specific Requirements

The application of nanochitosan in aquaculture is a dynamic field that offers various opportunities for optimization based on specific requirements. Whether it is employed for drug delivery to enhance fish health, disease prevention and control, formulation in fish feed, or water purification in aquaculture systems, tailoring nanochitosan is essential for achieving optimal performance and addressing the unique challenges within the aquaculture industry.

Nanochitosan can be optimized for drug delivery systems aimed at enhancing fish health. The size and surface properties of nanochitosan nanoparticles can be tailored to facilitate effective drug encapsulation, controlled release, and targeted delivery to fish tissues, ensuring improved therapeutic outcomes.

Nanochitosan's inherent antimicrobial properties can be harnessed for disease prevention and control in aquaculture. Optimization involves adjusting the size and surface characteristics of nanochitosan to maximize its antimicrobial activity, helping to mitigate the impact of bacterial, viral, or fungal infections in aquaculture settings. Furthermore, nanochitosan can be incorporated into fish feed formulations to improve nutrient delivery and enhance the nutritional value of the feed. The optimization process involves considering the particle size, stability, and bioavailability of nanochitosan to ensure its effective integration into fish feed, promoting growth, and overall fish health. Nanochitosan's adsorption properties can be optimized for water purification in aquaculture systems. By tailoring the size and surface chemistry of nanochitosan particles, it becomes more effective in adsorbing and removing contaminants such as heavy metals, pollutants, and toxins from the aquaculture water, contributing to a healthier environment for fish (El-Naggar et al., 2019).

Optimization also considers the biodegradability and environmental impact of nanochitosan in aquaculture practices. Green synthesis approaches and environmentally friendly post-treatment methods are explored to ensure that nanochitosan aligns with sustainable and eco-friendly aquaculture practices. Nanochitosan can be optimized for the encapsulation of bioactive compounds, such as probiotics or vitamins, in aquaculture. The goal is to enhance the stability and bioavailability of these compounds, ensuring their effective delivery to fish for improved immune response and overall health.

Nanochitosan can be designed as part of controlled release systems, particularly in drug delivery and feed formulations. Optimization involves achieving a balance between the controlled release of active ingredients and the specific needs of the aquaculture system, ensuring sustained benefits over time. Optimization takes into account the challenges presented by aquatic environments. Nanochitosan must be stable in water, resist degradation, and maintain its functionality in the dynamic conditions of aquaculture systems (El-Naggar et al., 2021).

For broader industrial applications in aquaculture, optimization includes considerations of cost-effectiveness and scalability. Nanochitosan production processes are optimized to ensure efficient and economically viable large-scale application in aquaculture practices.

The optimization of nanochitosan for aquaculture applications involves tailoring its properties to meet specific requirements, whether it be for drug delivery, disease prevention, fish feed formulation, or water purification (El-Naggar et al., 2019, 2021). The versatility of nanochitosan allows for a multifaceted approach to enhance fish health, promote sustainable aquaculture practices, and address challenges within the aquaculture industry. Careful experimentation and adjustment of these parameters allow researchers and industries to fine-tune nanochitosan to meet specific performance criteria.

4 Experimental Design for Optimization

The optimization of nanochitosan synthesis demands a systematic and scientific approach through experimental design. The initial phase involves screening experiments employing factorial designs to swiftly identify primary factors influencing nanochitosan properties. These factors, encompassing reaction time, temperature, and chemical concentrations, are systematically varied to assess their main effects on responses like particle size and stability. Following this, response surface methodology (RSM), specifically Central Composite Design (CCD), is applied to delve into the intricate interactions between multiple factors and responses. Mathematical models are derived from the experimental data to predict optimal conditions for nanochitosan synthesis. (Sousa et al., 2020) The subsequent optimization studies utilize desirability functions and algorithms to determine the combination of factors that concurrently maximize or minimize multiple responses, while also considering constraints such as cost and environmental impact. Verification experiments are then conducted to validate the predicted outcomes, and statistical analyses, including ANOVA, are employed to assess the reliability of the optimization process. Scalability considerations, economic feasibility, and an iterative approach further refine the optimization. Environmental impact is addressed through the application of green chemistry principles, and interdisciplinary collaboration ensures a holistic perspective. A comprehensive experimental design offers a rigorous and structured methodology for tailoring nanochitosan properties to meet specific application requirements (Rodolfo et al., 2021).

4.1 Factorial Design

In the context of nanochitosan optimization, the implementation of a factorial design represents a robust and systematic experimental approach aimed at refining the synthesis of nanochitosan. This design methodology is particularly powerful as it enables the simultaneous variation of multiple factors, providing researchers with a comprehensive understanding of how these factors individually and interactively influence key response variables. These critical response variables encompass essential characteristics of nanochitosan, including particle size, stability, and functionality. By systematically manipulating factors such as reactant concentrations, reaction time, and temperature at different levels, the factorial design facilitates a nuanced exploration of a wide range of experimental conditions.

The significance of this approach lies in its capacity to efficiently uncover the intricate relationships between various factors and the desired outcomes in nanochitosan synthesis. Through careful analysis of main effects and interactions, researchers gain insights into the specific contributions of each factor to the properties of nanochitosan. This comprehensive exploration not only aids in the identification of critical factors that significantly impact the synthesis process but also serves as a foundation for subsequent optimization strategies.

The factorial design, therefore, plays a pivotal role in streamlining the optimization process, allowing researchers to navigate through a multitude of experimental conditions in a structured manner. This systematic variation of factors not only enhances the understanding of the synthesis process but also empowers researchers to fine-tune conditions, ultimately optimizing nanochitosan for specific applications. The factorial design serves as a powerful tool, offering efficiency and depth in the exploration of conditions and contributing substantially to the advancement of nanochitosan synthesis for diverse applications.

4.2 Response Surface Methodology (RSM)

Response Surface Methodology (RSM) represents a sophisticated and powerful statistical tool essential for the intricate optimization of nanochitosan synthesis. In nanotechnology, the properties of nanomaterials like nanochitosan play a pivotal role in their applications; RSM stands out for its effectiveness in navigating the complexities of the synthesis process. This methodology excels in probing the intricate relationships among multiple variables and their impact on critical response variables such as particle size, stability, and functionality. RSM extends and refines the insights gained from factorial designs, offering a more detailed examination of the interactions between various factors (Naveen et al., 2020).

Central to the utility of RSM is its reliance on mathematical models, typically quadratic equations, which serve as predictive frameworks. These models facilitate the identification of optimal conditions for nanochitosan synthesis by capturing the multifaceted relationships between input variables and desired responses. Researchers systematically manipulate factors of interest within a well-defined experimental design, allowing for the construction of response surfaces that vividly portray the intricate relationship between these factors and the corresponding responses.

The analysis of these response surfaces becomes a key step in the optimization process. By scrutinizing the surfaces, researchers can precisely pinpoint the conditions that either maximize or minimize desired properties of nanochitosan. This not only streamlines the optimization process but also provides profound insights into the nuanced intricacies of nanochitosan synthesis.

Moreover, RSM's unique ability to navigate the multidimensional parameter space efficiently enhances the precision of optimization efforts. In doing so, it guides researchers toward optimal conditions for tailoring nanochitosan properties to meet the specific requirements of diverse applications. RSM, as an indispensable component of the nanomaterial optimization toolkit, contributes significantly to advancing our understanding and control over nanochitosan synthesis, ultimately paving the way for its enhanced efficacy in various scientific and industrial domains.

5 Characterization Techniques

Characterizing nanochitosan is essential for evaluating its properties and ensuring its suitability for various applications. Several characterization techniques (refer to chapter "Real-World Application of Nanochitosan in Refinery-Produced Water Treatment: A Case Study" for full details) are employed to analyze the structural, morphological, and physicochemical features of nanochitosan.

5.1 Transmission Electron Microscopy (TEM)

Transmission Electron Microscopy (TEM) stands out as a powerful imaging technique that plays a pivotal role in characterizing nanochitosan at the nanoscale. Operating on the principles of transmitted electrons through a thin specimen, TEM provides exceptionally high-resolution images that offer a detailed view of nanochitosan morphology. The capability to visualize nanochitosan at such fine resolutions enables researchers to precisely determine crucial parameters, including particle size, shape, and distribution.

In the context of nanochitosan, TEM becomes indispensable for elucidating the intricate details of its structure. The nanoscale features of chitosan particles, which are often in the range of a few nanometers to a hundred nanometers, can be accurately observed and analyzed. The resulting images not only showcase the external morphology of nanochitosan but also reveal internal structures, such as porosity or core-shell configurations if present.

The determination of particle size through TEM is particularly critical, as it provides insights into the uniformity or heterogeneity of the nanochitosan population. The precise measurement of particle dimensions aids in understanding the physical characteristics that influence the material's behavior and performance in various applications. TEM facilitates the observation of any aggregation or agglomeration of nanochitosan particles, information crucial for assessing the stability of the nanoparticles. The distribution of particles within a sample can be evaluated, offering valuable data on the homogeneity or dispersion of nanochitosan (Ghadi et al., 2014).

TEM serves as an indispensable tool in nanochitosan characterization, offering unparalleled resolution for visualizing the nanomaterial's morphology. The detailed insights gained through TEM contribute significantly to understanding the structure–property relationships of nanochitosan, guiding its optimization for diverse applications in fields such as medicine, materials science, and environmental remediation.

5.2 Scanning Electron Microscopy (SEM)

Scanning Electron Microscopy (SEM) is a powerful imaging technique employed to investigate the surface morphology of nanochitosan, complementing the insights obtained from Transmission Electron Microscopy (TEM). SEM operates by scanning the surface of a specimen with a focused electron beam, and the subsequent detection of secondary electrons produces detailed images of the sample's topography. In the context of nanochitosan, SEM serves as an invaluable tool for capturing high-resolution images that offer a wealth of information about the external features of the nanoparticles (Vladár & Hodoroaba, 2020).

One of the primary advantages of SEM lies in its ability to provide a three-dimensional view of nanochitosan surfaces. This allows researchers to discern not only particle size and shape but also surface characteristics, including roughness, texture, and any unique structural attributes. The information obtained through SEM is particularly valuable for understanding how nanochitosan particles interact with their surroundings and potential interfaces in various applications.

SEM's role in elucidating particle size and shape is crucial for comprehensively characterizing nanochitosan. While TEM excels in capturing nanoscale details, SEM offers a broader view, making it well-suited for observing larger areas and gaining a more representative understanding of the nanochitosan population. The combination of SEM and TEM data provides a comprehensive picture of both internal and external morphological features. Moreover, SEM enables the observation of nanochitosan aggregates or clusters, shedding light on the potential agglomeration behavior of the nanoparticles. This information is vital for assessing the stability and dispersibility of nanochitosan in different environments or formulations.

SEM is an indispensable tool in nanochitosan characterization, offering detailed and visually rich insights into the surface morphology of nanoparticles. The complementary use of SEM alongside TEM enhances the overall understanding of nanochitosan, guiding researchers in tailoring its properties for applications spanning nanomedicine, biomaterials, and environmental science.

5.3 *Dynamic Light Scattering (DLS)*

Dynamic Light Scattering (DLS) serves as a valuable technique for characterizing nanochitosan in solution by providing insights into its hydrodynamic size and size distribution. This non-invasive method capitalizes on the principles of light scattering to analyze the Brownian motion-induced fluctuations in the intensity of scattered light, enabling the determination of particle size in the nanometer range. In the context of nanochitosan, DLS is especially useful for assessing the hydrodynamic size, which accounts for the effective size of the particle in a liquid medium. Unlike techniques such as electron microscopy that measure the dry state size, DLS provides information about the size of nanoparticles as they exist in a solution, considering the hydration layer around the particles. This distinction is crucial for understanding how nanochitosan behaves in practical applications, such as drug delivery or environmental remediation, where interactions in liquid environments are prevalent (Ramos, 2017).

The size distribution data obtained through DLS is equally important, revealing the range of sizes present within the nanochitosan sample. A narrow size distribution indicates homogeneity, while a broader distribution suggests variability in particle sizes. This information is essential for optimizing nanochitosan properties, as applications often require a specific and consistent size range for optimal performance. One of the key applications of DLS in nanochitosan characterization is in assessing the stability of nanoparticles in solution. The technique can detect changes in size distribution over time, providing valuable information about the propensity of nanochitosan to aggregate or agglomerate. Stability is a critical factor, especially in biomedical applications, as it influences the effectiveness of drug delivery systems and the overall performance of nanochitosan-based materials.

Dynamic Light Scattering offers a non-destructive and efficient means of characterizing nanochitosan in solution, providing crucial information about its hydrodynamic size and size distribution. The insights gained through DLS are instrumental in optimizing the stability and performance of nanochitosan in liquid environments, guiding the development of applications in areas such as drug delivery, biomedicine, and nanomaterial science.

5.4 *Fourier Transform Infrared Spectroscopy (FTIR)*

Fourier Transform Infrared Spectroscopy (FTIR) is a powerful analytical technique widely used for the characterization of nanochitosan, providing valuable information about its chemical composition and molecular structure. This technique is based on the principle of measuring the absorption of infrared radiation by the sample, revealing the vibrational modes of its constituent molecules.

In the case of nanochitosan, FTIR is instrumental in identifying the specific chemical bonds and functional groups present in the material. Chitosan, the precursor to nanochitosan, is derived from chitin and undergoes deacetylation to form chitosan. FTIR allows researchers to distinguish between the functional groups associated with chitin (e.g., acetyl groups) and those characteristic of chitosan (e.g., amino groups). The presence and intensity of absorption bands in the infrared spectrum provide a fingerprint that can be used to identify the various chemical components in the nanochitosan structure (Eid, 2022).

One of the critical parameters assessed through FTIR is the degree of deacetylation (DD) of chitosan. The degree of deacetylation is a measure of the extent to which acetyl groups in chitin are removed to form chitosan. FTIR spectra of nanochitosan can be analyzed to quantify the ratio of acetyl groups to amino groups, providing a quantitative measure of the degree of deacetylation. This parameter is crucial because it influences the properties of nanochitosan, including its solubility, charge density, and reactivity. The information obtained from FTIR helps researchers assess the purity and quality of nanochitosan products. FTIR also allows for the identification of additional functional groups that may be introduced during the synthesis or modification of nanochitosan. For example, the presence of certain chemical modifications or coatings can be confirmed by characteristic peaks in the FTIR spectrum.

Fourier Transform Infrared Spectroscopy is a versatile tool for characterizing nanochitosan, providing detailed information about its chemical composition and structural features. The analysis of specific bonds and functional groups, along with the quantification of the degree of deacetylation, enhances our understanding of nanochitosan properties and aids in tailoring its characteristics for diverse applications, including drug delivery, wound healing, and environmental remediation.

5.5 X-Ray Diffraction (XRD)

X-ray diffraction (XRD) is a powerful analytical technique employed for investigating the crystalline structure and crystallinity of nanochitosan, providing essential insights into its molecular arrangement and structural characteristics. This non-destructive method relies on the interaction of X-rays with the crystal lattice of a material, resulting in a diffraction pattern that can be analyzed to determine the spatial arrangement of atoms within the nanochitosan sample.

In the context of nanochitosan, XRD is particularly valuable for discerning the degree of order and organization within the chitosan nanoparticles. Crystallinity refers to the extent to which the atoms in a material are arranged in a repeating, periodic fashion. Chitosan, derived from chitin, inherently possesses a semi-crystalline nature. The application of XRD allows researchers to quantify the level of crystallinity, providing information about the consistency and regularity of the molecular structure within the nanochitosan particles (Chhantyal, 2022).

The diffraction pattern generated by XRD provides information about the spacing between crystal planes, allowing for the determination of the crystallographic parameters of nanochitosan. Peaks in the XRD pattern correspond to specific crystal planes, and the intensity and position of these peaks offer insights into the arrangement of atoms within the crystalline lattice. For nanochitosan, the XRD analysis aids in identifying the crystal phases present and understanding how various processing techniques or modifications may impact the material's crystalline structure. Moreover, XRD is sensitive to changes in the nanochitosan structure induced by factors such as particle size reduction or modifications. As the particle size decreases to the nanoscale, the XRD pattern may exhibit broadening of diffraction peaks, indicating the presence of smaller crystallites or an increase in the amorphous portion of the material.

By employing X-ray Diffraction, researchers gain a deeper understanding of the molecular arrangement and structural characteristics of nanochitosan. This knowledge is pivotal for tailoring the properties of nanochitosan for specific applications, such as drug delivery, where the crystalline structure can influence the material's mechanical strength, stability, and release kinetics. Overall, XRD plays a crucial role in elucidating the nanoscale structure of chitosan nanoparticles, contributing to the optimization of nanochitosan for diverse technological and biomedical applications.

5.6 *Nuclear Magnetic Resonance (NMR) Spectroscopy*

Nuclear Magnetic Resonance (NMR) spectroscopy is a sophisticated analytical technique employed for studying the molecular structure of nanochitosan. Based on the principles of nuclear magnetic resonance, this method provides detailed insights into the chemical environment of atoms within the chitosan molecule, facilitating a comprehensive structural characterization.

In the context of nanochitosan, NMR spectroscopy is particularly valuable for unraveling the complex molecular architecture of chitosan at the atomic level. Chitosan, derived from chitin through deacetylation, consists of repeating units with various functional groups. NMR allows researchers to investigate these structural components, providing information about the arrangement of atoms, the types of chemical bonds present, and the overall configuration of the nanochitosan molecule (Agarwal et al., 2018).

One of the key advantages of NMR is its ability to provide detailed information about the chemical shifts of different nuclei in the chitosan structure. Each type of atom exhibits a characteristic resonance frequency, or chemical shift, in the NMR spectrum. By analyzing these chemical shifts, researchers can identify specific functional groups, such as acetyl and amino groups, and gain insights into the connectivity and arrangement of these groups within the nanochitosan molecule.

NMR spectroscopy also enables the determination of molecular conformation, including the spatial orientation of functional groups and any potential variations in the chitosan structure. This information is crucial for understanding how the nanochitosan structure may impact its properties, such as solubility, reactivity, and interaction with other molecules in different applications. Furthermore, NMR can be used to monitor changes in the chitosan structure due to modifications or processing steps involved in the synthesis of nanochitosan. Whether its alterations in the degree of deacetylation, introduction of surface modifications, or changes in molecular weight, NMR provides a powerful tool for tracking these structural modifications.

Nuclear magnetic resonance spectroscopy is an invaluable technique for elucidating the molecular structure of nanochitosan. Its ability to probe the chemical environment of atoms within the chitosan molecule offers a detailed and nuanced understanding of the nanochitosan structure, aiding in the optimization of its properties for various applications in biomedicine, materials science, and environmental technologies.

5.7 Zeta Potential Measurement

Zeta potential measurements play a crucial role in characterizing the surface charge of nanochitosan particles, providing valuable information about their stability and potential interactions in a solution. Zeta potential is the electric potential at the shear plane surrounding a charged particle in a colloidal suspension, and it serves as a key indicator of the surface charge of nanoparticles.

In the context of nanochitosan, understanding the surface charge is essential because it directly influences the particle's behavior in solution. Nanoparticles with a high zeta potential, whether positively or negatively charged, typically exhibit greater electrostatic repulsion between them. This repulsion prevents particles from aggregating or coalescing, contributing to the overall stability of the colloidal system. On the other hand, low zeta potential values may indicate a higher likelihood of particle aggregation, which can compromise the stability of the nanochitosan dispersion (Fatfat et al., 2023).

The zeta potential of nanochitosan is particularly important in applications where dispersion stability is critical, such as in drug delivery systems, biomedical formulations, or environmental remediation. A stable dispersion ensures that the nanoparticles remain well-dispersed in solution, preventing undesired particle interactions that could lead to precipitation or flocculation. Furthermore, the zeta potential of nanochitosan influences its interactions with other charged entities, such as cells or biomolecules. In drug delivery applications, for instance, the surface charge of nanochitosan particles can affect their cellular uptake and interactions with biological membranes. By controlling the zeta potential, researchers can tailor the surface properties of nanochitosan for optimal performance in specific biological or environmental contexts.

Zeta potential measurements are typically obtained through techniques like electrophoretic light scattering, where the velocity of charged particles under an applied electric field is measured. The resulting zeta potential values offer quantitative information about the surface charge, aiding researchers in fine-tuning the nanochitosan formulation for desired properties.

Zeta potential measurements provide critical insights into the surface charge of nanochitosan particles, guiding the optimization of colloidal stability and interactions in solution. This information is vital for designing nanochitosan-based materials with tailored properties for a wide range of applications, from drug delivery to water treatment.

5.8 UV-Visible Spectroscopy

UV-Visible spectroscopy is a powerful analytical technique employed to study the absorption characteristics of nanochitosan. This method relies on the absorption of ultraviolet (UV) and visible light by molecules, providing valuable information about electronic transitions within the material. In the context of nanochitosan, UV-Visible spectroscopy is utilized to gain insights into the electronic structure of the nanoparticles and to estimate their concentration in solution.

One of the primary applications of UV-Visible spectroscopy for nanochitosan involves the analysis of electronic transitions. When nanochitosan particles are exposed to UV or visible light, electrons within the material can absorb energy and transition to higher energy states. The absorption spectrum obtained from the UV-Visible spectroscopy reflects the wavelengths of light absorbed by the nanochitosan, providing information about the electronic transitions occurring within the material. This data aids in understanding the energy levels and band structure of nanochitosan, offering insights into its optical properties. Furthermore, UV-Visible spectroscopy is a useful tool for estimating the concentration of chitosan in a solution. By analyzing the absorption at a specific wavelength corresponding to a characteristic electronic transition of chitosan, researchers can establish a correlation between absorbance and concentration. This relationship allows for the quantification of chitosan concentration in the nanochitosan sample. Accurate concentration determination is crucial for various applications, particularly in fields like pharmaceuticals or materials science, where precise control of the chitosan concentration is necessary for optimal performance (Abbas, 2019).

UV-Visible spectroscopy is a non-destructive and relatively straightforward technique, making it a convenient method for routine analysis of nanochitosan samples. The obtained spectra can also be used to monitor changes in the electronic structure of nanochitosan induced by modifications or processing steps during synthesis. UV-Visible spectroscopy provides valuable information about the absorption characteristics of nanochitosan, helping researchers understand electronic transitions and estimate chitosan concentration in solution. This analytical tool contributes to the comprehensive characterization of nanochitosan and is essential for optimizing its properties for various applications, including drug delivery, biomaterials, and environmental technologies.

5.9 *Thermogravimetric Analysis (TGA)*

Thermogravimetric Analysis (TGA) is a powerful analytical technique used to investigate the thermal stability and degradation behavior of nanochitosan. This method involves measuring the weight changes of a sample as a function of temperature, providing valuable insights into the material's response to heat and its thermal decomposition characteristics.

In the context of nanochitosan, TGA is employed to understand how the material behaves under different temperature conditions. Chitosan, the precursor to nanochitosan, is known for its thermal stability, but as it undergoes various processing steps to form nanoparticles, its thermal properties may be altered. TGA allows researchers to pinpoint the temperature at which nanochitosan begins to degrade and quantify the extent of degradation, offering critical information about its thermal stability. The TGA curve typically shows weight loss as a function of temperature, with distinct steps corresponding to different thermal events. The initial weight loss often represents the removal of adsorbed or bound water. Subsequent steps may correspond to the decomposition of organic components in nanochitosan, such as the breakdown of chitosan polymer chains (Loganathan et al., 2017).

Analyzing the TGA data provides information about the thermal degradation profile of nanochitosan, including the onset temperature of degradation, the temperature range over which significant degradation occurs, and the residual weight after complete decomposition. This data is essential for understanding the temperature-dependent properties of nanochitosan and is valuable in the optimization of its processing conditions. Moreover, TGA can be employed to assess the purity and composition of nanochitosan samples. Differences in thermal stability between chitosan and potential impurities or additives can be identified through distinct weight loss patterns on the TGA curve.

The insights gained from TGA are crucial for tailoring nanochitosan for specific applications. For instance, in biomedical applications where nanochitosan may be exposed to elevated temperatures during sterilization processes, understanding its thermal stability is paramount. Similarly, in environmental remediation or materials science applications, knowledge of the thermal degradation behavior aids in predicting the material's performance under different temperature conditions.

Thermogravimetric Analysis is a valuable tool for investigating the thermal stability and degradation behavior of nanochitosan. The information obtained through TGA contributes to a comprehensive understanding of the material's thermal properties, guiding its optimization for diverse applications in areas such as medicine, materials science, and environmental technologies.

5.10 *Raman Spectroscopy*

Raman spectroscopy is a powerful analytical technique that provides valuable information about molecular vibrations in nanochitosan, offering insights into structural changes and interactions at the nanoscale. This non-destructive method is based on the inelastic scattering of monochromatic light, revealing the vibrational modes of chemical bonds within a material.

In the context of nanochitosan, Raman spectroscopy serves as a sensitive tool for probing the molecular structure of the nanoparticles. It relies on the phenomenon of Raman scattering, where incident photons interact with the material and undergo energy changes corresponding to the vibrational modes of the chemical bonds. The resulting Raman spectrum provides a unique fingerprint that can be used to identify specific molecular vibrations and gain information about the composition and structural arrangement of nanochitosan.

One of the key advantages of Raman spectroscopy is its ability to provide detailed information about the functional groups and chemical bonds present in nanochitosan. Different vibrational modes correspond to specific molecular configurations, allowing researchers to discern between various components within the nanoparticles. This is particularly useful for monitoring structural changes induced by processing steps or modifications during nanochitosan synthesis. Additionally, Raman spectroscopy is sensitive to the crystallinity of nanochitosan. The presence of crystalline regions can result in distinct Raman peaks, and changes in peak intensities or positions can be indicative of alterations in the crystalline structure. This is crucial for understanding the nanoscale organization of chitosan within the nanoparticles.

Furthermore, Raman spectroscopy is effective in studying interactions at the nanoscale, such as those between nanochitosan and other substances or surfaces. It can provide information about the bonding environment and the influence of external factors on the molecular structure of nanochitosan. The non-destructive nature of Raman spectroscopy makes it particularly valuable for characterizing nanochitosan without altering its properties. This is essential for applications where maintaining the integrity of the nanomaterial is critical (Gouadec & Colomban, 2007).

Raman spectroscopy is a versatile technique that offers detailed insights into the molecular vibrations, structural changes, and interactions within nanochitosan. Its ability to provide information at the nanoscale contributes to the comprehensive characterization of nanochitosan, guiding its optimization for applications in fields such as medicine, materials science, and environmental technologies.

5.11 Atomic Force Microscopy (AFM)

Atomic Force Microscopy (AFM) is a powerful imaging technique that provides high-resolution images of nanochitosan surfaces, offering detailed insights into the nanoscale structure of chitosan particles. AFM utilizes a sharp tip mounted on a flexible cantilever to scan the surface of a sample, measuring the interactions between the tip and the material at the atomic or molecular level.

In the context of nanochitosan, AFM allows researchers to visualize the three-dimensional topography of individual particles with exceptional resolution. The technique provides not only qualitative information about the surface morphology but also quantitative data, such as particle height, size, and distribution. This level of detail is crucial for understanding the nanoscale structure of chitosan nanoparticles and tailoring their properties for specific applications (Klapetek et al., 2011).

One of the primary advantages of AFM is its ability to capture high-resolution images of nanochitosan surfaces, revealing features that may not be easily discernible through other microscopy techniques. AFM can distinguish individual nanoparticles and characterize their surface roughness, providing information about the overall texture of the nanochitosan material. Moreover, AFM is a versatile tool that can be employed in various modes, such as tapping mode or contact mode, depending on the specific requirements of the study. Tapping mode, for instance, minimizes the interaction forces between the tip and the sample, reducing the potential for damage or alteration of the nanochitosan surface during imaging.

AFM's quantitative capabilities extend to measuring mechanical properties, such as stiffness or elasticity, of nanochitosan particles. This information is valuable for understanding the mechanical behavior of the nanoparticles and can be essential in applications where the mechanical properties of nanochitosan play a crucial role, such as in biomaterials or tissue engineering. Furthermore, AFM can be employed to study dynamic processes in real-time, providing insights into changes in nanochitosan structure or behavior under different environmental conditions. This dynamic imaging capability is especially useful for tracking interactions between nanochitosan particles and other substances.

Atomic Force Microscopy is a versatile and powerful tool for characterizing nanochitosan at the nanoscale. Its ability to provide high-resolution images, quantitative data on particle morphology, and insights into mechanical properties contributes to a comprehensive understanding of the nanoscale structure of chitosan nanoparticles. AFM's versatility makes it an invaluable technique in optimizing nanochitosan for diverse applications, from drug delivery to materials science and beyond.

5.12 *Brunauer–Emmett–Teller (BET) Surface Area Analysis*

Brunauer–Emmett–Teller (BET) Surface Area Analysis is a widely used technique to determine the specific surface area of nanochitosan, offering valuable insights into its reactivity, especially in applications involving adsorption processes. The BET method is based on the principle of physical adsorption of gas molecules onto the surface of a solid material, and it is particularly effective for porous materials like nanochitosan.

In the context of nanochitosan, the specific surface area is a critical parameter that reflects the amount of surface area available for interactions with other substances. The BET analysis involves exposing the nanochitosan sample to a known gas, typically nitrogen, at various pressures. As the gas molecules adsorb onto the surface of the nanochitosan particles, the resulting isotherm is used to calculate the specific surface area based on the BET equation. The specific surface area determined through BET analysis provides quantitative information about the nanochitosan's porosity and the extent of its exposed surface. This information is crucial for understanding the material's reactivity, particularly in processes where a high surface area is desirable, such as in adsorption-based applications (Nasrollahzadeh et al., 2019).

Nanochitosan, due to its porous nature and large surface area, exhibits enhanced reactivity and adsorption capacity. This makes it valuable in applications such as water treatment, where nanochitosan can adsorb contaminants like heavy metals or organic pollutants. The specific surface area obtained from BET analysis serves as a key parameter in predicting and optimizing the adsorption efficiency of nanochitosan for these applications.

Furthermore, the BET analysis is often used to assess the impact of various factors, including particle size, synthesis methods, or surface modifications, on the specific surface area of nanochitosan. This allows researchers to tailor the nanochitosan's properties for specific applications by optimizing its surface characteristics.

The quantitative data provided by BET surface area analysis contributes to a comprehensive understanding of nanochitosan's physical properties and aids in the design and optimization of nanomaterials for a wide range of applications, including environmental remediation, drug delivery, and catalysis.

The practical applications of some of these described techniques in the optimization of nanochitosan are discussed later in chapter "Real-World Application of Nanochitosan in Refinery-Produced Water Treatment: A Case Study"..

6 Challenges and future prospects

Nanochitosan optimization techniques face several challenges, and addressing these challenges is crucial for the successful development and application of nanochitosan-based materials. Achieving consistent and reproducible synthesis methods for nanochitosan is challenging. Variations in raw materials, reaction conditions, and equipment can lead to differences in the properties of the nanochitosan produced. Translating laboratory-scale synthesis to large-scale production is a common hurdle. Maintaining the same properties and performance of nanochitosan at a larger scale requires careful consideration of the production process and conditions. Ensuring the purity of nanochitosan is essential for its safe use in various applications. Contaminants from the starting materials or by-products of the synthesis process can affect the performance and biocompatibility of nanochitosan. Understanding the biological interactions and potential toxicity of nanochitosan is crucial, especially in biomedical applications. The impact on cells, tissues, and organs must be thoroughly investigated to ensure safety. Nanochitosan may undergo agglomeration or degradation over time, affecting its stability and shelf life. Developing methods to enhance the stability of nanochitosan and extend its shelf life is an ongoing challenge (El-Sayed, 2019). Accurate characterization of nanochitosan is essential for understanding its properties and optimizing its performance. However, conventional characterization techniques may not be suitable for nanoscale materials, requiring the development of new and advanced analytical methods. Tailoring the surface properties of nanochitosan to meet specific application requirements often involves functionalization and surface modification. Achieving controlled and well-defined modifications poses challenges in terms of reproducibility and maintaining the core properties of nanochitosan. The cost of raw materials and the complexity of the synthesis process can impact the economic viability of nanochitosan production. Finding cost-effective and sustainable sources of chitosan and optimizing production processes are ongoing challenges. Nanochitosan-based products may face regulatory challenges due to the lack of standardized testing protocols and guidelines. Addressing these issues is essential for the successful commercialization and widespread use of nanochitosan. Different applications of nanochitosan, such as drug delivery, wound healing, and water treatment, have unique challenges. Optimizing nanochitosan for specific applications requires a thorough understanding of the application requirements and potential hurdles. Addressing these challenges requires interdisciplinary collaboration between researchers in materials science, chemistry, biology, and engineering to develop robust nanochitosan optimization techniques that can unlock the full potential of this versatile material.

6.1 Conclusion

The optimization of nanochitosan involves careful consideration of various parameters, including acid concentration, temperature, and reaction time. These factors collectively impact the physicochemical properties and performance of nanochitosan in diverse applications. Achieving the desired particle size, degree of deacetylation, surface charge, crystallinity, and other characteristics requires a nuanced approach to parameter optimization. Researchers and manufacturers must navigate challenges such as standardization, scalability, biocompatibility, and environmental sustainability. The effects of acid concentration, temperature, and reaction time on nanochitosan properties are interconnected, and a holistic understanding is crucial for tailoring nanochitosan to specific applications. Optimizing nanochitosan involves striking a balance between achieving the desired properties and addressing practical considerations such as cost-effectiveness and environmental impact. Advances in nanochitosan optimization techniques hold promise for applications ranging from drug delivery and biomedical devices to water treatment and beyond. Continued research and interdisciplinary collaboration are essential to unlock the full potential of nanochitosan, ensuring its successful integration into various industrial and biomedical settings.

References

Abbas, Q. (2019). Understanding the UV-vis spectroscopy for nanoparticles. *Journal of Nanomaterials & Molecular Nanotechnology, 8*, 3. https://doi.org/10.4172/2324-8777.1000268

Agarwal, N., Nair, M. S., Mazumder, A., & Poluri, K. M. (2018). Chapter 3 – Characterization of nanomaterials using nuclear magnetic resonance spectroscopy. In S. M. Bhagyaraj, O. S. Oluwafemi, N. Kalarikkal, & S. Thomas (Eds.), *Micro and nano technologies, characterization of nanomaterials* (pp. 61–102). Woodhead Publishing. ISBN 9780081019733.

Bhatia, A., Shard, P., Chopra, D., & Mishra, T. (2011). Chitosan nanoparticles as carrier of immunorestoratory plant extract: Synthesis, characterization and immunorestoratory efficacy. *International Journal of Drug Delivery, 3*, 381–385.

Cardoso, V. M., Ferreira, L. M. B., Comparetti, E. J., Sampaio, I., Ferreira, N. N., Miranda, R. R., & Zucolotto, V. (2022). Chapter 4 – Stimuli-responsive polymeric nanoparticles as controlled drug delivery systems. In V. Gajbhiye, K. R. Gajbhiye, & S. Hong (Eds.), *Stimuli-responsive nanocarriers* (pp. 87–117). Academic Press. ISBN 9780128244562.

Chhantyal, P. (2022). *The use of X-ray diffraction for nanoparticle characterization*. The State of Scientific Marketing '23 Market Report. https://www.azooptics.com/Article.aspx?ArticleID=2180

Cho, Y., Jun, T. K., & Park, H. J. (2012). Preparation, characterization, and protein loading properties of N-acyl chitosan nanoparticles. *Journal of Applied Polymer Science, 124*, 1366–1371. https://doi.org/10.1002/app.34931

Ding, Y., & Kan, J. (2017). Optimization and characterization of high pressure homogenization produced chemically modified starch nanoparticles. *Journal of Food Science and Technology, 54*(13), 4501–4509. https://doi.org/10.1007/s13197-017-2934-8. Epub 2017 Oct 26. PMID: 29184257; PMCID: PMC5686031.

dos Santos, A. M., Carvalho, S. G., Ferreira, L. M. B., Chorilli, M., & Gremião, M. P. D. (2022). Understanding the role of electrostatic interactions on the association of 5-fluorouracil to chitosan-TPP nanoparticles. *Colloids and Surfaces A: Physicochemical and Engineering Aspects, 640*, 128417. ISSN 0927-7757.

Dubey, S., Mody, N., Sharma, R., Agrawal, U., & Vyas, S. P. (2016). Chapter 4 – Nanobiomaterials: Novel nanoplatforms for protein and peptide delivery. In A. M. Grumezescu (Ed.), *Nanobiomaterials in drug delivery* (pp. 111–146). William Andrew Publishing. ISBN 9780323428668.

Eid, M. M. (2022). Characterization of nanoparticles by FTIR and FTIR-microscopy. In *Handbook of consumer nanoproducts*. Springer. https://doi.org/10.1007/978-981-16-8698-6_89

El-Naggar, M. M., Abou-Elmagd, W. S. I., Suloma, A. M., El-Shabaka, H. A., Khalil, M. T., & Abd El-Rahman, F. A. A. (2019). Optimization and physicochemical characterization of chitosan and chitosan nanoparticles extracted from the crayfish Procambarus clarkii wastes. *Journal of Shellfish Research, 38*(2), 385–395. https://doi.org/10.2983/035.038.0220

El-Naggar, M. M., Haneen, D. S. A., Mehany, A. B. M., & Khalil, M. T. (2020). New synthetic chitosan hybrids bearing some heterocyclic moieties with potential activity as anticancer and apoptosis inducers. *International Journal of Biological Macromolecules, 150*, 1323–1330. https://doi.org/10.1016/j.ijbiomac.2019.10.142

El-Naggar, M. M., Salah, S., El-Shabaka, H. A., Abd El-Rahman, F. A. A., Khalil, M. T., & Suloma, A. (2021). Efficacy of dietary chitosan and chitosan nanoparticles supplementation on health status of Nile tilapia, Oreochromis niloticus (L.). *Aquaculture Reports, 19*, 100628. https://doi.org/10.1016/j.aqrep.2021.100628

El-Naggar, M. M., Medhat, F., & Taha, A. (2022). Applications of chitosan and chitosan nanoparticles in fish aquaculture. *Egyptian Journal Of Aquatic Biology And Fisheries, 26*(1), 23–43. https://doi.org/10.21608/ejabf.2022.213365

Elsawy, M. A., Saad, G. A., & Sayed, A. M. (2016). Mechanical, thermal, and dielectric properties of poly(lactic acid)/chitosan nanocomposites. *Polymer Engineering and Science, 56*, 987–994. https://doi.org/10.1002/pen.24328

El-Sayed, A.-F. M. (2019). *Tilapia culture*. Academic Press. https://doi.org/10.1016/C2017-0-04085-5

Fadlaoui, S., El Asri, O., Mohammed, L., Sihame, A., Omari, A., & Melhaoui, M. (2019). Isolation and characterization of chitin from shells of the freshwater crab Potamon algeriense. *Progress on Chemistry and Application of Chitin and its Derivatives, XXIV*, 23–35. https://doi.org/10.15259/PCACD.24.002

Fatfat, Z., Karam, M., Maatouk, B., Fahs, D., & Gali-Muhtasib, H. (2023). Chapter 7 – Nanoliposomes as safe and efficient drug delivery nanovesicles. In A. K. Nayak, M. S. Hasnain, B. Laha, & S. Maiti (Eds.), *Advanced and modern approaches for drug delivery* (pp. 159–197). Academic Press. ISBN 9780323916684.

Ferreira, L. M. B., dos Santos, A. M., Boni, F. I., dos Santos, K. C., Robusti, L. M. G., de Souza, M. P. C., Ferreira, N. N., Carvalho, S. G., Cardoso, V. M. O., Chorilli, M., Cury, B. S. F., de Godoi, D. R. M., & Gremião, M. P. D. (2020). Design of chitosan-based particle systems: A review of the physicochemical foundations for tailored properties. *Carbohydrate Polymers, 250*, 116968. ISSN 0144-8617.

Fonseca, L. R., Santos, T. P., Czaikoski, A., & Cunha, R. L. (2020). Modulating properties of polysaccharides nanocomplexes from enzymatic hydrolysis of chitosan. *Food Research International, 137*, 109642. ISSN 0963-9969.

Ghadi, A., Mahjoub, S., Tabandeh, F., & Talebnia, F. (2014). Synthesis and optimization of chitosan nanoparticles: Potential applications in nanomedicine and biomedical engineering. *Caspian Journal of Internal Medicine, 5*(3), 156–161. PMID: 25202443; PMCID: PMC4143737.

Ghannam, H., Talab, A. S., Dolganova, N. V., Hussein, A. M. S., & Abdelmaguid, N. M. (2016). Characterization of chitosan extracted from different crustacean shell wastes. *Journal of Applied Sciences, 16*, 454–461.

Gouadec, G., & Colomban, P. (2007). Raman spectroscopy of nanomaterials: How spectra relate to disorder, particle size and mechanical properties. *Progress in Crystal Growth and Characterization of Materials, 53*(1), 1–56. ISSN 0960-8974.

Hao, R., Xing, R., Xu, Z., Hou, Y., Gao, S., & Sun, S. (2010). Synthesis, functionalization, and biomedical applications of multifunctional magnetic nanoparticles. *Advanced Materials, 22*, 2729–2742.

Hejjaji, E. M. A., Smith, A. M., & Morris, G. A. (2018). Evaluation of the mucoadhesive properties of chitosan nanoparticles prepared using different chitosan to tripolyphosphate (CS:TPP) ratios. *International Journal of Biological Macromolecules, 120*, 1610–1617. https://doi.org/10.1016/j.ijbiomac.2018.09.185

Hijazi, N., Le Moigne, N., Rodier, E., Sauceau, M., Vincent, T., Benezet, J. C., & Fages, J. (2019). Biocomposite films based on poly(lactic acid) and chitosan nanoparticles: Elaboration, microstructural and thermal characterization. *Polymer Engineering and Science, 59*, E350–E360. https://doi.org/10.1002/pen.24983

Hu, Y., Jiang, X., Ding, Y., Ge, H., Yuan, Y., & Yang, C. (2002). Synthesis and characterization of chitosan-poly(acrylic acid) nanoparticles. *Biomaterials, 23*, 3193–3201. https://doi.org/10.1016/S0142-9612(02)00071-6

Huang, G., Lu, C., & Yang, H. (2019). Chapter 3 – Magnetic nanomaterials for magnetic bioanalysis. In X. Wang & X. Chen (Eds.), *Micro and nano technologies, novel nanomaterials for biomedical, environmental and energy applications* (pp. 89–109). Elsevier. ISBN 9780128144978.

Jang, M. K., Kong, B. G., Jeong, Y. I., Lee, C. H., & Nah, J. W. (2004). Physicochemical characterization of α-chitin, β-chitin, and γ-chitin separated from natural resources. *Journal of Polymer Science Part A: Polymer Chemistry, 42*, 3423–3432.

Kaya, M., Sargin, I., Tozak, K. O., Baran, T., Erdogan, S., & Sezen, G. (2013). Chitin extraction and characterization from *Daphnia magna* resting eggs. *International Journal of Biological Macromolecules, 61*, 459–464.

Kaya, M., Baran, T., Mentes, A., Asaroglu, M., Sezen, G., & Tozak, J. O. (2014). Extraction and characterization of α-chitin and chitosan from six different aquatic invertebrates. *Food Biophysics, 9*, 145–157.

Klapetek, P., Valtr, M., Nečas, D., Salyk, O., & Dzik, P. (2011). Atomic force microscopy analysis of nanoparticles in non-ideal conditions. *Nanoscale Research Letters, 6*, 514. http://www.nanoscalereslett.com/content/6/1/514

Kumar, A., Vimal, A., & Kumar, A. (2016). Why chitosan? From properties to perspective of mucosal drug delivery. *International Journal of Biological Macromolecules, 91*, 615–622.

Lavall, R. L., Assis, O. B. G., & Campana-Filho, S. P. (2007). β-Chitin from the pens of *Loligo* sp.: Extraction and characterization. *Bioresource Technology, 98*, 2465–2472.

Levitin, S. V., Gal'braikh, L. S., Grunin, Y. B., & Masas, D. S. (2014). Supramolecular structure of chitosan acid-hydrolysis products. *Fibre Chemistry, 46*(3), 147–150. https://doi.org/10.1007/s10692-014-9579-4

Liu, H., & Gao, C. (2009). Preparation and properties of ionically cross-linked chitosan nanoparticles. *Polymers for Advanced Technologies, 20*, 613–619. https://doi.org/10.1002/pat.1306

Loganathan, S., Valapa, R. B., Mishra, R. K., Pugazhenthi, G., & Thomas, S. (2017). Chapter 4 – Thermogravimetric analysis for characterization of nanomaterials. In S. Thomas, R. Thomas, A. K. Zachariah, & R. K. Mishra (Eds.), *Micro and nano technologies, thermal and rheological measurement techniques for nanomaterials characterization* (pp. 67–108). Elsevier. ISBN 9780323461399.

Murakami, Y. & Shimoyama, Y. (2017). Production of nanosuspension functionalized by chitosan using supercritical fluid extraction of emulsion;The Journal of Supercritical Fluids, 128, 121-127, ISSN 0896-8446, https://doi.org/10.1016/j.supflu.2017.05.014.

Nasrollahzadeh, M., Atarod, M., Sajjadi, M., Sajadi, S. M., & Issaabadi, Z. (2019). Chapter 6 – Plant-mediated green synthesis of nanostructures: Mechanisms, characterization, and applications. In M. Nasrollahzadeh, S. Mohammad Sajadi, M. Sajjadi, Z. Issaabadi, & M. Atarod (Eds.), *Interface science and technology* (Vol. 28, pp. 199–322). Elsevier. ISSN 1573-4285, ISBN 9780128135860.

Naveen, N. R., Kurakula, M., & Gowthami, B. (2020). Process optimization by response surface methodology for preparation and evaluation of methotrexate loaded chitosan nanoparticles. *Materials Today: Proceedings, 33*(Part 7), 2716–2724. ISSN 2214-7853.

Ngan, L. T. K., Wang, S. L., Hiep, D. M., Luong, P. M., Vui, N. T., Dinh, T. M., & Dzung, N. A. (2014). Preparation of chitosan nanoparticles by spray drying, and their antibacterial activity. *Research on Chemical Intermediates, 40*, 2165–2175.

Ostolska, I., & Wisniewska, M. (2014). Application of the zeta potential measurements to explanation of colloidal Cr_2O_3 stability mechanism in the presence of the ionic polyamino acids. *Colloid & Polymer Science, 292*, 2453–2464.

Paulino, A. T., Simionato, J. I., Garcia, A. J., & Nozaki, J. (2006). Characterization of chitosan and chitin produced from silkworm crys-talides. *Carbohydrate Polymers, 64*, 98–103.

Ramos, A. P. (2017). Dynamic light scattering applied to nanoparticle characterization. In A. L. Da Róz, M. Ferreira, F. de Lima Leite, & O. N. Oliveira (Eds.), *Micro and nano technologies, nanocharacterization techniques* (pp. 99–110). William Andrew Publishing. ISBN 9780323497787.

Ribeiro, E.F.. São Paulo State University (Unesp), Institute of Biosciences, Humanities and Exact Sciences (Ibilce), Campus São José do Rio Preto, SP, 15054-000, Brazil. *et al.* Chitosan and crosslinked chitosan nanoparticles: Synthesis, characterization and their role as Pickering emulsifiers, https://doi.org/10.1016/j.carbpol.2020.116878 (2020).

Riegger, B. R., Baurer, B., Mirzayeva, A., Tovar, G. E. M., & Bach, M. (2018). Systematic approach for preparation of chitosan nanoparticles via emulsion crosslinking as potential adsorbent in waste-water treatment. *Carbohydrate Polymers, 180*, 46–54.

Rodolfo, C., Eusébio, D., Ventura, C., Nunes, R., Florindo, H. F., Costa, D., & Sousa, Â. (2021). Design of experiments to achieve an efficient chitosan-based DNA vaccine delivery system. *Pharmaceutics, 13*(9), 1369. https://doi.org/10.3390/pharmaceutics13091369. PMID: 34575445; PMCID: PMC8471690.

Roncal, T., Oviedo, A., de Armentia, I. L., Fernández, L., & Villarán, M. C. (2007). High yield production of monomer-free chitosan oligosaccharides by pepsin catalyzed hydrolysis of a high deacetylation degree chitosan. *Carbohydrate Research, 342*(18), 2750–2756. https://doi.org/10.1016/j.carres.2007.08.023

Sakuma, S., Hayashi, M., & Akashi, M. (2011). Design of nanoparticles composed of graft copolymers for oral peptide delivery. *Advanced Drug Delivery Reviews, 47*, 21–37.

Shard, P., Bhatia, A., & Sharma, D. (2014). Optimization and physicochemical parameters on synthesis of chitosan nanoparticles by ionic gelation technique. *International Journal of Drug Delivery, 6*, 58–63.

Shi, A. M., Li, D., Wang, L. J., Li, B. Z., & Adhikari, B. (2011). Preparation of starch-based nanoparticles through high-pressure homogenization and miniemulsion cross-linking: Influence of various process parameters on particle size and stability. *Carbohydrate Polymers, 83*, 1604–1610.

Shukla, S. K., Mishra, A. K., Arotiba, O. A., & Mamba, B. B. (2013). Chitosan-based nanomaterials: A state-of-the-art review. *International Journal of Biological Macromolecules, 59*, 46–58. https://doi.org/10.1016/j.ijbiomac.2013.04.043

Sivashankari, P. R., & Prabaharan, M. (2017). *Deacetylation modification techniques of chitin and chitosan*. Elsevier. https://doi.org/10.1016/B978-0-08-100230-8.00005-4

Sousa, Â., Faria, R., Albuquerque, T., Bhatt, H., Biswas, S., Queiroz, J. A., & Costa, D. (2020). Design of experiments to select triphenylphosphonium-polyplexes with suitable physicochemical properties for mitochondrial gene therapy. *Journal of Molecular Liquids, 302*, 112488. https://doi.org/10.1016/j.molliq.2020.112488

Thomasin, C., Merkle, H. P., & Gander, B. A. (1997). Physico-chemical parameters governing protein microencapsulation into biodegradable polyesters by coacervation. *International Journal of Pharmaceutics, 147*(2), 173–186.

Vladár, A. E., & Hodoroaba, V. (2020). Chapter 2.1.1 – Characterization of nanoparticles by scanning electron microscopy. In V.-D. Hodoroaba, W. E. S. Unger, & A. G. Shard (Eds.), *Micro and nano technologies, characterization of nanoparticles* (pp. 7–27). Elsevier. ISBN 9780128141823.

Wang, Y., & Li, J. (2011). Effects of chitosan nanoparticles on survival, growth and meat quality of tilapia, *Oreochromis nilotica. Nanotoxicology, 5*, 425–431.

Wang, Y., Chang, Y., Yu, L., Zhang, C., Xu, X., Xue, Y., Li, Z., & Xue, C. (2013). Crystalline structure and thermal property characterization of chitin from Antarctic krill (*Euphausia superba*). *Carbohydrate Polymers, 92*, 90–97.

Wijesena, R. N., Tissera, N., Kannangara, Y. Y., Lin, Y., Amaratunga, G. A. J., & De Silva, K. M. N. (2015). A method for top-down preparation of chitosan nanoparticles and nanofibers. *Carbohydrate Polymers, 117*, 731–738. https://doi.org/10.1016/j.carbpol.2014.10.055

Wong, C. Y., Al-Salami, H., & Dass, C. R. (2020). Formulation and characterisation of insulinloaded chitosan nanoparticles capable of inducing glucose uptake in skeletal muscle cells in vitro. *Journal of Drug Delivery Science and Technology, 57*, 101738. https://doi.org/10.1016/j.jddst.2020.101738

Yanat, M., & Schroën, K. (2021). Preparation methods and applications of chitosan nanoparticles; with an outlook toward reinforcement of biodegradable packaging. *Reactive and Functional Polymers, 161*, 104849.

Yin, Y., Dang, Q., Liu, C., Yan, J., Cha, D., Yu, Z., Cao, Y., Wang, Y., & Fan, B. (2017). Itaconic acid grafted carboxymethyl chitosan and its nanoparticles: Preparation, characterization and evaluation. *International Journal of Biological Macromolecules, 102*, 10–18. https://doi.org/10.1016/j.ijbiomac.2017.04.005

Zhang, H. L., Wu, S. H., Tao, Y., Zang, L. Q., & Su, Z. Q. (2010). Preparation and characterization of water-soluble chitosan nanoparticles as protein delivery system. *Journal of Nanomaterials, 2010*, 898910.

Zhao, D., Yu, S., Sun, B., Gao, S., Guo, S., & Zhao, K. (2018). Biomedical applications of chitosan and its derivative nanoparticles. *Polymer, 10*, 462.

Nanochitosan-Based Fish Disease Prevention and Control

Margaret Ikhiwili Oniha, Olusola Luke Oyesola, Olugbenga Samson Taiwo, Stephen Oluwanifise Oyejide, Seyi Akinbayowa Akindana, Christiana Oluwatoyin Ajanaku, and Patrick Omoregie Isibor

Contents

M. I. Oniha (✉)
Department of Biological Sciences, Covenant University, Ota, Ogun State, Nigeria
e-mail: margaret.oniha@covenantuniversity.edu.ng

O. L. Oyesola · O. S. Taiwo · C. O. Ajanaku
Department of Chemistry, Landmark University, Omu aran, Kwara State, Nigeria

S. O. Oyejide
Department of Cell Biology and Genetics, University of Lagos, Lagos, Nigeria

S. A. Akindana
Department of Botany, University of Ibadan, Oyo, Nigeria

P. O. Isibor
Department of Biological Sciences, College of Science and Technology,
Covenant University, Ota, Ogun State, Nigeria

P. O. Isibor et al. (eds.), *Nanochitosan-Based Enhancement of Fisheries and Aquaculture*, https://doi.org/10.1007/978-3-031-52261-1_4

1 Introduction

Currently, infections and diseases account for one of the primary causes of human mortality with regard to diverse routes. These maladies are engendered by bacteria, fungi, parasites and viruses. Since days beyond recall, the prevention and treatment of maladies have gained pronounced focus dating back to the origination of penicillin. Undeterred by copious remedial approaches, challenges abound that incorporate infectious disease dynamics and emergence of antimicrobial-resistant microorganisms, which have validated the desideratum to construct more efficacious modes of action and drug delivery schemes to achieve effectual prevention and treatment goal. Studies have posited that drug molecules can derive ideal drug stacking capability in particles due to the high facet of nanoparticles (Kim et al., 2010). Researchers' interest has been centered on vectors characterised with stubby cost, perpetual biocompatibility and minimal side effects have attracted the interest of researchers. One of the current unique scientific procedures, employing different functional materials, is nanoencapsulation for revamping the bioavailability, solvability and retention time of biologically active composites. Nanotechnology, being one of the most active areas of modern data research, is a prominent technique possessing significance in economic, social and ecological sectors (El-Saadony, et al., 2021a, 2021b). Nanocarriers protect the efficacious ingredient from degradation precipitated by photolysis and hydrolysis (Sathiyabama et al., 2019). Nanoparticles exert their activity either directly (He et al., 2011) or as carrier systems for synthetics utilised in the field (Ihegwuagu et al., 2016; Kashyap et al., 2015) thus providing more enhanced efficiency and revamped environmental safety (Sekhon, 2014). These compositions constitute of a totally novel cum enhanced attribute premised on specific properties such as size, distribution and shape (Saad et al., 2021; El-Saadony, et al., 2021c). Additionally, there have been results to validate the significant improvement of anti-infection capacity obtained through the synergy between chitosan and orthodox medicine (Meng et al., 2021). Polymer nanoparticles are broadly utilised in the biomedical field as implements in the diagnosis and treatment of diseases (Uthaman et al., 2015). Due to their role as a delivery carrier, polymer nanoparticles can incorporate into or be loaded with multitudinous drugs with subsequent increased effective discharge of these medicaments. Furthermore, synthetic resin nanoparticles can encapsulate medications on their facets. These polymer-based nanoparticles are capacitated to target molecules with specific receptors on the cell facet as well as to invade cells which will facilitate a supplementary secured plus efficacious distribution of targeted medicaments and in gene therapy (De Jong & Borm, 2008). Hydrophilic faceted polymer nanoparticles are extensively applied as vectors based on their very diminished nonspecific peptide adsorption features. Moreover, these polymer composites can be employed to diagnose and treat complicated maladies. Based on its good physicochemical plus distinct biological features, chitosan obtains applications in numerous industries that include medical, edibles, synthetic, cosmetics, aqueous treatment, metal extrication and recovery, biochemical as well as biomedical engineering economies. However, chitosan is not soluble in aqueous solutions, a major disadvantage that limits its widespread application in living systems (Ngo et al., 2015). However, chitosan possesses some functional types that permit graft modification that confers the modified chitosan with special characteristics. These modifications can be employed to synthetically modify chitosan for solubility enhancement and subsequently extend its applications. These chemical modifications produce many kinds of chitosan derivatives that have sustained-release properties and are non-toxic, biocompatible and biodegradable (Chua et al., 2012). Chitosan nanoparticles are used as drug carriers facilitated by their good biological-compatibility and biological-degradation (Wang et al., 2011). In furtherance, chitosan nanoparticles possess broad application in medicament and vaccine conveyance, as vaccine adjuvant, as an antimicrobial, in tissue engineering including other implementations.

Chitosan (CS) is a natural cationic biopolymer configured of N-acetyl-D-glucosamine and Dglucosamine units connected by β-1,4-glycosidic linkages (Elieh-Ali-Komi & Hamblin, 2016). Chitosan and chitosan oligosaccharides have become well-known biological control agents due to their non-toxic, biodegradable and biocompatible properties (Singh & Chaudhary, 2010). In agriculture, chitosan has been validated to be the most abundant natural polymer with dual functions, which includes the capability to control pathogenic microorganisms by preventing growth, sporulation, spore viability, germination and cell destruction, inducing different defense responses in host plants. While the second function is to inhibit various biochemical processes during the phytopathogenic interaction. Chitosan is a cationic polyose acquired from the basic chitin acetylation procedure of the cell mass of parasites and crustal outer shell, which is predominantly associated with wound healing, cell polymorphism activation, fibroblast initiation, cytokine generation amongst others (Khairy et al., 2022). The biological-polymer obtained as the free product of chitin acetylation is known as nanochitosan, and it is composed of glucosamine and N-acetyl residues. It is generally obtained in immense quantities from lobster exoskeleton residuals and shrimp walls at an inexpensive cost (Sharif et al., 2018).

In physiological environments, chitosan presents positive charges consequence by the presence of quaternary ammonium salt groups. With colossal effectual groups on the molecular sequence, chitosan possesses the capacity to be structurally and synthetically reformed to effect immune stimulation (Gorbach et al., 1994), the boosting of wound healing (Madhumathi et al., 2010), and antibacterial and antifungal effects (Qi et al., 2004). Its versatility and adaptability provide a distinctive opportunity for the generation of new antibacterial remedials and prohibition of infectious maladies. Composites of chitosan systems employ chitosan's features to obtain immense remedial outturns such as the utilisation of chitosan's adhesive capacity to achieve non-invasive mucosal vaccine vectors. By combining chitosan with additional wound dressings such as hydrogels, its antibacterial activities plus wound healing is presented. Carriers bearing chitosan show perdurable in vivo biological compatibility (Meng et al., 2021).

Chitosan shows its mechanism of antibacterial by binding to the negatively charged bacterial cell wall, thus leading to a change in the permeability of the cell coat and then adheres to the DNA to inhibit its replication. All physical and biological characteristics of chitosan are based on the degree of deacetylation and molecular weight. The number and distribution of acetyl components in chitosan effectuate its biodegradability and cytotoxicity, and the intensity of deacetylation of chitosan directly affects the efficacy of antigen delivery and the function of chitosan as an adjuvant (Meng et al., 2021). All amine and hydroxyl operative groups available in chitosan can be non-synthetically modified to introduce other composites including hydroxyalkyl (Kurita, 2006; Sashiwa et al., 2003), carboxyalkyl (Abreu & Campana-Filho, 2005), succinyl (Kato et al., 2004), thiol (Bernkop-Schnürch et al., 2004; Roldo et al., 2004), and sulfate (Holme & Perlin, 1997), to compose copious chitosan by-products. The adhesion and porosity enhancement of chitosan by-products depend on the intensity of substitution or quaternisation of chitosan. Every configuration of chitosan-composed nanoscale carriers utilises the advantage of the size to obtain lofty remedial effects (Meng et al., 2021). Varying disbursement schemes and associated requirements for discharge systems exist as well such as nanoparticles as drug delivery systems require responsive discharge features and good biocompatibility, while wound dressings emphasise the mechanical activities of the material to achieve sustained drug release. The response to differential infectious maladies requires diversified treatment configurations such as hydrogels that need good biocompatibility for the promotion of wound-healing ability. Hydrogels encased with therapeutic drugs have been proven to promote wound repair or improve antibacterial and anti-infective properties (Meng et al., 2021).

2 Mechanism of Chitosan in Disease Prevention and Treatment

Chitosan and its derivatives possess antibacterial effects on Gram-negative bacteria and Gram-positive bacteria (Sadeghi et al., 2008). Inhibition of bacterial growth can be revamped by incorporating the antibacterial agents into the chitosan-composed nanoparticles with evidence that chitosan functions concertedly with other nanocrystals (Sobhani et al., 2017). The lofty specific facet yields higher charge density with subsequent higher interaction with the elements of microorganisms (Qi et al., 2004; Regiel-Futyra et al., 2015) with immense positive results but minimal negative effects including the achievement of enhanced antibacterial properties through the addition of chitosan films onto gold nanoparticles (Fig. 1). Chitosan-composited nanoparticles are more efficacious to pure chitosan polymers and doxycycline in bacterial growth inhibition as obtained with *Escherichia coli* and copious *Staphylococcus* (Qi et al., 2004). Findings of Friedman et al. situated the efficacy of antibacterial and immunological activities of chitosan-sodium alginate nanoparticles in the restriction of *Pseudomonas acnes* growth (Friedman et al., 2013). In furtherance, the even dispersion of chitosan-combined nanoparticles in the sample engenders consequential cell membrane damage (Holban et al., 2014). This similar mechanism is also employed in fungal growth inhibition but with some resistance to chitosan as observed in some species such as *Aspergillus niger* due to the presence of its 10% cell wall comprised chitin (Ma & Lim, 2003; Ing et al., 2012). Precise choices are made with respect to chitosan-contained nanoparticles against fungal growths for absolute remedial results.

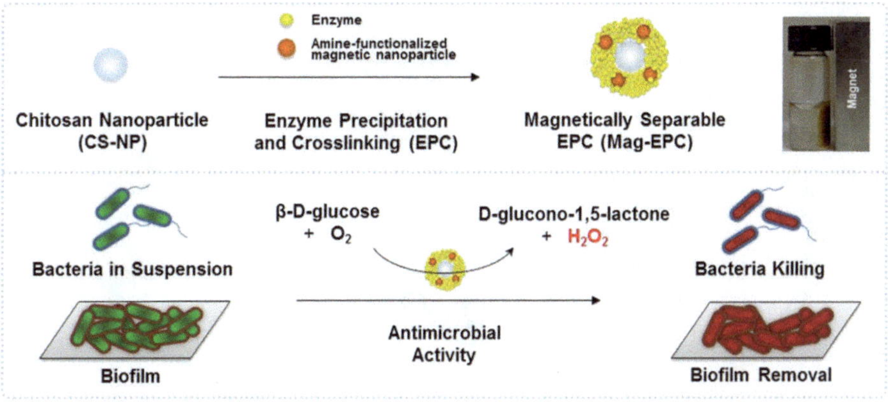

Fig. 1 Diagram of enzyme-immobilised chitosan nanoparticles as effective antibacterial agents (source: Kyung-Min et al., 2019)

Helicobacter pylori can induce gastroesophageal reflux disease and chronic atrophic gastritis (Luo et al., 2018). There has been an intense elevation in the drug resistance of *H. pylori* with consequent gross diminished efficacy of conventional antibiotics. Chitosan nanoparticles can be utilised for gastric delivery and continuous application (Modi et al., 2013), and this process can circumvent drug malabsorption in the stomach. These chitosan-based nanoparticles can also be employed for ocular remedial. Another research by Zhou et al. revealed enhanced corneal permeation that is void of increasing corneal irritation through their formulated chitosan-encased polylactic acid nanocrystals for ocular drug discharge (Zhou et al., 2013). These are some of the benefits that aggrandised the strategic configuration of chitosan-constituted nanoparticles in drug disbursement schemes. In addition, the biodegradable feature of nanodelivery scheme encapsulating polyoses and peptides has additionally received focus with researchers. Reports have posited the compositions of nanoparticles by electrostatic complexation of proteins/polyoses (Raei et al., 2017; Huang et al., 2016). The perdurability of implanting agent and nanocrystals can be revamped by accumulating the spatial and electrostatic repulsion within peptide and polysaccharide. The lucidity of nutrient components can be loftily revamped through encasing hydrophobic and hydrophilic nutritive constituents into peptide/dextrose electrostatic combinations (Fig. 2) (Joye et al., 2015).

In infectious disease management, drug treatments often yield adverse side effects and contribute to the growing resistance of bacteria and viruses to medications. Vaccination stands as a pivotal measure in preventing and treating infectious diseases effectively. Research indicates that initiating the immune system response can substantially curb the recurrence of infectious diseases (Look et al., 2010).

The mucosal surface, encompassing areas like the nose, respiratory tract, oropharynx, gastrointestinal tract, and genitourinary system, serves as the primary entry point for pathogens, including viruses and bacteria (Kanner et al., 2019).

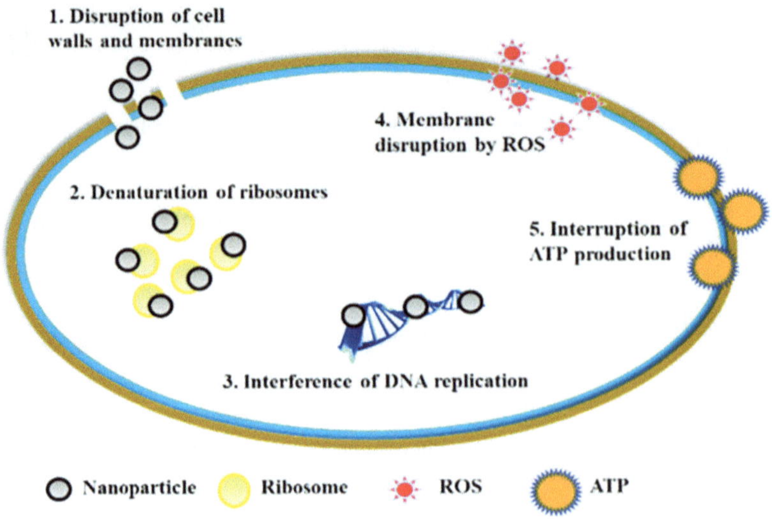

Fig. 2 Antimicrobial mechanism of nanochitosan (*Source*: Chen et al., 2017)

Beyond providing humoural and cellular immune defenses in these mucosal regions, mucosal immunity also extends to systemic immunity (Holmgren & Czerkinsky, 2005). Polymer-based vectors have emerged as a promising avenue for mucosal vaccine delivery due to their ability to target specific sites for antigen delivery (des Rieux et al., 2006). Moreover, these carriers shield antigens from harsh environmental conditions such as pH variations, bile, and enzymes in the gastrointestinal tract, while regulating the release of antigens. Studies indicate that polymer-based composites can augment the immune response when delivering antigens through mucosal routes (Andrianov & Payne, 1998).

Upon encountering an antigen, B cells transform into antibody-secreting plasma cells, generating antibodies vital for eliminating pathogens from mucosal surfaces, as depicted in Fig. 2. Meanwhile, dendritic cells (DCs) present the antigen via major histocompatibility complex (MHC) class I and class II molecules to CD8 + and CD4 + T cells. Activation of CD8 + T cells and CD4 + Th1 cells triggers the production of cytotoxic T lymphocytes (CTL) and activated macrophages, enabling the elimination of intracellular pathogens or infected cells (cellular response). In contrast, the activation pathway of CD4 + Th2 cells stimulates activated B lymphocytes to secrete antibodies for neutralising extracellular pathogens (humoural response).

The use of polymer-based carriers, particularly chitosan and its derivatives, has garnered significant attention for their efficacy in delivering antigens through mucosal routes. The adhesive properties of chitosan primarily stem from the electrostatic interaction between its positively charged molecular structure and the negatively charged surfaces of cells and mucus (Illum et al., 2001).

Studies, such as the one conducted by Subbiah et al. (2012), have focused on loading hepatitis B virus surface antigen (HBsAg) into N,N,N-trimethyl chitosan nanoparticles (TMC NPs) for controlled intranasal administration. In vivo immune studies have demonstrated that the adjuvant capability of antigen-loaded TMC NPs maintains high stability over an extended period. Enhancing the distribution efficiency of antigens, vaccine vectors modified with targeting ligands have shown promising outcomes (Phanse et al., 2013). For instance, chitosan modified with mannose has been utilised for nasal vaccine delivery (Macri et al., 2016). Similarly, the targeting capability of vaccine vectors can be augmented by attaching targeting peptides to chitosan (Jung et al., 2015). Chitosan, owing to its positively charged nature, holds substantial promise in configuring vaccine vectors, showcasing its potential in the field of antigen delivery.

Wounds, burns, surgical incisions, and related tissue trauma represent common sites for localised infections, often leading to prolonged wound healing, abscess formation, wound reopening (wound dehiscence), and, in severe cases, life-threatening complications (Sun et al., 2020). Developing novel biological substances to prevent infections has become a critical focus, and chitosan stands out as a promising candidate for wound dressing due to its inherent antibacterial properties. Chitosan, by virtue of its interaction with polyvalent electronegative molecules or anions, forms an ion network through coordination and secondary interactions, adeptly creating a gel-like network (Dash et al., 2011). Various substances with three-dimensional network compositions have been engineered as novel platforms, including three-dimensional scaffolds, drug reservoirs, bandages, and wound dressings (Wu et al., 2015; Cheng et al., 2013; Park et al., 2015). Emphasis has been placed on natural biomedical hydrogels, particularly chitosan, due to its favourable biocompatibility and biomimetic properties (Xu et al., 2020). These chitosan-based gels can encapsulate biologically active compounds within their network through physical interactions or chemical bonding. For instance, a hydrogel composed of carboxymethyl chitosan (CMC) and oxidised dextran, designed for loading anti-infective drugs like ceftriaxone sodium, has exhibited good biocompatibility and promising anti-infective outcomes in vivo (Li et al., 2017). In another study, Hu et al. (2018) developed double cross-linked amorphous hydrogels (CMC-ALG) by combining CMC, alginate (ALG), and calcium chloride. These hydrogels, promoting wound healing while demonstrating antibacterial properties, were utilised to store epidermal growth factor (EGF) through electrostatic interaction and divalent chelation. Similarly, Yang et al. (2018) crafted a high-performance hybrid chitosan-polyacrylamide (CS-pam) ion-covalent double-network hydrogel, dynamically adjusting its structure and mechanics in situ for improved performance. Furthermore, Chen et al. (2017) developed an antibacterial alginate/chitosan hydrogel dressing integrated with gelatin microspheres, enhancing the biodegradability and mechanical properties of the dressing. This formulation, utilising chitosan acetate, displayed evident antibacterial activities.

In comparative studies against alginate sponge bandages and silver sulfadiazine cream, chitosan acetate bandages exhibited superior efficacy in reducing bacterial load on the skin, notably in preventing systemic infections in animal models induced by *P. aeruginosa* and *P. mirabilis*.

3 Application of Nanochitosan in Controlling Bacterial, Viral, and Fungal Infections

3.1 Chitosan's Role in the Control of Bacterial Infections

Numerous nano-polymeric systems, comprising both synthetic and natural polymers, have been developed and explored, each possessing distinct advantages and drawbacks. Natural polymers like alginate, chitosan, and hyaluronic acid have been investigated extensively for their potential in nanoparticle methodologies. Despite advancements in drug delivery, oral administration remains a preferred route (Sharifi-Rad et al., 2021). The focus on antibacterial and antimicrobial agents and their disinfection systems remains crucial in daily research endeavours.

The antimicrobial efficacy of nanochitosan against pathogens arises from its intricate interactions with microbial cells, involving a complex mechanism that disrupts several fundamental cellular processes. Nanochitosan intervenes in the structural integrity of microbial cell walls and membranes, resulting in their disruption or increased permeability (refer to Fig. 2). This interference compromises the microorganism's protective shield, leading to cell lysis or dysfunction. Moreover, nanochitosan induces denaturation or modification of ribosomal structures within microbial cells. This disruption in ribosomal function, pivotal for protein synthesis, disturbs crucial cellular processes and impedes the microbe's capacity to produce essential proteins.

Furthermore, nanochitosan inhibits the replication of microbial DNA, disrupting the accurate duplication of genetic material in the microorganism. This interference obstructs the cell's ability to reproduce effectively.

Reactive oxygen species (ROS) generated by nanochitosan prompt destabilisation of microbial cell membranes, leading to oxidative stress and damage to cellular components, further compromising cell integrity. Additionally, nanochitosan disrupts the production of adenosine triphosphate (ATP), a crucial cellular energy source. This inhibition undermines the microbe's energy generation, resulting in cellular dysfunction and impairment of essential metabolic processes.

These collective actions orchestrated by nanochitosan contribute significantly to its potent antimicrobial efficacy against a broad spectrum of pathogens. By targeting multiple cellular components and vital processes essential for microbial survival, nanochitosan demonstrates robust antimicrobial properties, presenting promising applications across diverse sectors, including health care, agriculture, and food preservation.

Extensive studies have explored the applicability of various materials across diverse sectors, including healthcare facilities, industrial settings, marine laboratories, and residential environments (Ali et al., 2015; Hosseinnejad & Jafari, 2016). Within the realm of antibacterial materials, two primary categories exist: inorganic and organic constituents. Inorganics encompass metals, metal oxides and metal phosphates (Tuncer, 2007).

Amongst these inorganic materials, metal oxides such as TiO2, ZnO, MgO and CaO have garnered significant attention due to their stability under adverse conditions and recognised safety for both animal and human integration (Teli & Kale, 2011). On the organic front, while compounds like phenols, halogenated compounds, and quaternary ammonium salts have historically been in focus, recent attention has shifted towards chitosan (CTS) and chitin (Hosseinnejad & Jafari, 2016).

Polymer/metal nanocomposites present a potent option, allowing controlled release of metal species into living organism environments, offering intriguing prospects for various biotechnological applications. Studies have validated the efficacy of polymer-based nanocomposites loaded with stabilised copper nanoparticles, proposed as biostatic coatings, establishing systematic correlations between material properties and biological effects.

In a separate study, researchers in Japan (Chung et al., 2004) discovered a nickel-alloy coating exhibiting antibacterial properties, showing promise in reducing the SARS coronavirus. Furthermore, the assessment of silver-encased materials and titanium dioxide photocatalyst components revealed their potential as inorganic environmental purification functional materials, as evaluated through in vitro tests. The antibacterial activity was determined by measuring the zone of inhibition on nutrient agar plates generated by the silver-loaded polymer coating against bacteria (antibacterial effect).

The inclusion of Ag nanoparticle-reinforced polymer composites exhibited significant inhibition zones against Staphylococcus aureus and Escherichia coli. This effect is attributed to the interaction between silver and thiol groups present in bacterial proteins (Yilmaz Atay, 2020). While the exact mechanism underlying this antibacterial activity remains incompletely understood, the antimicrobial effectiveness of chitosan is influenced by diverse factors acting in an organised and independent manner.

The proposed antibacterial activity of chitosan involves several mechanisms. The most widely recognised mechanism suggests that chitosan binds to the negatively charged bacterial cell wall, precipitating cell disruption and altering membrane permeability. This disruption is followed by chitosan attaching to DNA, inhibiting DNA replication and ultimately leading to cell death (Nagy et al., 2011). Another potential mechanism proposes that chitosan acts as a chelating agent, selectively binding to trace metal elements, inducing toxin production, and inhibiting microbial growth (Divya et al., 2017).

The polycationic structure of chitosan is imperative for its antibacterial activity. Under lower environmental pH conditions (below the pKa of chitosan and its derivatives), the electrostatic interaction between the polycationic structure and the predominantly anionic components on the surface of microorganisms plays a pivotal role in its antibacterial action (Kong et al., 2010).

The formation of the polycationic structure becomes prominent under acidic conditions, with specific derivatives altering the pKa of chitosan and leading to protonation at higher pH values (Yang et al., 2005). As the positive charge density of chitosan intensifies, its antibacterial properties are enhanced, as seen with quaternised chitosan and chitosan metal complexes (Xie et al., 2007). Conversely, if the polycationic nature of chitosan is diminished or reversed, its corresponding antibacterial capacity decreases or is lost.

In addition to protonation, the quantity of amino groups linked to C-2 on chitosan backbones plays a crucial role in electrostatic interactions, contributing to increased antibacterial activity with a higher number of amino groups.

3.2 Chitosan's Role in the Control of Viral Infections

Viruses have the capacity to invade host cells, hijack their machinery, and replicate using their own RNA or DNA, often resulting in the destruction of the host cell. The intricate viral life cycle poses challenges in developing effective antiviral drugs. To address limitations in antiviral drug development, nanotechnology-based approaches have emerged as promising solutions. These methods aim to overcome challenges such as drug solubility and toxicity, enhancing the selectivity of antiviral drugs towards viruses and infected cells while preserving the integrity of healthy host cells.

Chitosan (CH) is a bio-polymer obtained through partial de-acetylation of chitin, a naturally occurring polysaccharide consisting of randomly distributed (1 → 4)-linked N-acetyl glucosamine and glucosamine units (Rashki et al., 2021). It is commonly obtained as a white powder, characterised by rigid, inflexible and nitrogen-containing glucose sequences of varying length and molecular weight (Badawy & Rabea, 2011). Moreover, chitosan possesses multifaceted applications owing to its non-toxicity, biodegradability and intrinsic antimicrobial properties. Its utilisation spans various fields including biomedical compositions, genetic engineering, agriculture, environmental pollution control, food and paper manufacturing, photography and water treatment (Cheba, 2011).

Chitosan, synthesised by the partial deacetylation of chitin through alkaline hydrolysis, commonly refers to cationic co-polymers comprised predominantly of 2-amino-2-deoxy-β-D-glucose residues (60% to 100%) and 2-acetamino-2-deoxy-β-D-glucoside residues (0% to 50%), linked together by ß (1 → 4) bonds (Nasrollahzadeh et al., 2020). The degree of deacetylation, usually >60%, dictates the total number of amide and primary amine residues, influencing chitosan's solubility and its chemical, biological and physical characteristics (Boroumand et al., 2021).

Chitosan, a naturally derived polymer, has been employed to create nanoparticles (NPs). These NPs are biocompatible, biodegradable, less toxic, easily formulated and serve as effective drug delivery systems (DDSs). Their versatility makes them promising candidates for various biomedical applications.

Infectious diseases caused by bacteria, fungi, viruses and parasites collectively contribute to an estimated 15 million fatalities globally. Amongst these, significant mortality is attributed to major infections like HIV, malaria, tuberculosis and acute respiratory infections, including the recent COVID-19 pandemic (Boroumand et al., 2021). Viral infections pose substantial health concerns worldwide, impacting human health and socio-economic progress. The effectiveness of treating viral infections is hindered by the growing incidence of drug resistance, notably observed in cases like HIV. The emergence of drug-resistant viral infections presents a significant public health challenge, leading to widespread morbidity and mortality, along with the need for costly medications (Boroumand et al., 2021).

Nanotechnology pertains to particles sized within the nanometer range, typically measuring 10^–9 or one-billionth of a meter (Parboosing et al., 2012). Nanobiotechnology focuses on the interactions between nanoscience and biological systems, whereas nanomedicine involves the utilisation of nanostructured materials for diagnosing, preventing or treating diseases (Medepalli, 2008). The initial application of nanoparticles (NPs) in medicine aimed to enhance the solubility and stability of drugs with low bioavailability (Schütz et al., 2013).

Nanoparticles exhibit antiviral activity through various mechanisms. Their unique characteristics, including small size, high surface area-to-volume ratio and modifiable surface charge, enable their application in viral treatment. Additionally, some nanoparticles exhibit intrinsic antiviral activity, such as silver nanoparticles (AgNPs) or dendrimers, possibly owing to their biomimetic features (Gagliardi, 2017; Lara et al., 2010; Mallipeddi & Rohan, 2010). Another mechanism involves the drug encapsulation capacity of nanoparticles, ensuring enhanced stability, optimised dosage and controlled release of the loaded drug (Chiappetta et al., 2011; Kumar et al., 2012).

Nanoparticles can be tailored to form stable structures or modified by attaching polyethylene glycol (PEG) to further improve their characteristics (Goldberg et al., 2007; McNeil, 2011; Santos-Martinez et al., 2014). Moreover, engineered nanoparticles can significantly enhance drug delivery by employing targeting molecules that identify specific receptors or biomarkers, thereby enhancing specificity for target cells, tissues or subcellular compartments (McNeil, 2011).

Chitosan, the second most abundant polysaccharide in nature, has found widespread use in drug formulation and pharmaceuticals due to its non-toxicity, biodegradability, biocompatibility and ability to traverse tight epithelial junctions (Ilium, 1998).

Chitosan (CH) demonstrates the capability to open tight intercellular junctions, facilitating improved penetration of loaded medications into tissues for better absorption by target cells. CH nanoparticles (NPs) hold promise as biomaterials to enhance antigen distribution and the performance of vaccines. Associating antigens with chitosan-based nanoparticle formulations has shown enhanced uptake of antigens by mucosal lymphatic tissues, resulting in more potent mucosal and systemic immune responses to diverse antigens (Bramwell & Perrie, 2006).

The mucoadhesive properties of chitosan contribute to prolonged adhesion of CH NPs, enabling an extended contact time with bloodstream capillaries, thereby enhancing the uptake of antigen proteins or plasmid DNA. The respiratory mucosal surface serves as a primary barrier for immune defense and a principal site for influenza virus infection. Studies have underscored that mucosal vaccines can effectively induce both systemic and local mucosal immune responses (Yuki & Kiyono, 2003). Combining vaccines with adjuvants enhances the induction of mucosal immunity, with some adjuvants, like *Escherichia coli* heat-labile toxin or cholera toxin, demonstrating the potential to induce acute diarrhoea and damage to the central nervous system (Glück et al., 2000; van Ginkel et al., 2000). Hence, the development of more effective and safer adjuvants is crucial for improving mucosal immunisation.

The interaction between nanoparticles and antigens enhances antigen-specific acquired immune reactions by boosting uptake by antigen-presenting cells, such as dendritic cells (DCs) and macrophages. NP uptake by DCs triggers the up-regulation of costimulatory molecules, stimulates cytokine production, and increases T-cell stimulation (Uto et al., 2007), rendering NP distribution systems efficient adjuvants for subunit vaccines. Mucosal immunisation can evoke a mucosal immune response with or without a systemic immune response, and oral vaccination is more widely accepted due to its higher efficiency. This efficacy stems from the degradation of antigens in the acidic stomach environment and by enzymes in the intestinal tract.

Chitosan significantly modulates the functional behaviour of numerous immune cells, including granulocytes and macrophages. Upon subcutaneous implantation, chitosan initiated chemotaxis of macrophages in Canis familiaris L., leading to increased nitric oxide production by these macrophages in vitro. Additionally, it prompted leukocytosis in the peripheral blood of laboratory dogs. The secretion of nitric oxide was primarily attributed to the N-acetylglucosamine residues present in chitosan, which demonstrated higher efficacy compared to N-acetylmannosamine or N-acetylgalactosamine (Boroumand et al., 2021).

Macrophages, being antigen-presenting cells, play a pivotal role in triggering cellular and humoural immune responses upon interaction with T-helper cells. Augmenting the functional activity of macrophages with chitosan might hold significance in suppressing viral infections in animals. Upon absorption of chitosan or chitin nanoparticles through phagocytosis, mouse alveolar macrophages exhibited increased production of reactive oxygen species. Similarly, mouse splenocytes demonstrated elevated secretion of γ-interferon, which aids in suppressing viral replication by impeding the translation ability of genomic RNAs or early viral mRNAs (Boroumand et al., 2021).

Sulphated derivatives of chitosan have been formulated specifically to inhibit retrovirus replication. Research has indicated that N-carboxymethyl chitosan-N,O-sulphate could impede the generation of virus-specific polypeptides, decrease HIV-1 replication in cultured T-cells, and inhibit Rausher murine leukaemia virus in cultured mouse fibroblasts.

3.3 Chitosan's Role in the Control of Fungal Infections

The clinical management of fungal diseases primarily relies on four classes of drugs: nucleoside analogues, azoles, echinocandins and polyenes (Robbins et al., 2016). However, the limited selection of available treatments, combined with the widespread use of antifungal medications, poses a potential risk of exacerbating drug resistance. Research has indicated a concerning global surge in the emergence of human fungal pathogens, resulting in reduced efficacy against fungal infections (Yang et al., 2010; Ford et al., 2015). Consequently, there is a critical need to develop novel therapeutic strategies or new antifungal agents (Shih et al., 2019).

Chitosan has been recognised for its broad-spectrum antimicrobial activity against Gram-positive bacteria, Gram-negative bacteria and fungi (Shih et al., 2019). Several review articles corroborate that the antimicrobial efficacy of chitosan is significantly correlated with its degree of deacetylation and pH levels (Cheung et al., 2015; Hosseinnejad & Jafari, 2016). Higher degrees of deacetylation enhance the antimicrobial activity of chitosan, while pH levels influence its antifungal and antimicrobial effects, with greater antimicrobial activity observed at lower pH values. Chitosan exhibits antimicrobial activity as a cationic polymer when the pH is below 6.5 (Shih et al., 2019).

The cationic nature of chitosan allows it to interact with the negatively charged surfaces of microbial cells, disturbing the balance between anions and cations and leading to an inhibitory response (Martinez-Camacho et al., 2010). The antimicrobial efficacy of chitosan is highly reliant on its inherent characteristics and the specific type of bacteria or fungi involved (Hosseinnejad & Jafari, 2016). In their study, Shih et al. (2019) investigated the potential antifungal effects of chitosan against Candida albicans. Their findings indicated that chitosan's antifungal action was mediated by the inhibition of SAGA (Spt-Ada-Gcn5-acetyltransferase) complex component expression, subsequently altering cell surface integrity. Additionally, chitosan treatment reduced the levels of chitin and β-glucan in C. albicans cells and modified the ultrastructure of the cell wall and membrane by suppressing SAGA complex component expression. Similar to its impact on bacteria, chitosan's activity against fungal growth is believed to be fungistatic rather than fungicidal, potentially inducing regulatory changes in both the host and fungus.

Overall, studies have highlighted chitosan's effectiveness in restraining spore germination, germ tube elongation, and radial growth, with numerous investigations conducted on yeasts and molds associated with food and plant spoilage.

4 Mechanisms of Action and Effectiveness against Common Aquatic Pathogens

4.1 Chitosan as an Antimicrobial Agent

Chitosan is a versatile antipathogenic chemical, particularly used as a natural fungicide. Several investigations have been conducted to study its methods of action. Chitosan influences the development and morphology of economically significant aquatic diseases such as *Rhizopus stolonifer* and *Botrytis cinerea*. Many other pathogenic fungi, such as *Alternaria* spp., *Colletotrichium* spp. and *Trichoderma* spp., are similarly inhibited by this polymer (Lopez-Moya et al., 2019). Chitosan causes energy-dependent permeabilisation of plasma membranes in sensitive fungi. This polymer has an antibacterial effect against harmful microorganisms (bacteria). Chitosan, like fungus, permeabilises bacterial plasma membranes (Palma-Guerrero et al., 2010). Chitosan suppresses the production of biofilms in *A. fumigatus* (Kvasničková et al., 2016). The inhibition of nutrients such as carbon and nitrogen improve chitosan's antifungal effectiveness against pathogenic microorganisms (Lopez-Moya et al., 2019). Nutrient deprivation causes alterations in cell wall architecture, affecting fungal development. In this regard, fungal cell walls with limited branching as a result of glucan deposition are more sensitive to chitosan. Since plasma membrane-associated synthase complexes synthesise important cell wall components (glucans and chitin), there is a direct relationship between cell wall and membrane. Chitosan inhibits the development of pathogenic fungi especially in a carbon nutritional state. Chitosan considerably lowers *C. albicans* pathogenicity in *Galleria mellonella* L. under the required circumstances (Aranda-Martinez et al., 2016).

4.2 Chitosan Alters Gene Expression in Aquatic Pathogens and Fungi

Aquatic pathogens involve bacteria, viruses, fungi, and protozoa. These are known to pose a risk to both animal and human health. The diversity of these pathogenic microorganisms and the infectious diseases they cause are expanding. Some frequently occurring aquatic bacterial pathogens include *Escherichia coli, Campylobacter* spp., *Salmonella* spp. and *Cryptosporidium* (Stec et al., 2022). The major categories of aquatic viruses include megavirales (Pithovirus, Tupanvirus, Mimivirus), virophages (these belong to the family Lavidaviridae), polintons (Virus-Like Transposons) amongst others. Others include emerging and re-emerging viruses such as Iridovirus (Marine et al., 2017; Weinheimer & Aylward, 2022). Aquatic pathogenic fungi include *Ichthyophonus* spp., *Fusarium* spp., *Aphanomyces* spp., *Saprolegnia* spp., *Branchiomyces* spp. and *Achlya* spp. amongst others (Purabi Sarkar et al., 2022). The protozoan parasites most frequently linked with aquatic environments are *Giardia* and *Cryptosporidium*, the causative agents of giardiasis and cryptosporidiosis, respectively (Sánchez et al., 2018). Their pathogen–host interactions lead to the genesis of infectious diseases which are connected to incredibly complicated genetic alterations observed in plants, animals and humans.

Aquatic pathogens are susceptible to chitosan in a variety of ways due to variations in the components of the cell wall (Yan et al., 2021). Investigations have been carried out to determine the role of plasma membranes in the sensitivity of aquatic pathogens to chitosan. A study highlights the role of Ca^{2+} in plasma membrane remodelling during cell fusion which points to chitosan's method of action on sensitive aquatic microorganisms (Zilly et al., 2011). Aquatic pathogenic microorganisms' first reaction to chitosan comprises its partial membrane permeabilisation and the commencement of ROS generation (Ing et al., 2012).

Several investigations have been carried out to determine the role of plasma membranes in the sensitivity of fungi to chitosan. The membranes of chitosan-sensitive fungi are highly fluid. On the contrary, chitosan-resistant fungi have low-fluidity membranes (Valenzuela-Ortiz et al., 2022). Using flow cytometry, it was recently revealed that chitosan permeabilises the plasma membrane of *N. crassa*. This results in the generation of reactive oxygen species (ROS) within the cell and cell death (Palma-Guerrero et al., 2009). According to RNAseq data and gene ontology (GO) analysis, the primary categories stimulated by chitosan are oxidoreductase activity, plasma membrane and transport (Lopez-Moya et al., 2016). Chitosan also improves yeast oxidative metabolism, respiration and transport GO activities. Chitosan induces the plasma membrane response to stress and cell wall integrity genes in aquatic microorganisms (Jaime et al., 2012). Lipase Class III, Monosaccharide transporter and Glutathione transferase genes are confirmed as primary chitosan targets in aquatic pathogenic fungi by testing deletion strain mutants (Lopez-Moya & Lopez-Llorca, 2016).

Chitosan provides useful intervention in membrane healing, catabolite assimilation and buffering ROS surplus derived from the breakdown of chitosan. Antifungal proteins, such as PAF from *Penicillium chrysogenum*, work similarly to chitosan by permeabilising plasma membranes and inducing ROS generation. The synthesis of these oxidative by-products of metabolism reflects the energetic state of the cell. This would explain why chitosan-induced plasma membrane permeabilisation is an energy-dependent process (Arnold et al., 2023). Chitosan's antifungal action is abolished by a chemical or physiological obstruction in the electron transport chain. The presence of peroxisomes and, more specifically, mitochondria, the primary organelles involved in ROS formation, supports the importance of ROS metabolism's reaction to chitosan. It is understood that chitosan generates an intracellular ROS burst, which initiates the oxidation of FFA from cell membranes. Increased membrane oxidation eventually leads to complete plasma membrane permeabilisation, which may be responsible for chitosan's antifungal activity (Guarnieri et al., 2022). In some fungi, glucose deprivation induces ROS production and inhibits development. This again explains the connection between nutritional content, ROS and antimicrobial effect (Dazhong et al., 2021).

4.3 *Chitosan as Gene Modulator*

Chitosan can be coupled with tolerant fungi such as biocontrol fungi (BCF). BCF may break down chitosan by utilising it as a food source. *P. chlamydosporia*, a nematophagous fungus, can survive large dosages of chitosan (Lopez-Nuñez et al., 2022). The genome of *P. chlamydosporia* demonstrates an increase in hydrolases. This may be due to the fungus's multitrophic (saprotrophic, endophytic and nematophagous) behaviour. *P. chlamydosporia* encodes enzymes that produce and break down chitosan, such as chitin deacetylases or chitosanases (Abd El-Hack et al., 2020). Investigations show that chitosan promotes proteases and stimulates the production of the vcp1 serine protease, which is also implicated in the infection of nematode eggs by *P. chlamydosporia*. Chitosan also increases virulence by inducing the accumulation of vcp1 and scp1 (a serine carboxypeptidase) proteases in the appressoria of *P. chlamydosporia* infecting root-knot nematode eggs. Chitosan also promotes BCF sporulation (*P. chlamydosporia* and *Beauveria bassiana* (Subhoshmita et al., 2016).

Other research indicates that chitosan promotes development and sporulation in mycoparasitic biocontrol fungus *Trichoderma* spp. (Javier et al., 2008; Kappel et al., 2022). Other *Trichoderma* spp. (*T. harzianun* and *T. neocrassum*) are hypersensitive to chitosan (Zavala-González et al., 2016). *T. koningiopsis* has a plasma membrane high in saturated FFA and is resistant to chitosan. Other *Trichoderma* spp. that are extremely sensitive to chitosan while having a high amount of polyunsaturated FFAs. The use of chitosan in conjunction with BCF brings up new ecologically acceptable options for pest and disease management caused by insects, nematodes or fungi (Lopez-Moya et al., 2019).

Chitosan is a versatile chemical with several uses. It has been shown in clinical studies to suppress the growth of several microorganisms (Ke et al., 2021). Chitosan application under low nutritional (carbon and nitrogen) status favours the antifungal mechanism of action of this polymer on its hosts (Peter et al., 2023). Chitosan also inhibits the growth of the model fungus *Neisseria crassa*. Chitosan significantly inhibits *N. crassa* spore germination and growth development. Chitosan stimulates the expression of genes producing lipase Class III, monosaccharide transporter and glutathione transferase in *N. crassa* spores. These are the primary chitosan targets in this filamentous fungus (Fig. 3). In the model yeast (*S. cerevisiae*), the major chitosan target is the membrane protein ARL1 (Lopez-Moya et al., 2019; Lopez-Nuñez et al., 2022).

5 Chelation of Nutrients by Chitosan

When the chelation effect outweighs the electrostatic force, which occurs when the mixture's pH is greater than the pKa of chitosan, metal ions (such as Ni^{2+}, Zn^{2+}, Co^{2+}, Fe^{2+} and Cu^{2+}) present on the surface of microorganisms' cell walls can be chelated by the amino groups of chitosan (Wickham et al., 2009; Yan et al., 2021). The cell membrane can be stabilised by divalent cations. The divalent metal ions that bind to wall teichoic acids (WTAs) in Gram-positive bacteria might reduce the attraction between nearby phosphate groups, improving the stability of the polymer structure and the strength of the cell wall (Fig. 4).

Divalent cations that bind to WTAs can reduce osmotic pressure variations on either side of the microbial cell. LPS or polyanionic molecules made up of diverse negatively charged phosphate groups is found in the outer leaflet of the OM in the case of Gram-negative bacteria. The stability of the bacterial OM can be maintained by the divalent metal cations, which can reduce the repulsive forces amongst

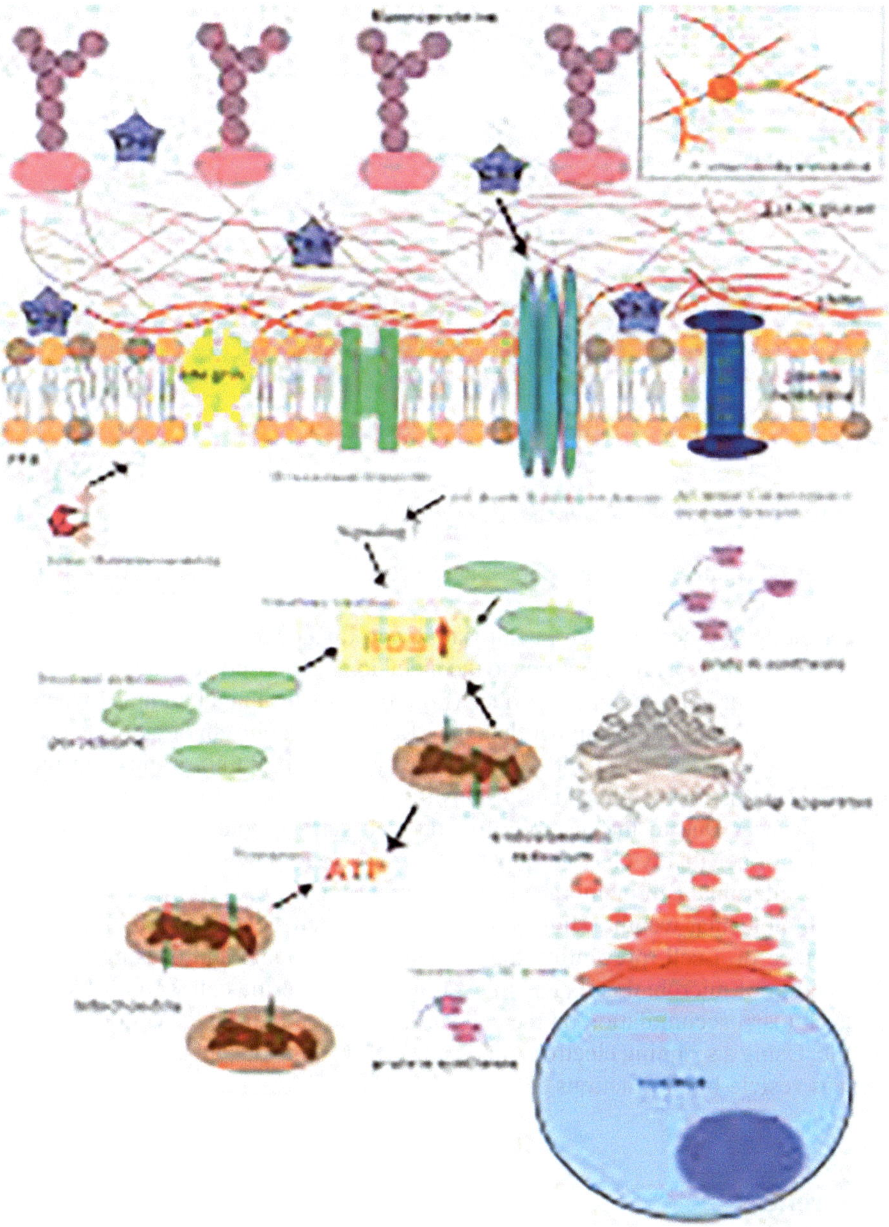

Fig. 3 Mode of action of chitosan as a gene modulator. (*Source*: Lopez-Nuñez et al., 2022)

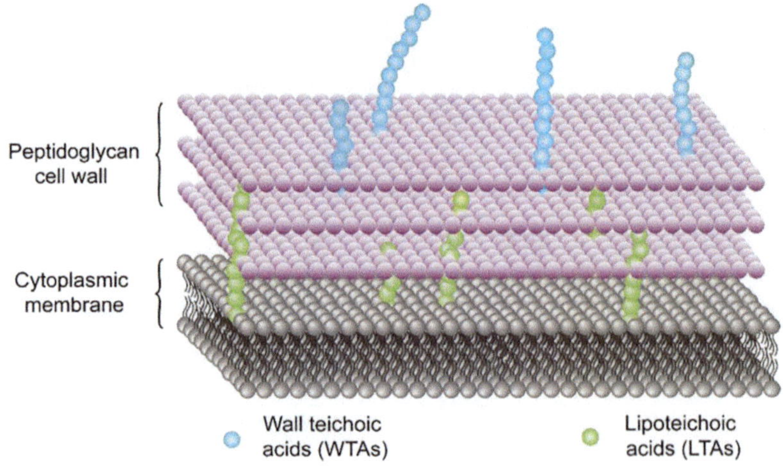

Fig. 4 Binding of metal ions to wall teichoic acids (WTAs). (*Source*: Feng et al., 2021)

aggregated negatively charged phosphate groups (Swoboda et al., 2010; Clifton et al., 2015).

Chelating abilities are seen in chitosan. Unprotonated amino groups of chitosan can give their lone pair of electrons to the metal ions of phosphate groups in the LPS or WTAs on the cell membrane surface (Fig. 3) to create a metal complex when the pH value of the medium is higher than the pKa value of chitosan or chitosan derivatives (Kong et al., 2010). Chitosan's positively charged amino groups can outbid divalent cations for the phosphate groups found in LPS or WTAs on the cell membrane surface. Due to the instability of the cell surface potential and the mutual attraction of negatively charged phosphate groups caused by such a chelation reaction, the microbial cell membrane may rupture (Feng et al., 2021).

High molecular weight (High-MW) chitosan can create a thick polymer film on the surface of a cell, preventing the exchange of nutrients and causing complete destruction of microbial cell. The thicker appearance of the cell walls, which suggests chitosan accumulation on the cell surface, allowed for the detection of such a profile. Using a scanning electron microscope (SEM), the flocculation effect may be seen as vesicle-like formations (Feng et al., 2021).

References

Abd El-Hack, M. E., El-Saadony, M. T., Shafi, M. E., Zabermawi, N. M., Arif, M., Batiha, G. E., et al. (2020). Antimicrobial and antioxidant properties of chitosan and its derivatives and their applications: A review. *International Journal of Biology and Macromolecules., 164*, 2726–2744. https://doi.org/10.1016/j.ijbiomac.2020.08.153

Abreu, F. R., & Campana-Filho, S. P. (2005). Preparation and characterization of carboxymethylchitosan. *Polímerosology., 15*, 79–83. https://doi.org/10.1590/S0104-14282005000200004

Ali, A., Noh, N. M., & Mustafa, M. A. (2015). Antimicrobial activity of chitosan enrichedwith lemongrass oil against anthracnose of bell pepper. *Food Packaging and Shelf Life., 3*, 56–61. https://doi.org/10.1016/j.fpsl.2014.10.003

Andrianov, A. K., & Payne, L. G. (1998). Polymeric carriers for oral uptake of microparticulates. *Advanced Drug Delivery Review., 34*(2), 155–170. https://doi.org/10.1016/S0169-409X(98)00038-6

Aranda-Martinez, A., Lenfant, N., Escudero, N., Zavala-Gonzalez, E. A., Henrissat, B., & Lopez-Llorca, L. V. (2016). CAZyme content of Pochonia chlamydosporia reflects that chitin and chitosan modification are involved in nematode parasitism. *Environmental Microbiology, 18*(11), 4200–4215. https://doi.org/10.1111/1462-2920.13544

Arnold, N. D., Garbe, D., & Brück, T. B. (2023). Proteomic and transcriptomic analyses to decipher the Chitinolytic response of *Jeongeupia* spp. *Marine Drugs, 21*(8), 448. https://doi.org/10.3390/md21080448

Badawy, M. E. I., & Rabea, E. I. (2011). A biopolymer chitosan and its derivatives as promising antimicrobial agents against plant pathogens and their applications in crop protection. *International Journal of Carbohydrate Chemistry, 2011.*

Bernkop-Schnürch, A., Guggi, D., & Pinter, Y. (2004). Thiolated chitosans: Development and in vitro evaluation of a mucoadhesive, permeation enhancing oral drug delivery system. *Journal of Controlled Release, 94*(1), 177–186. https://doi.org/10.1016/j.jconrel.2003.10.005

Boroumand, H., Badie, F., Mazaheri, S., Seyedi, Z. S., Nahand, J. S., Nejati, M., Baghi, H. B., Abbasi-Kolli, M., Badehnoosh, B., Ghandali, M., Hamblin, M. R., & Mirzaei, H. (2021). Chitosan-based nanoparticles against viral infections. *Frontiers in Cellular Infection Microbiology., 11*, 1–22.

Bramwell, V. W., & Perrie, Y. (2006). Particulate delivery systems for vaccines: What can we expect? *Journal of Pharmacy and Pharmacology., 58*, 717–728. https://doi.org/10.1211/jpp.58.6.0002

Cheba, B. A. (2011). Chitin and chitosan: Marine biopolymers with unique properties and versatile applications. *Global Journal of Biotechnology and Biochemistry., 6*, 149–153.

Chen, H., Xing, X., Tan, H., Jia, Y., Zhou, T., Chen, Y., et al. (2017). Covalently antibacterial alginate-chitosan hydrogel dressing integrated gelatin microspheres containing tetracycline hydrochloride for wound healing. *Material Science and Engineering Communication., 70*, 287–295. https://doi.org/10.1016/j.msec.2016.08.086

Cheng, Y., He, C., Ding, J., Xiao, C., Zhuang, X., & Chen, X. (2013). Thermosensitive hydrogels based on polypeptides for localized and sustained delivery of anticancer drugs. *Biomaterials, 34*(38), 10338–10347. https://doi.org/10.1016/j.biomaterials.2013.09.064

Cheung, R. C., Ng, T. B., Wong, J. H., & Chan, W. Y. (2015). Chitosan: An update on potential biomedical and pharmaceutical applications. *Marine Drugs, 13*, 5156–5186. https://doi.org/10.3390/md13085156

Chiappetta, D. A., Facorro, G., de Celis, E. R., & Sosnik, A. (2011). Synergistic encapsulation of the anti-HIV agent efavirenz within mixed poloxamine/poloxamer polymeric micelles. *Nanomedicine, 7*, 624–637. https://doi.org/10.1016/j.nano.2011.01.017

Chua, B. Y., Al Kobasi, M., Zeng, W., Mainwaring, D., & Jackson, D. C. (2012). Chitoson microparticles and nanoparticles as biocompatible delivery vehicles for peptide and protein-based immunocontraceptive vaccines. *Molecular Pharmaceuticals., 9*, 81–90. https://doi.org/10.1021/mp200264m

Chung, Y. C., Su, C. C., Chen, Y., Jia, G., Wang, J. C. G., Wu, H., & Lin, J. G. (2004). Relationship between antibacterial activity of chitosan and surface characteristics of cell wall. *Acta Pharmacology Singapore., 25*, 932–936.

Clifton, L. A., Skoda, M. W., Le Brun, A. P., Ciesielski, F., Kuzmenko, I., Holt, S. A., & Lakey, J. H. (2015). Effect of divalent cation removal on the structure of gram-negative bacterial outer membrane models. *Langmuir, 31*, 404–412. https://doi.org/10.1021/la504407v

Dash, M., Chiellini, F., Ottenbrite, R. M., & Chiellini, E. (2011). Chitosan-A versatile semi-synthetic polymer in biomedical applications. *Progressive Polymer Science., 36*(8), 981–1014. https://doi.org/10.1016/j.progpolymsci.2011.02.001

De Jong, W. H., & Borm, P. J. (2008). Drug delivery and nanoparticles: Applications and hazards. *International Journal of Nanomedicine, 3*, 133–149. https://doi.org/10.2147/IJN.S596
des Rieux, A., Fievez, V., Garinot, M., Schneider, Y. J., & Préat, V. (2006). Nanoparticles as potential oral delivery systems of proteins and vaccines: A mechanistic approach. *Journal of Controlled Release, 116*(1), 1–27. https://doi.org/10.1016/j.jconrel.2006.08.013
Divya, K., Vijayan, S., Tijith, K. G., & Jisha, M. S. (2017). Antimicrobial properties of chitosan nanoparticles: Mode of action and factors affecting activity. *Fibers Polymers., 18*, 221–230. https://doi.org/10.1007/s12221-017-6690-1
Elieh-Ali-Komi, D., & Hamblin, M. R. (2016). Chitin and chitosan: production and application of versatile biomedical nanomaterials. *International Journal of Advanced Research., 4*(3), 411–427.
El-Saadony, M. T., Alkhatib, F. M., Alzahrani, S. O., Shafi, M. E., El. Abdel-Hamid, S., Taha, T. F., Aboelenin, S. M., Soliman, M. M., & Ahmed, N. H. (2021a). Impact of mycogenic zinc nanoparticles on performance, behavior, immune response, and microbial load in *Oreochromis niloticus*. *Saudi Journal of Biological Science., 28*(8), 4592–4604.
El-Saadony, M. T., Desoky, E. S. M., Saad, A. M., Eid, R. S., Selem, E., & Elrys, A. S. (2021b). Biological silicon nanoparticles improve Phaseolus vulgaris L. yield and minimize its contaminant contents on a heavy metals-contaminated saline soil. *Journal of Environmental Science., 106*, 1–14.
El-Saadony, M. T., Sitohy, M. Z., Ramadan, M. F., & Saad, A. M. (2021c). Green nanotechnology for preserving and enriching yogurt with biologically available iron (II). *Innovative Food Science and Emerging Technology., 69*, 102645.
Feng, P., Luo, Y., Ke, C., Qiu, H., Wang, W., Zhu, Y., Hou, R., Xu, L., & Wu, S. (2021). Chitosan-based functional materials for skin wound repair: Mechanisms and applications. *Frontiers in Bioengineering and Biotechnology., 9*, 650598. https://doi.org/10.3389/fbioe.2021.650598
Ford, C. B., Funt, J. M., Abbey, D., Issi, L., Guiducci, C., Martinez, D. A., et al. (2015). The evolution of drug resistance in clinical isolates of *Candida albicans*. *eLife, 4*, e00662. https://doi.org/10.7554/eLife.00662
Friedman, A. J., Jenny, P., Schairer, D. O., Champer, J., Qin, M., Pirouz, A., et al. (2013). Antimicrobial and anti-inflammatory activity of chitosan alginate nanoparticles: A targeted therapy for cutaneous pathogens. *Journal of Investigative Dermatology., 133*(5), 1231–1239. https://doi.org/10.1038/jid.2012.399
Gagliardi, M. (2017). Biomimetic and bioinspired nanoparticles for targeted drug delivery. *Therapeutic Delivery, 8*, 289–299. https://doi.org/10.4155/tde-2017-0013
Glück, R., Mischler, R., Durrer, P., Fürer, E., Lang, A. B., Herzog, C., et al. (2000). Safety and immunogenicity of intranasally administered inactivated trivalent virosome-formulated influenza vaccine containing Eschevichia coli heat-labile toxin as a mucosal adjuvant. *Journal of Infectious Diseases., 181*, 1129–1132. https://doi.org/10.1086/315337
Goldberg, M., Langer, R., & Jia, X. (2007). Nanostructured materials for applications in drug delivery and tissue engineering. *Journal of Biomaterial and Science Polymer Editorial., 18*, 241–268. https://doi.org/10.1163/156856207779996931
Gorbach, V. I., Krasikova, I. N., Luk'yanov, P. A., Loenko, Y. N., Solov'eva, T. F., Ovodov, Y. S., et al. (1994). New glycolipids (chitooligosaccharide derivatives) possessing immunostimulating and antitumor activities. *Carbohydrate Research, 260*(1), 73–82. https://doi.org/10.1016/0008-6215(94)80023-5
Guarnieri, A., Triunfo, M., Scieuzo, C., et al. (2022). Antimicrobial properties of chitosan from different developmental stages of the bioconverter insect *Hermetia illucens*. *Scientific Reports, 12*, 8084. https://doi.org/10.1038/s41598-022-12150-3
He, L., Liu, Y., Mustapha, A., & Lin, M. (2011). Antifungal activity of zinc oxide nanoparticles against *Botrytis cinerea* and *Penicillium expansum*. *Microbiological Research, 1666*, 207–215.
Holban, A. M., Grumezescu, V., Grumezescu, A. M., Vasile, B. S., Trusca, R., Cristescu, R., et al. (2014). Antimicrobial nanospheres thin coatings prepared by advanced pulsed laser technique. *Beilstein Journal of Nanotechnology., 5*, 872–880. https://doi.org/10.3762/bjnano.5.99

Holme, K. R., & Perlin, A. S. (1997). Chitosan N-sulfate. *A water-soluble polyelectrolyte Carbohydrate Research., 302*(1), 7–12. https://doi.org/10.1016/S0008-6215(97)00117-1

Holmgren, J., & Czerkinsky, C. (2005). Mucosal immunity and vaccines. *Natural Medicine., 11*(4), S45–S53. https://doi.org/10.1038/nm1213

Hosseinnejad, M., & Jafari, S. M. (2016). Evaluation of different factors affecting antimicrobial properties of chitosan. *International Journal of Biology and Macromolecules., 85*, 467–475. https://doi.org/10.1016/j.ijbiomac.2016.01.022

Hu, Y., Zhang, Z., Li, Y., Ding, X., Li, D., Shen, C., et al. (2018). Dual-crosslinked amorphous polysaccharide hydrogels based on chitosan/alginate for wound healing applications. *Macromolecules Rapid Communication., 39*(20), 1800069. https://doi.org/10.1002/marc.201800069

Huang, X., Huang, X., Gong, Y., Xiao, H., McClements, D. J., & Hu, K. (2016). Enhancement of curcumin water dispersibility and antioxidant activity using core-shell protein-polysaccharide nanoparticles. *Food Research International., 87*, 1–9. https://doi.org/10.1016/j.foodres.2016.06.009

Ihegwuagu, N. E., Sha'Ato, R., Tor-Anyin, T. A., Nnamonu, I. A., Buekes, P., Sone, B., et al. (2016). Facile formulation of starch-silver-nanoparticle encapsulated dichlorovos and chlorpyrifos for enhanced insecticide delivery. *New Journal of Chemistry., 40*, 1777–1784.

Ilium, L. (1998). Chitosan and its use as a pharmaceutical excipient. *Pharmaceutical Research, 15*, 1326–1331. https://doi.org/10.1023/A:1011929016601

Illum, L., Jabbal-Gill, I., Hinchcliffe, M., Fisher, A. N., & Davis, S. S. (2001). Chitosan as a novel nasal delivery system for vaccines. *Advanced Drug Delivery Review., 51*(1), 81–96. https://doi.org/10.1016/S0169-409X(01)00171-5

Ing, L. Y., Zin, N. M., Sarwar, A., & Katas, H. (2012). Antifungal activity of chitosan nanoparticles and correlation with their physical properties. *International Journal of Biomaterials, 2012*, 632698. https://doi.org/10.1155/2012/632698

Jaime, M. D. L. A., Lopez-Llorca, L. V., Conesa, A., Lee, A. Y., Proctor, M., Heisler, L. E., Gebbia, M., Giaever, G., Westwood, J. T., & Nislow, C. (2012). Identification of yeast genes that confer resistance to chitosan oligosaccharide (COS) using chemogenomics. *BMC Genomics, 13*, 267. https://doi.org/10.1186/1471-2164-13-267

Joye, I. J., Nelis, V. A., & McClements, D. J. (2015). Gliadin-based nanoparticles: Stabilization by post-production polysaccharide coating. *Food Hydrocolloids, 43*, 236–242. https://doi.org/10.1016/j.foodhyd.2014.05.021

Jung, S. N., Kang, S. K., Yeo, G. H., Li, H. Y., Jiang, T., Nah, J. W., et al. (2015). Targeted delivery of vaccine to dendritic cells by chitosan nanoparticles conjugated with a targeting peptide ligand selected by phage display technique. *Macromolecules and Biosciences., 15*(3), 395–404. https://doi.org/10.1002/mabi.201400352

Kanner, E., Gorelov, A. V., Pechkurov, D., Gorelova, E., Maksimov, M., & Ermolaeva, A. (2019). Mucosal immune system of digestive and respiratory tracts: Possibilities of prevention and treatment of infectious diseases. *Medical Council., 2019*, 100–107. https://doi.org/10.21518/2079-701X-2019-11-100-107

Kappel, L., Kosa, N., & Gruber, S. (2022). The multilateral efficacy of chitosan and Trichoderma on sugar beet. *Journal of Fungi, 8*(2). https://doi.org/10.3390/jof8020137

Kashyap, P. L., Xu, X., & Heiden, P. (2015). Chitosan nanoparticle-based delivery systems for sustainable agriculture. *International Journal of Biological Macromolecules., 77*, 36–51.

Kato, Y., Onishi, H., & Machida, Y. (2004). N-succinyl-chitosan as a drug carrier: Water-insoluble and water-soluble conjugates. *Biomaterials, 25*(5), 907–915. https://doi.org/10.1016/S0142-9612(03)00598-2

Ke, C.-L., Deng, F.-S., Chuang, C.-Y., & Lin, C.-H. (2021). Antimicrobial actions and applications of chitosan. *Polymers, 13*(6), 904. https://doi.org/10.3390/polym13060904

Khairy, A. M., Tohamy, M. A. R., Zayed, M. A., Mahmoud, S. F., El-Tahan, A. M., El-Saadony, M. T., & Mesiha, P. K. (2022). Eco-friendly application of nano-chitosan for controlling potato and tomato bacterial wilt. *Saudi Journal of Biological Sciences., 29*(4), 2199–2209.

Kim, B. Y., Rutka, J. T., & Chan, W. C. (2010). Nanomedicine. *New England Journal of Medicine., 363*(25), 2434–2443. https://doi.org/10.1056/NEJMra0912273

Kong, M., Chen, X. G., Xing, K., & Park, H. J. (2010). Antimicrobial properties of chitosan and mode of action: A state of the art review. *International Journal of Food Microbiology., 144*, 51–63. https://doi.org/10.1016/j.ijfoodmicro.2010.09.012

Kumar, A., Ma, H., Zhang, X., Huang, K., Jin, S., Liu, J., et al. (2012). Gold nanoparticles functionalized with therapeutic and targeted peptides for cancer treatment. *Biomaterials, 33*, 1180–1189. https://doi.org/10.1016/j.biomaterials.2011.10.058

Kurita, K. (2006). Chitin and chitosan: Functional biopolymers from marine crustaceans. *Marine Biotechnology, 8*(3), 203. https://doi.org/10.1007/s10126-005-0097-5

Kvasničková, E., Paulíček, V., Paldrychová, M., Ježdík, R., Maťátková, O., & Masák, J. (2016). Aspergillus fumigatus DBM 4057 biofilm formation is inhibited by chitosan, in contrast to baicalein and rhamnolipid. *World Journal of Microbiology & Biotechnology, 32*(11), 187. https://doi.org/10.1007/s11274-016-2146-9

Kyung-Min, Y., Jisung, Y., Manab, D. A., Sung-Gil, H., Inseon, L., Han, S. K., Li, N. K., Jahyun, N., Seok-Joon, K., Moon Il, K., Warayuth, S., Jonathan, S. D., & Jungbae, K. (2019). Enzyme-immobilized chitosan nanoparticles as environmentally friendly and highly effective antimicrobial agents. *Biomacromolecules, 20*(7), 2477–2485.

Lara, H. H., Ayala-Nuñez, N. V., Ixtepan-Turrent, L., & Rodriguez-Padilla, C. (2010). Mode of antiviral action of silver nanoparticles against HIV-1. *Journal of Nanobiotechnology, 8*, 1. https://doi.org/10.1186/1477-3155-8-1

Li, Z., He, C., Yuan, B., Dong, X., & Chen, X. (2017). Injectable polysaccharide hydrogels as biocompatible platforms for localized and sustained delivery of antibiotics for preventing local infections. *Macromolecular Bioscience, 17*(4). https://doi.org/10.1002/mabi.201600347

Look, M., Bandyopadhyay, A., Blum, J. S., & Fahmy, T. M. (2010). Application of nanotechnologies for improved immune response against infectious diseases in the developing world. *Adanced Drug Delivery Review., 62*(4), 378–393. https://doi.org/10.1016/j.addr.2009.11.011

Lopez-Moya, F., & Lopez-Llorca, L. V. (2016). Omics for investigating chitosan as an antifungal and gene modulator. *Journal of Fungi, 2*(1), 11. https://doi.org/10.3390/jof2010011

Lopez-Moya, F., Kowbel, D., Nueda, M. J., Palma-Guerrero, J., Glass, N. L., & Lopez-Llorca, L. V. (2016). Neurospora crassa transcriptomics reveals oxidative stress and plasma membrane homeostasis biology genes as key targets in response to chitosan. *Molecular BioSystems, 12*(2), 391–403. https://doi.org/10.1039/c5mb00649j

Lopez-Moya, F., Suarez-Fernandez, M., & Lopez-Llorca, L. V. (2019). Molecular mechanisms of chitosan interactions with fungi and plants. *International Journal of Molecular Sciences, 20*(2), 332. https://doi.org/10.3390/ijms20020332

Lopez-Nuñez, R., Suarez-Fernandez, M., Lopez-Moya, F., & Lopez-Llorca, L. V. (2022). Chitosan and nematophagous fungi for sustainable management of nematode pests. *Frontiers in Fungal Biology, 3*, 980341. https://www.frontiersin.org/articles/10.3389/ffunb.2022.980341

Luo, M., Jia, Y. Y., Jing, Z. W., Li, C., Zhou, S. Y., Mei, Q. B., et al. (2018). Construction and optimization of pH-sensitive nanoparticle delivery system containing PLGA and UCCs-2 for targeted treatment of *helicobacter pylori*. *Colloids Surface Biology., 164*, 11–19. https://doi.org/10.1016/j.colsurfb.2018.01.008

Ma, Z. S., & Lim, L. Y. (2003). Uptake of chitosan and associated insulin in Caco-2 cell monolayers: A comparison between chitosan molecules and chitosan nanoparticles. *Pharmaceutical Research, 20*(11), 1812–1819. https://doi.org/10.1023/B:PHAM.0000003379.76417.3e

Macri, C., Dumont, C., Johnston, A. P., & Mintern, J. D. (2016). Targeting dendritic cells: A promising strategy to improve vaccine effectiveness. *Clinical Translational Immunology., 5*(3), 66. https://doi.org/10.1038/cti.2016.6

Madhumathi, K., Sudheesh Kumar, P. T., Abhilash, S., Sreeja, V., Tamura, H., Manzoor, K., et al. (2010). Development of novel chitin/nanosilver composite scafolds for wound dressing applications. *Journal of Material Science and Material Medicine., 21*(2), 807–813. https://doi.org/10.1007/s10856-009-3877-z

Mallipeddi, R., & Rohan, L. C. (2010). Progress in antiretroviral drug delivery using nanotechnology. *International Journal of Nanomedicine, 5*, 533–547.

Marine, R. L., Nasko, D. J., Wray, J., Polson, S. W., & Wommack, K. E. (2017). Novel chaperonins are prevalent in the virioplankton and demonstrate links to viral biology and ecology. *ISME Journal., 11*, 2479–2491.

Martinez-Camacho, A. P., Cortez-Rocha, M. O., Ezquerra-Brauer, J. M., Graciano-Verdugo, A. Z., Rodriguez-Felix, F., Castillo-Ortega, M. M., et al. (2010). Chitosan composite films: Thermal, structural, mechanical and antifungal properties. *Carbohyd.rate Polymers., 82*, 305–315. https://doi.org/10.1016/j.carbpol.2010.04.069

McNeil, S. E. (2011). Unique benefits of nanotechnology to drug delivery and diagnostics. *Methods in Molecular Biology., 697*, 3–8. https://doi.org/10.1007/978-1-60327-198-1_1

Modi, J., Joshi, G., & Sawant, K. (2013). Chitosan based mucoadhesive nanoparticles of ketoconazole for bioavailability enhancement: Formulation, optimization, in vitro and ex vivo evaluation. *Drug Development and Industrial Pharmacy., 39*(4), 540–547. https://doi.org/10.3109/03639045.2012.666978

Nagy, A., Harrison, A., Sabbani, S., Munson, R. S., Dutta, P. K., Jr., & Waldman, W. J. (2011). *International Journal of Nanomedicine, 6*, 1833.

Nasrollahzadeh, M., Shafiei, N., Nezafat, Z., Bidgoli, N. S. S., & Soleimani, F. (2020). Recent progresses in the application of cellulose, starch, alginate, gum, pectin, chitin and chitosan based (nano) catalysts in sustainable and selective oxidation reactions: A review. *Carbohydrate Polymers., 116353*. https://doi.org/10.1016/j.carbpol.2020.116353

Ngo, D. H., Vo, T. S., Ngo, D. N., Kang, K. H., Je, G. Y., Pham, H. N., Byun, H. G., & Kim, S. K. (2015). Biological effects of chitosan and its derivatives. *Food Hydrocolloides, 51*, 200–216. https://doi.org/10.1016/j.foodhyd.2015.05.023

Palma-Guerrero, J., Huang, I.-C., Jansson, H.-B., Salinas, J., Lopez-Llorca, L. V., & Read, N. D. (2009). Chitosan permeabilizes the plasma membrane and kills cells of Neurospora crassa in an energy dependent manner. *Fungal Genetics and Biology: FG & B, 46*(8), 585–594. https://doi.org/10.1016/j.fgb.2009.02.010

Palma-Guerrero, J., Lopez-Jimenez, J. A., Pérez-Berná, A. J., Huang, I.-C., Jansson, H.-B., Salinas, J., Villalaín, J., Read, N. D., & Lopez-Llorca, L. V. (2010). Membrane fluidity determines sensitivity of filamentous fungi to chitosan. *Molecular Microbiology, 75*(4), 1021–1032. https://doi.org/10.1111/j.1365-2958.2009.07039.x

Parboosing, R., Maguire, G. E., Govender, P., & Kruger, H. G. (2012). Nanotechnology and the treatment of HIV infection. *Viruses, 4*, 488–520. https://doi.org/10.3390/v4040488

Park, J., Kim, I. Y., Patel, M., Moon, H. J., Hwang, S. J., & Jeong, B. (2015). 2D and 3D hybrid systems for enhancement of chondrogenic differentiation of tonsil-derived mesenchymal stem cells. *Advanced Functional Material., 25*(17), 2573–2582. https://doi.org/10.1002/adfm.201500299

Phanse, Y., Carrillo-Conde, B. R., Ramer-Tait, A. E., Roychoudhury, R., Pohl, N. L. B., Narasimhan, B., et al. (2013). Functionalization of polyanhydride microparticles with di-mannose influences uptake by and intracellular fate within dendritic cells. *Acta Biomaterials., 9*(11), 8902–8909. https://doi.org/10.1016/j.actbio.2013.06.024

Purabi Sarkar, V., Stefi, R., Giva Kuppusamy, M., Aminur, R., Preetham, E., Ramasamy, H., Aziz, A., & Jesu, A. (2022). Pathogenic fungi affecting fishes through their virulence molecules. *Aquaculture, 548*(1), 737553. https://doi.org/10.1016/j.aquaculture.2021.737553

Qi, L., Xu, Z., Jiang, X., Hu, C., & Zou, X. (2004). Preparation and antibacterial activity of chitosan nanoparticles. *Carbohydrate Research, 339*(16), 2693–2700. https://doi.org/10.1016/j.carres.2004.09.007

Raei, M., Shahidi, F., Farhoodi, M., Jafari, S. M., & Rafe, A. (2017). Application of whey protein-pectin nano-complex carriers for loading of lactoferrin. *International Journal of Biology and Macromolecules., 105*, 281–291. https://doi.org/10.1016/j.ijbiomac.2017.07.037

Rashki, S., Asgarpour, K., Tarrahimofrad, H., Hashemipour, M., Ebrahimi, M. S., Fathizadeh, H., et al. (2021). Chitosan-based nanoparticles against bacterial infections. *Carbohydrate Polymers., 251*, 117108. https://doi.org/10.1016/j.carbpol.2020.117108

Regiel-Futyra, A., Kus-Liśkiewicz, M., Sebastian, V., Irusta, S., Arruebo, M., Stochel, G., et al. (2015). Development of noncytotoxic chitosan-gold nanocomposites as efficient antibacterial materials. *ACS Applied Material Interfaces., 7*(2), 1087–1099. https://doi.org/10.1021/am508094e

Robbins, N., Wright, G. D., & Cowen, L. E. (2016). Antifungal drugs: The current armamentarium and development of new agents. *Microbiology Spectrum., 4*. https://doi.org/10.1128/microbiolspec.FUNK-0002-2016

Roldo, M., Hornof, M., Caliceti, P., & Bernkop-Schnürch, A. (2004). Mucoadhesive thiolated chitosans as platforms for oral controlled drug delivery: Synthesis and in vitro evaluation. *European Journal of Pharmacy and Biopharmaceuticals., 57*(1), 115–121. https://doi.org/10.1016/S0939-6411(03)00157-7

Saad, A. M., El-Saadony, M. T., El-Tahan, A. M., Sayed, S., Moustafa, M. A., Taha, A. E., & Ramadan, M. M. (2021). Polyphenolic extracts from pomegranate and watermelon wastes as substrate to fabricate sustainable silver nanoparticles with larvicidal effect against *Spodoptera littoralis. Saudi Journal of Biological Science., 28*(10), 5674–5683.

Sadeghi, A. M. M., Dorkoosh, F. A., Avadi, M. R., Saadat, P., Rafiee-Tehrani, M., & Junginger, H. E. (2008). Preparation, characterization and antibacterial activities of chitosan, N-trimethyl chitosan (TMC) and N-diethylmethyl chitosan (DEMC) nanoparticles loaded with insulin using both the ionotropic gelation and polyelectrolyte complexation methods. *International Journal of Pharmacy., 355*(1), 299–306. https://doi.org/10.1016/j.ijpharm.2007.11.052

Sánchez, C., López, M. C., Galeano, L. A., et al. (2018). Molecular detection and genotyping of pathogenic protozoan parasites in raw and treated water samples from Southwest Colombia. *Parasites and Vectors, 11*, 563. https://doi.org/10.1186/s13071-018-3147-3

Santos-Martinez, M. J., Rahme, K., Corbalan, J. J., Faulkner, C., Holmes, J. D., Tajber, L., et al. (2014). Pegylation increases platelet biocompatibility of gold nanoparticles. *Journal of BioMedicine and Nanotechnology., 10*, 1004–1015. https://doi.org/10.1166/jbn.2014.1813

Sashiwa, H., Kawasaki, N., Nakayama, A., Muraki, E., Yajima, H., Yamamori, N., et al. (2003). Chemical modification of chitosan. Part 15: Synthesis of novel chitosan derivatives by substitution of hydrophilic amine using N-carboxyethylchitosan ethyl ester as an intermediate. *Carbohydrate Research, 338*(6), 557–561. https://doi.org/10.1016/S0008-6215(02)00492-5

Sathiyabama, M., Indhumathi, M., & Muthukumar, S. (2019). Chitosan nanoparticles loaded with thiamine stimulate growth and enhances protection against wilt disease in chickpea. *Carbohydrate Polymers., 212*, 169–177.

Schütz, C. A., Juillerat-Jeanneret, L., Mueller, H., Lynch, I., & Riediker, M. (2013). Therapeutic nanoparticles in clinics and under clinical evaluation. *Nanomedicine (London, England), 8*, 449–467. https://doi.org/10.2217/nnm.13.8

Sekhon, B. S. (2014). Nanotechnology in Agri-food production: An overview. *Nanotechnology, Science and Applications., 7*, 31–53.

Sharif, R., Mujtaba, M., Ur Rahman, M., Shalmani, A., Ahmad, H., Anwar, T., & Wang, X. (2018). The multifunctional role of chitosan in horticultural crops; a review. *Molecules, 23*(4), 872–891.

Sharifi-Rad, J., Quispe, C., Butnariu, M., Rotariu, L. S., Sytar, O., Sestito, S., Rapposelli, S., Akram, M., Iqbal, M., Krishna, A., Kumar, N. V. A., Braga, S. S., Cardoso, S. M., Jafernik, K., Ekiert, H., Cruz-Martins, N., Szopa, A., Villagran, M., Mardones, L., Martorell, M., Docea, A. O., & Calina, D. (2021). Chitosan nanoparticles as a promising tool in nanomedicine with particular emphasis on oncological treatment. *Cancer Cell International, 21*, 1–31.

Shih, P.-Y., Liao, Y.-T., Tseng, Y.-K., Deng, F.-S., & Lin, C.-H. (2019). A potential antifungal effect of chitosan against *Candida albicans* is mediated via the inhibition of SAGA complex component expression and the subsequent alteration of cell surface integrity. *Frontiers in Microbiology, 10*, 602. https://doi.org/10.3389/fmicb.2019.00602

Singh, A., & Chaudhary, B. (2010). Preliminary phyco-chemical analysis and in vitro antibacterial screening of *Pithophora oedogonia* (Mont.) Wittrock: A freshwater green alga forming mats in the water bodies. *Journal of Algal Biomass Utililization., 1*(2), 33–41.

Sobhani, Z., Mohammadi, S. S., Montaseri, H., & Khezri, E. (2017). Nanoparticles of chitosan loaded ciprofloxacin: Fabrication and antimicrobial activity. *Advanced Pharmaceutical Bulletin., 7*(3), 427–432. https://doi.org/10.15171/apb.2017.051

Stec, J., Kosikowska, U., Mendrycka, M., Stępień-Pyśniak, D., Niedźwiedzka-Rystwej, P., Bębnowska, D., Hrynkiewicz, R., Ziętara-Wysocka, J., & Grywalska, E. (2022). Opportunistic pathogens of recreational waters with emphasis on antimicrobial resistance-a possible subject of human health concern. *International Journal of Environmental Research and Public Health., 19*(12), 7308. https://doi.org/10.3390/ijerph19127308

Subbiah, R., Ramalingam, P., Ramasundaram, S., Kim, D. Y., Park, K., Ramasamy, M. K., et al. (2012). N, N, N-Trimethyl chitosan nanoparticles for controlled intranasal delivery of HBV surface antigen. *Carbohydrate Polymers., 89*(4), 1289–1297. https://doi.org/10.1016/j.carbpol.2012.04.056

Sun, L., Song, L., Zhang, X., Zhou, R., Yin, J., & Luan, S. (2020). Poly(γ-glutamic acid)-based electrospun nanofibrous mats with photodynamic therapy for effectively combating wound infection. *Material Science and Engineering Communication., 113*, 110936. https://doi.org/10.1016/j.msec.2020.110936

Swoboda, J. G., Campbell, J., Meredith, T. C., & Walker, S. (2010). Teichoic acid function, biosynthesis, and inhibition. *Chembiochem, 11*, 35–45. https://doi.org/10.1002/cbic.200900557

Teli, M. D., & Kale, R. D. (2011). Polyester nanocomposite fibers with antibacterial properties. *Advanced and Applied Scientific Research., 2*, 491–502.

Tuncer, M. (2007). *Effects of chloride ion and the types of oxides on the antibacterial activities of inorganic oxide supported ag materials* (MSc. Thesis). IzmirInstitute of Technology.

Uthaman, S., Lee, S. J., Cherukula, K., Cho, C., & Park, I. K. (2015). Polysaccharide-coated magnetic nanoparticles for imaging and gene therapy. *BioMed Research International., 2015*, 1–14. https://doi.org/10.1155/2015/959175

Uto, T., Wang, X., Sato, K., Haraguchi, M., Akagi, T., Akashi, M., et al. (2007). Targeting of antigen to dendritic cells with poly (γ-glutamic acid) nanoparticles induces antigen-specific humoral and cellular immunity. *Journal of Immunology., 178*, 2979–2986. https://doi.org/10.4049/jimmunol.178.5.2979

Valenzuela-Ortiz, G., Gaxiola-Camacho, S. M., San-Martín-Hernández, C., Martínez-Téllez, M. Á., Aispuro-Hernández, E., Lizardi-Mendoza, J., & Quintana-Obregón, E. A. (2022). Chitosan sensitivity of fungi isolated from mango (*Mangifera indica* L.) with anthracnose. *Molecules (Basel, Switzerland), 27*(4), 1244. https://doi.org/10.3390/molecules27041244

van Ginkel, F. W., Jackson, R. J., Yuki, Y., & McGhee, J. R. (2000). Cutting edge: The mucosal adjuvant cholera toxin redirects vaccine proteins into olfactory tissues. *Journal of Immunology., 165*, 4778–4782. https://doi.org/10.4049/jimmunol.165.9.4778

Wang, J. J., Zeng, Z. W., Xiao, R. Z., Xie, T., Zhou, G. L., Zhan, X. R., & Wang, S. L. (2011). Recent advances of chitosan nanoparticles as drug carriers. *International Journal of Nanomedicine, 6*, 765–774. https://doi.org/10.2147/IJN.S17296

Weinheimer, A. R., & Aylward, F. O. (2022). Infection strategy and biogeography distinguish cosmopolitan groups of marine jumbo bacteriophages. *ISME Journal., 16*, 1657. https://doi.org/10.1038/s41396-022-01214-x

Wickham, J. R., Halye, J. L., Kashtanov, S., Khandogin, J., & Rice, C. V. (2009). Revisiting magnesium chelation by teichoic acid with phosphorus solid-state NMR and theoretical calculations. *The Journal of Physical Chemistry, 113*, 2177–2183. https://doi.org/10.1021/jp809313j

Wu, X., He, C., Wu, Y., Chen, X., & Cheng, J. (2015). Nanogel-incorporated physical and chemical hybrid gels for highly effective chemo-protein combination therapy. *Advanced Functional Materials., 25*(43), 6744–6755. https://doi.org/10.1002/adfm.201502742

Xie, Y. J., Liu, X. F., & Chen, Q. (2007). Synthesis and characterization of water-soluble chitosan derivate and its antibacterial activity. *Carbohydrate Polymers., 69*, 142–147. https://doi.org/10.1016/j.carbpol.2006.09.010

Xu, M., Hao, X., Hu, Z., & Yan, Q. (2020). Palladium-bridged polymers as CO-biosignal-responsive self-healing hydrogels. *Polymerization Chemistry., 11*(4), 779–783. https://doi.org/10.1039/C9PY01660K

Yan, D., Li, Y., Liu, Y., Li, N., Zhang, X., & Yan, C. (2021). Antimicrobial properties of chitosan and chitosan derivatives in the treatment of enteric infections. *Molecules, 26*(23), 7136. https://doi.org/10.3390/molecules26237136

Yang, T. C., Chou, C. C., & Li, C. F. (2005). Antibacterial activity of N-alkylated dissaccharide chitosan derivatives. *International Journal of Food Microbiology., 97*, 237–245. https://doi.org/10.1016/S0168-1605(03)00083-7

Yang, Y. L., Cheng, M. F., Wang, C. W., Wang, A. H., Cheng, W. T., Lo, H. J., et al. (2010). The distribution of species and susceptibility of amphotericin B and fluconazole of yeast pathogens isolated from sterile sites in Taiwan. *Medical Mycology., 48*, 328–334. https://doi.org/10.3109/13693780903154070

Yang, Y., Wang, X., Yang, F., Wang, L., & Wu, D. (2018). Highly elastic and ultratough hybrid ionic–covalent hydrogels with tunable structures and mechanics. *Advanced Materials, 30*(18), 1707071. https://doi.org/10.1002/adma.201707071

Yuki, Y., & Kiyono, H. (2003). New generation of mucosal adjuvants for the induction of protective immunity. *Review of Medicine and Virology., 13*, 293–310. https://doi.org/10.1002/rmv.398

Zavala-González, E. A., Lopez-Moya, F., Aranda-Martinez, A., Cruz-Valerio, M., Lopez-Llorca, L. V., & Ramírez-Lepe, M. (2016). Tolerance to chitosan by Trichoderma species is associated with low membrane fluidity. *Journal of Basic Microbiology, 56*(7), 792–800. https://doi.org/10.1002/jobm.201500758

Zhou, W., Wang, Y., Jian, J., & Song, S. (2013). Self-aggregated nanoparticles based on amphiphilic poly(lactic acid)-grafted-chitosan copolymer for ocular delivery of amphotericin B. *International Journal of Nanomedicine, 8*, 3715–3728. https://doi.org/10.2147/ijn.S51186

Zilly, F. E., Halemani, N. D., Walrafen, D., Spitta, L., Schreiber, A., Jahn, R., & Lang, T. (2011). Ca2+ induces clustering of membrane proteins in the plasma membrane via electrostatic interactions. *The EMBO Journal., 30*, 1209.

Applications of Nanochitosan in Fish Disease Management

Franklyn Nonso Iheagwam, Doris Nnenna Amuji, and Collins Ojonugwa Mamudu

Contents

F. N. Iheagwam (✉)
Department of Biochemistry, Covenant University, Ota, Nigeria

Covenant University Public Health and Wellness Research Cluster, Ota, Nigeria
e-mail: franklyn.iheagwam@covenantuniversity.edu.ng

D. N. Amuji
Department of Biochemistry, Covenant University, Ota, Nigeria

Covenant Applied Informatics and Communication Africa Centre of Excellence, Ota, Nigeria

C. O. Mamudu
Biochemistry Program, City University of New York Graduate Center, New York, USA

P. O. Isibor et al. (eds.), *Nanochitosan-Based Enhancement of Fisheries and Aquaculture*, https://doi.org/10.1007/978-3-031-52261-1_5

1 Introduction

Chitosan, a naturally occurring polysaccharide derived from crustacean shells, holds significant importance for therapeutic purposes. It is the deacetylated form of chitin and serves as a crucial material in the medical field due to its unique properties. Chitosan exhibits remarkable biocompatibility, lacks characteristics related to antigens, and is highly compatible with living tissues (De Sousa Victor et al., 2020). Its haemostatic and anti-thrombogenic characteristics make it a pivotal tool for medical utilization, including controlled drug release, drug encapsulation, enzyme and immobilised cells, as well as serving as a carrier for genes. Chitosan is also biodegradable, breaking down into oligo-products through enzymatic processes, which are subsequently metabolised. Many chitosan derivatives share their biocompatibility and non-toxic nature when interacting with living tissues.

Chitosan's diverse range of potential uses in the medical and biomedical fields stems from its exceptional properties, including biodegradability, immunocompatibility, low toxicity, and biocompatibility. It finds numerous applications across various medical sectors, including pharmaceutical development and drug delivery for a wide array of substances such as antibiotics, anti-inflammatory agents, vaccines, proteins, peptides, and growth factors. Additionally, chitosan can be employed in antimicrobial contexts, gene delivery, and gene therapy, in addition to promoting wound healing and treating burns. It holds promise in regenerative medicine and tissue engineering, spanning from the regeneration of bone, ligaments, cartilage, and tendons to aiding in the regeneration of liver, neural, and skin tissues. Furthermore, chitosan has potential applications in the field of cancer, playing roles in treatment, therapy, and diagnostic strategies. Its versatility extends to dermatology, ophthalmology, dentistry, biosensors, bioimaging, assistance for enzymes in a fixed state, as well as in veterinary medicine (Sivanesan et al., 2021). In the realm of medical utilization, chitosan and its derivatives, including chitooligosaccharides, can be readily transformed into various formats such as solutions, hydrogels, sponges, nanoparticles, membranes and films (including pure films or blends), adhesives, as well as nanofibers (Morin-Crini et al., 2019).

Chitosan faces a significant limitation, in that it lacks solubility in aqueous solutions, which restricts its broad utilization in biological contexts (Qin et al., 2006; Aranaz et al., 2021; Desai et al., 2021). Nevertheless, chitosan possesses functional groups that offer the potential for graft modification, granting modified chitosan unique characteristics. These alterations can be harnessed to chemically alter chitosan, enhancing its solubility and thereby expanding its range of utilisation. Such chemical alterations give rise to various chitosan derivatives characterized by sustained-release properties, non-toxicity, biocompatibility, and biodegradability. Numerous standard methods exist for the fabrication of chitosan-based nanoparticles, including processes like ionic gelation, covalent cross-linking, precipitation, polymerisation, self-assembly, creation of chitosan-drug complexes, and spray-drying. Moreover, nanochitosan, in the form of chitosan nanoparticles, can enhance the body's immune response, leading to antitumor activity. These nanoparticles are highly biocompatible, biodegradable, and amenable to modification, making them valuable as drug carriers. Chitosan nanoparticles find widespread utilization in the transportation of drugs and vaccines, vaccine adjuvants, antimicrobial function, biosensors, tissue creation, and numerous other utilizations.

The effectiveness of nanochitosan in these applications depends on the consistency and dimensions of the produced microspheres. The size of particles impacts the adsorption of antigens and their distribution, ultimately influencing the immune response. Factors such as microsphere structure, surface micropore size, and antigen release rate also play crucial roles in microsphere functionality. Nanochitosan is typically obtained through methods such as emulsion crosslinking, ionic crosslinking, solvent evaporation, spray drying, precipitation, or flocculation, and chitosan solution coating (Ahuekwe et al., 2023a, b).

The use of nanochitosan as a drug carrier for controlled release has led to numerous successes in the diagnosis, detection, and treatment of various diseases. Nanochitosan positive surface charge and mucoadhesive features enable it to attach to mucous membranes and facilitate the gradual and extended release of therapeutic substances, making it particularly suitable for mucosal delivery. Chitosan's ability to inhibit the growth of bacteria and fungi is especially valuable in wound treatment, effectively harnessing its antimicrobial properties and compatibility with biological systems (Ahuekwe et al., 2023a, b).

2 Role of Nanochitosan in Disease Prevention and Treatment

Nanoparticle technology is increasingly being utilized as a formulation approach to address the challenges associated with oral drug delivery. Nanoparticles offer several advantages, including their small particle size, large surface area, and potential for surface modification. The reduced particle size is well-known for its ability to improve the rate at which drugs dissolve. Moreover, nanoparticles can enhance the durability of pharmaceuticals that are sensitive to acids in the gastrointestinal tract, a feature that sets them apart from other drug delivery systems.

Chitosan, a versatile material, can be transformed into polymeric nanoparticles referred to as nanochitosan. These nanoparticles have gained extensive application in the biomedical field, serving as valuable tools for both the detection and treatment of diseases. When used as a drug delivery carrier, nanochitosan exhibits the capability to adsorb or load multiple drugs, allowing for more precise control over drug release. Furthermore, nanochitosan can coat drugs on its outer surface. Their unique capacity to focus on molecules by binding to particular receptors located on the surfaces of cells and facilitating cellular entry enhances the secure and effective administration of drugs and gene therapy to specific targets. Particularly noteworthy are hydrophilic-surfaced nanochitosan variants, which are greatly favoured as carriers because of their minimal non-specific protein adsorption features. These nanochitosan carriers find extensive use in the diagnosis and treatment of complex illnesses.

2.1 Drug Delivery

One of the most significant applications of nanochitosan is in drug delivery systems (Jana & Jana, 2019). Nanochitosan-based drug carriers can encapsulate therapeutic agents, protecting them from degradation and enabling their targeted delivery to specific tissues or cells. This enhances drug efficacy while reducing side effects and the need for frequent dosing. This is particularly valuable in cancer treatment, where nanochitosan allows for the precise delivery of chemotherapeutic agents to cancer cells while sparing healthy tissue (Sadoughi et al., 2020). Nanochitosan is employed also as a carrier for gene delivery, enabling the introduction of therapeutic genes into cells. This is valuable in gene therapy due to its ability to form complexes, as well as its biodegradability and biocompatibility. Its involvement in gene delivery relies on its capacity to undergo protonation in acidic conditions, resulting in the formation of complexes with DNA through electrostatic interactions.

These chitosan–DNA complexes are straightforward to assemble and demonstrate higher effectiveness in comparison to more prevalent systems. They have been reported to successfully transfect different cell types, including human embryonic kidney cells, cervical cancer cells, primary chondrocytes, and fibroblast cells. Targeted therapy minimizes collateral damage and enhances the therapeutic index. Nanochitosan-based drug delivery systems offer controlled and sustained release of medications. This ensures that therapeutic levels of drugs are maintained over an extended period, improving patient compliance and treatment outcomes, as exemplified below with several instances of oral drug delivery. For instance, catechin and epigallocatechin, which are antioxidants found in green tea, tend to degrade in solution of intestinal fluid and have limited absorption across intestinal membranes. Encapsulating them within chitosan nanoparticles can significantly improve their intestinal absorption.

Tamoxifen, an anti-cancer medication characterized by its low water solubility, presents a viable option for the delivery of cancer drugs through the oral route. The formulation of tamoxifen into nanoparticles composed of lecithin and chitosan serves to improve its ability to permeate the intestinal epithelium. These nanoparticles exhibit mucoadhesive properties and enhance tamoxifen permeation through the paracellular pathway. Researchers, such as Feng et al. (2009) have explored efficient oral delivery strategies for drug use in treating cancer. They have developed nanoparticles of doxorubicin hydrochloride (DOX) using chitosan and carboxymethyl chitosan. These nanostructures have been found to improve the DOX absorption in the small intestine. Alendronate sodium, used in osteoporosis treatment, faces challenges related to limited absorption in the digestive system and potential gastrointestinal adverse effects. By formulating nanochitosan through an ion gelation technique, researchers have achieved high encapsulation efficiency for alendronate sodium. Additionally, the release of the drug from these nanoparticles is pH-dependent.

Nasal delivery serves as a non-invasive method for administering drugs to target the respiratory system, and the brain, or achieve circulation that encompasses the entire body. Effective nasal absorption is crucial for the drugs to exert their therapeutic effects. Nanochitosan, owing to its distinctive properties, finds applications in nasal drug delivery. Nasal absorption primarily occurs through three pathways: transcellular, paracellular, and via the trigeminal nerves. For instance, carbamazepine, commonly used in epilepsy treatment, needs to cross the blood–brain barrier (BBB) to be effective. When administered nasally, carboxymethyl nanochitosan containing carbamazepine demonstrated improved absorption and better targeting in the brain (Zhao et al., 2018).

Several mechanisms govern the release of drugs from nanochitosan, including polymer swelling, dispersion of the adsorbed medication, drug permeation through the polymeric structure, breakdown of the polymer, and a combination of both erosion and degradation. The initial rapid release of drugs from nanochitosan can occur due to polymer swelling. Nanochitosan formation can result from polymer swelling, leading to pore formation, or from drug diffusion originating from the surface of the polymer. Additionally, nanochitosan demonstrates pH-dependent drug release due to the solubility characteristics of chitosan. Nanochitosan can modify the drug release pattern, allowing for adjustable drug release rates and influencing the pharmacokinetic characteristics of the encapsulated medicine.

In regulated drug delivery through diffusion, the drug penetrates the inner structure of the polymeric matrix toward the external environment. Within the polymer, the polymer chains form a diffusion barrier, impeding the drug's passage and acting as the rate-controlling membrane for drug release. This diffusion mechanism can also be connected to polymer expansion or deterioration. Polymer swelling involves the absorption of water into the polymer until the polymer ultimately dissolves. This release mechanism is contingent on the polymer's solubility in either water or the adjacent biological medium. When the polymer interacts with the surrounding medium and undergoes swelling, the polymer chains disentangle, leading to drug release from that specific section of the polymer matrix (Mohammed et al., 2017).

Erosion and polymer breakdown are interconnected processes. Polymer breakdown can sometimes induce subsequent physical wear and tear as chemical bonds are severed. Polymer erosion is a multifaceted process involving swelling, diffusion, and dissolution. Erosion can occur in two ways: homogenous erosion, where the polymer erodes uniformly throughout the matrix, and heterogeneous erosion, where the erosion progresses from the outer part toward the inner structure of the polymer. The breakdown of polymers can be triggered by various factors, including the surrounding environment and the presence of enzymes. The release of drugs is contingent upon various factors, including the specific polymer type, internal bonding, any additives like chitosan derivatives, and the characteristics of the nanoparticles, including their size and shape, which determine the surface area and free energy.

2.2 *Wound Healing and Tissue Regeneration*

Nanochitosan has been used in wound dressings and tissue engineering. In regenerative medicine and tissue creation, nanochitosan plays a crucial role. It can be used to create scaffolds for tissue regeneration, supporting the repair of damaged tissues such as bone, cartilage, skin, and neural tissue. Its antibacterial properties help prevent infections, while its ability to promote cell adhesion and proliferation accelerates wound healing and tissue regeneration. This is crucial in managing chronic wounds, burns, and tissue injuries. It finds application in crafting artificial skin for grafting onto severe burn wounds and plays a role in surgical procedures, such as suture threads. Local infections are a common occurrence in wounds, burns, diseases, and surgical sites, and they pose significant risks to patients' health. These infections can result in delayed wound healing, wound reopening the formation of pus-filled pockets, systemic infection, and in severe instances, potentially fatal outcomes. Consequently, there is an urgent and crucial need for the advancement of novel biomaterials that can effectively prevent such infections. Chitosan, owing to its inherent antibacterial properties, presents a promising solution for wound dressings. Nanochitosan hydrogel can entrap bioactive compounds via physical interactions or chemical bonding. For instance, a gel made from carboxymethyl chitosan (CMC) and dextran that underwent oxidation was developed to load anti-infective medications like ceftriaxone sodium. This formulated gel has demonstrated excellent biocompatibility and has shown significant anti-infective effects in vivo studies, including subcutaneous infection and caecal ligation and perforation models.

Furthermore, Chen et al. (2017) created an antibacterial alginate/chitosan hydrogel dressing that incorporates gelatine microspheres. This innovative dressing serves as an effective solution for preventing and treating infections in wounds and surgical sites.

Nanochitosan serves as a 3D scaffold for tissue growth, activating macrophage function and promoting cell proliferation. It enhances the activity of pronuclear leukocytes and triggers the activation of macrophage fibroblasts, leading to improved granulation and tissue repair. The gradual breakdown of N-acetyl-β-d-glucosamine encourages fibroblast multiplication, resulting in the deposition of collagen and synthesis of hyaluronic acid within the wound. This increases the wound-healing process and mitigates scar formation. Kavitha Sankar et al. (2017) devised a lyophilized glutaraldehyde-crosslinked chitosan sponge for achieving haemostasis in blood. The sponge functioned as a physical obstruction, causing rapid blood coagulation. Another form of composite particle, tricalcium phosphate-chitosan, has found application as a replacement for bone and as a support in tissue creation, exhibiting high efficacy in bone formation. These nanoparticles possess the ability to fill specific defect sites, potentially serving as bone substitutes, enhancing drug release capacity, and acting as scaffolds for osteoblast cell cultures.

Diagnostics Nanochitosan finds valuable applications in biosensors and bioimaging for the early diagnosis of diseases, particularly in the detection of specific biomarkers and pathogens. One illustrative example is the development of magnetic nanoparticles coated with chitosan (CS-MNPs) for the identification of *Escherichia coli* and *Staphylococcus aureus* in specimens. These CS-MNPs proved highly effective in quantifying both gram-negative *E. coli* and gram-positive *S. aureus* using 2,2′-casino-bis (3-ethylbenzothiazoline-6-sulfonic acid chemistry), providing a visible, unaided-eye detection method. This innovation suggests that CS-MNPs hold promise for a novel approach to detecting a wide range of bacteria in diverse samples.

2.3 Neurological and Ophthalmic Applications

In neurological disorders and ocular infections of aquatic organisms, nanochitosan can be formulated to cross the blood–brain barrier making it relevant in the treatment of neurological disorders and effectively delivering drugs to the eyes, offering new possibilities for treatment (Alajangi et al., 2022; Mikušová & Mikuš, 2021).

2.4 Immunomodulation

Nanochitosan can modulate the immune response, making it useful in autoimmune disease management. It can help regulate immune responses, reducing inflammation and tissue damage. It can help regulate the immune system's activity, potentially reducing inflammation and tissue damage in conditions like rheumatoid arthritis (Janakiraman et al., 2018).

2.5 Infectious Disease Management

Various concentrations of nanochitosan have been assessed as an alternative approach for disease control, both in laboratory settings (in vitro) and in living organisms (in vivo) (Khairy et al., 2022). Nanochitosan can manage pathogenic microorganisms by inhibiting their growth, sporulation, spore viability, and germination ultimately causing cell destruction by exhibiting antimicrobial properties, which can be harnessed in the development of antimicrobial agents and wound care products. These products can be applied topically for wound care, where they help to inhibit bacterial growth and promote wound healing. It also has potential as a vaccine adjuvant, enhancing the body's immune response to vaccines. It can enhance the antigenicity of vaccines, leading to better protection against infectious diseases.

Infectious diseases pose significant global health threats in aquaculture. Emerging viruses and drug-resistant microorganisms continuously strain the effectiveness of available treatments. In response to this crisis, scientists have introduced a range of novel antiviral and antibacterial medications. Concurrently, advancements in functional materials have enhanced therapeutic results. Evidence demonstrates that the synergy between chitosan and conventional medicine substantially enhances their anti-infective capabilities.

In the context of preventing and treating infectious diseases, drug therapies often come with unwanted side effects and can lead to increased drug resistance in pathogens. Vaccination stands as an effective approach for both prevention and treatment, leveraging immune stimulation to effectively thwart the recurrence of infectious diseases (Ghattas et al., 2021). Notably, mucosal surfaces like the nasal, respiratory, oropharyngeal, gastrointestinal, and genitourinary systems serve as the primary points of entry for pathogens, including viruses and bacteria. Nanochitosan has emerged as a promising tool in mucosal vaccine delivery due to its ability to precisely target antigens (Gaglio et al., 2023; Rhee, 2020). Additionally, nanochitosan offers protection to antigens against harsh environmental conditions such as pH levels, bile, and digestive enzymes in the gastrointestinal tract, while also allowing for controlled and gradual release of the antigen. Studies have indicated that nanochitosan can heighten the immune response when delivering antigens through mucosal routes.

2.5.1 The Mechanism Through Which Nanochitosan Acts Against Antimicrobial Agents

Nanochitosan plays a widely recognized role as an antimicrobial agent, exhibiting high effectiveness against infectious agents. This heightened effectiveness arises from its superior capacity to engage with bacterial cell membranes. This antimicrobial activity is primarily attributed to the electrostatic interplay amongst the amino groups' charge of chitosan's glucosamine carrying a positive and the cell membrane of bacteria possessing a negative charge. This interaction induces alterations in membrane permeability and surface characteristics, ultimately leading to cellular demise and disruption of osmotic equilibrium. It also facilitates the entry of nanochitosan through the bacterial cell wall, resulting in a more robust engagement with charged molecules and a higher aggregation of nanochitosan at the site of interaction. Once penetrated, nanochitosan binds with the DNA, impeding DNA replication and ultimately causing bacterial cell death. Studies indicate that chitosan's binding to teichoic acid in bacteria with a positive cell wall charge, as well as lipopolysaccharide in a negative cell wall charge, carries out a crucial function in inducing mutations that disrupt cell membrane function, leading to disturbances in cell wall dynamics. Furthermore, nanochitosan demonstrates the capability to alter the electron transport chain of bacteria, potentially contributing to its effectiveness against microbes (Dilnawaz et al., 2023).

2.5.2 Application of Nanochitosan in Controlling Bacterial Infection

The mechanism behind chitosan's antibacterial action involves its binding to bacterial cell walls with a negative electrical charge. This interaction results in changes in the permeability of the bacterial cell envelope and subsequent attachment to bacterial DNA, inhibiting its replication (Meng et al., 2021). The presence and arrangement of acetyl groups within chitosan influence its biodegradability and its impact on cell health, with the extent of deacetylation impacting its potency in antigen delivery and its adjuvant properties. Chitosan and its modified forms demonstrate antibacterial properties against both Gram-negative and Gram-positive bacteria. Enhancing the efficacy of antibacterial agents can be achieved by loading them into nanochitosan. The intrinsic antibacterial properties of chitosan-based nanoparticles are improved at the nanoscale, with minimal increase in adverse effects. These nanoparticles demonstrate noticeable antibacterial effects surpassing the antibacterial efficacy of pure chitosan polymers and clinical standards (MubarakAli et al., 2018; Yilmaz Atay, 2019).

Studies have demonstrated that nanochitosan, especially when derived from low molecular weight chitosan, can prevent the development of *Streptococcus mutans* biofilm in laboratory settings, as it can be evenly distributed in samples and inflict substantial harm to cell membranes. In the context of addressing *H. pylori*-induced gastroesophageal reflux disease and chronic atrophic gastritis, where traditional antibiotics are becoming less effective due to drug resistance, nanochitosan can be utilised for gastric delivery. This method overcomes drug malabsorption in the stomach. Nanochitosan can also serve as an effective agent for ocular administration, with nanochitosan-based drug delivery systems exhibiting enhanced corneal penetration without increased corneal irritation (Onugwu et al., 2023; Zamboulis et al., 2020). In mucosal vaccine carriers, chitosan enhances the adhesive properties of vaccine carriers, enabling prolonged retention on mucosal surfaces. In the treatment of dental diseases, where traditional antibacterial agents often face challenges related to absorption and biofilm penetration, chitosan's adhesive properties are leveraged. Chitosan has been employed in the treatment of periodontitis, resulting in a rapid healing of alveolar bone and periodontal epithelium. Nanochitosan can be attached to oral cavity surfaces like teeth, tongue, and buccal mucosa, releasing antibacterial agents in response to oral environment changes, effectively penetrating bacterial biofilms responsible for dental infections for therapeutic purposes.

Due to its remarkable versatility, nanochitosan has emerged as a favoured alternative or adjunctive approach to antibiotics. It provides unique advantages, including precise and controlled drug delivery, as well as safeguarding the potency of the drug. To illustrate, in the case of brucellosis, a significant contributor being *Brucella* spp., an infectious disease affecting the digestive and respiratory tracts in humans and animals, the induction of mucosal immunity is crucial. This was investigated through the encapsulation of three recombinant proteins (rMdh, rOmp10, and rOmp19) from *Brucella abortus* in mucoadhesive nanochitosan, and subsequently studied for its immunogenicity following nasal administration in BALB/c mice (Soh et al., 2019; Dilnawaz et al., 2023).

Inflammatory bowel disease (IBD) poses a persistent challenge for physicians due to its recurrent nature and the difficulty in containment. Rifaximin, commonly prescribed for IBD, has been harnessed to create a controlled, colon-targeted delivery system using chitosan, aimed at enhancing drug solubility for an overall improvement in therapeutic efficacy. In regions where gastrointestinal diseases are prevalent, particularly in developing nations, there is an urgent need for efficient antimicrobial treatments targeting the responsible pathogens. An independent study conducted by Covarrubias et al. (2018) showcased the potent antibacterial effects of hybrid material comprised of copper and nanochitosan against cariogenic *S. mutans*, the culprit behind tooth decay. Furthermore, nanochitosan has exhibited promise in targeted therapy against cutaneous pathogens like *Propionibacterium acnes*, which is responsible for the development of acne. In addition, it has demonstrated effectiveness in decreasing the release of *E. coli* O157:H7 in a bovine model with uterine conditions. Likewise, an in vitro study, contingent on time and dosage, utilizing amphotericin-loaded nanochitosan, affirmed its anti-leishmanial activity. Sohail et al. (2021) proposed the application of these amphotericin-loaded nanochitosan particles for localized treatment against Leishmania following in-depth in vivo efficacy assessments.

A novel chitosan-coated human albumin nanoparticle formulation for the delivery of Colistin (Col/haNPs) has exhibited a notable antibacterial effect has been reported. This formulation led to a considerable decrease in minimum inhibitory concentration (MIC) values, and over time, it demonstrated a substantial decrease in the growth of bacteria, especially against Colistin-resistant strains, in comparison to free Colistin (Scutera et al., 2021). Cpl-1 loaded nanochitosan has also been employed against antibiotic-resistant *S. pneumoniae*, showcasing its potential in combatting drug-resistant strains.

2.5.3 Application of Nanochitosan in Controlling Viral Infection

Nanochitosan has emerged as a promising platform for delivering therapeutics against various viral diseases. To enhance HIV-1 treatment accessibility in resource-limited areas, scientists have created PLGA nanochitosan carriers containing hydrophilic drugs such as lamivudine (LMV) or nevirapine (NVP), and varying concentrations of PEGylated chitosan-loaded lamivudine (LPC) nanoparticles have demonstrated robust inhibition of HIV replication by reducing virion production (Asl et al., 2023; Mamo et al., 2010). To enhance the anti-HIV properties of Atripla, Shohani et al. (2017) designed a nanoconjugate using trimethyl chitosan. This approach was aimed at targeting HIV-infected brain astrocytes, which require the delivery of siRNA across BBB. By using small interfering RNA (siRNA) encapsulated in dual antibody-modified nanochitosan, they successfully inhibited HIV replication. The two antibodies, bradykinin B2 antibody and transferrin antibody, were utilized as targeting ligands, binding to bradykinin B2 receptor (B2R) and transferrin receptor (TfR) to facilitate siRNA delivery into astrocytes. In the context of interfering with HIV transmission, Timur et al. developed tenofovir-loaded nanochitosan vaginal gels for in vitro systems. They demonstrated the potential antiviral effect of chitosan-expressed siRNA against influenza nucleoprotein when delivered through the nasal passage. This approach protected lethal influenza challenges in BALB/c mice. A new nanoparticle system was created, involving chitosan combined with a recombinant plasmid depicting two short hairpin RNA (shRNA) sequences. This system targeted the RSV NS1 and P genes, effectively inhibiting the pathogenesis and replication of the respiratory syncytial virus (RSV) inside cells that are infected. A liposomal layer positioned between the nucleus and the outer layer guaranteed the stability and robustness of the plasmid DNA.

Furthermore, curcumin-encapsulated nanochitosan demonstrated antiviral activity against the hepatitis C virus, as confirmed in human hepatoma cell lines. These applications showcase the versatility and potential of chitosan nanoparticles in combating various viral diseases (Dilnawaz et al., 2023; Ng et al., 2020).

Given the high transmission rate of COVID-19, especially through aerosols released by infected individuals containing SARS-CoV-2 viruses, there is a critical need to protect healthcare workers. A proposed solution involves utilizing positively charged polymers like chitosan to create nanofibers that can be incorporated into specialized clothing for healthcare providers. These nanofibers would generate an electrostatic force of repulsion between the fabric surface and SARS-CoV-2, reducing the viral load around the wearer and mitigating the risk of virus transmission (Safer & Leporatti, 2021). Chitosan-based compounds have demonstrated effectiveness in inhibiting coronavirus infections in vitro and ex vivo. This is attributed to chitosan derivatives binding to the virus's S protein, thereby masking the spike protein and preventing it from interacting with cellular receptors. Nanochitosan, as a vaccine adjuvant or carrier, has been explored for delivering drugs, siRNA, and peptides through the intranasal pathway. Chitosan's capacity to activate the immune system and traverse mucosal epithelial cell tight junctions offers the potential for enhancing the immunogenicity of antigen molecules, making it a promising option for COVID-19 therapy.

For targeted delivery to dendritic cells (DCs), biotinylated nanochitosan was employed to deliver plasmid DNA encoding the SARS-CoV N protein. This approach resulted in elevated mucosal IgA and Increased systemic IgG response to the N protein following intranasal delivery, indicating its potential for immune response modulation. Innovatively, nanochitosan was employed as a diagnostic agent for identifying SARS-CoV-2. A voltammetric genosensor was designed for rapid COVID-19 diagnosis, with the RNA-dependent RNA polymerase (RdRP) sequence utilized as the focus for detecting the virus in sputum samples. RdRP plays a crucial role in the virus's genome replication.

2.5.4 Application of Nanochitosan in Controlling Fungal Infection

The antifungal mechanism of nanochitosan also involves close interaction with fungal cell membranes, leading to the rupture of these membranes (Lopez-Moya et al., 2019; Ing et al., 2012). However, it is important to note that nanochitosan may not be effective against all fungi, as some, like *Aspergillus niger*, exhibit resistance to it. *A. niger's* resistance is attributed to its cell wall, which contains a substantial amount of chitin. Therefore, when considering the use of nanochitosan for antifungal purposes, specific choices and strategies should be employed to achieve the desired therapeutic effect (Rozman et al., 2019).

In general, chitosan nanoparticles have demonstrated potency in hindering the sprouting of spores and the outward expansion of mycelium in various fungal species (Nami et al., 2021). Numerous studies have explored their impact on yeasts and fungi linked to food and plant contamination. For instance, Saharan et al. (2015) utilized the ionic gelation method to create nanochitosan loaded with copper (Cu) ions, testing its antifungal effectiveness against pathogenic fungi affecting tomatoes. The nanoparticles displayed a zeta potential of +22.6 mV, indicating a positive charge, enhancing particle stability and facilitating stronger electrostatic interactions with the cell membranes of pathogenic fungi. Cota-Arriola et al. (2016) proposed that the effectiveness of nanochitosan against *Aspergillus parasiticus* depends on particle size and the chitosan–sodium tripolyphosphate matrix. Meanwhile, Pilon et al. (2015) explored nanochitosan as a coating for fresh-cut apples and found that 110 nm nanoparticles were the most efficient in inhibiting the growth of various microorganisms, both mesophilic and psychrotrophic.

In a recent study conducted by Kheiri et al. (2017), they compared the antifungal effects of chitosan and nanochitosan on *Fusarium graminearum*. The study revealed that nanochitosan exhibited stronger antifungal activity, as evidenced by its lower minimal inhibitory concentration 50% (MIC50) value. This enhanced activity can be attributed to nanochitosan's small size, which allows for easier penetration of microorganism cell membranes. Additionally, Hernandez- Hernández-Lauzardo et al. (2008) conducted a comparison of chitosan with varying molecular weights for their antifungal effects on *Rhizopus stolonifer*. The findings showed that chitosan with lower molecular weight demonstrated the most pronounced suppression of mycelial growth, whereas chitosan with higher molecular weight caused disruptions in spore morphology, sporulation, and germination. These findings collectively emphasize the potential of nanochitosan as a valuable tool in combatting fungal growth and contamination, with its efficacy influenced by factors such as particle size, charge, and molecular weight (Ahmed et al., 2020).

3 Use of Nanochitosan in Aquaculture

Aquaculture has played a vital role in meeting the growing demand for animal protein, and consequently, in ensuring food security. Nonetheless, the aquaculture sector faces significant challenges related to environmental contamination and disease outbreaks. To address these challenges effectively, innovative technological approaches have emerged. Amongst these, nanotechnology stands out as a novel and promising tool with a wide array of applications in aquaculture and seafood preservation. Nanotechnology offers diverse possibilities, including the detection and management of pathogens, the promotion of fish and shellfish growth, the sterilization of aquaculture ponds, efficient delivery of nutrients and medications, as well as advancements in seafood processing and preservation, and water treatment (Fajardo et al., 2022).

3.1 Water Quality Management

While aquaculture plays a vital role in the food industry, it has raised concerns about water pollution resulting from wastewater discharge. Moreover, water intended for aquaculture must undergo treatment, especially when sourced from surface water, to ensure it does not harbour pathogenic microorganisms and other physiochemical pollutants. Nanochitosan has emerged as an effective tool in water and wastewater treatment, serving to eliminate contaminants (Osarenotor & Adetunji, 2023). Its application involves the removal of impurities and suspended solids from water, thereby enhancing the quality of water within aquaculture systems. For instance, nanochitosan is employed as an adsorbent for heavy metal removal due to its high reactivity and expansive surface area. Recent research has focused on the extraction of heavy metals from clays like kaolinite, bentonite, and montmorillonite using nanochitosan, taking advantage of the inherent ability of clays, chitosan, and chitin to bind heavy metals. Studies have also explored nanochitosan–clay composites for the removal of metal ions in recent years (Nasr-Eldahan et al., 2021). This application contributes to improved living conditions for aquatic organisms.

3.2 Nutrient Delivery and Immune System Enhancement

Nanochitosan nanoparticles have the potential to serve as carriers for delivering essential nutrients and supplements to aquatic organisms. This application can significantly boost the nutritional content of their diet, thereby fostering the growth of fish and other species cultivated in aquaculture.

In recent years, fish, crab, and shrimp farming have experienced substantial growth and become integral components of the aquaculture industry. Their contribution to seafood production is noteworthy and plays a pivotal role in shaping a country's economy, particularly within the seafood sector. However, various environmental factors, including water and soil quality, fluctuations in pH and salinity, mineral deficiencies, and disease outbreaks, can pose significant challenges to seafood production, particularly in fish and shrimp farming (Venugopal & Sasidharan, 2021). To address these challenges, the introduction of antibiotic-fortified and synthetically formulated fish feeds has been explored. These specialized feeds are designed to help aquatic organisms cope with both biotic and abiotic stresses, as mentioned earlier. Recent studies have demonstrated the effectiveness of nanotechnologically synthesized, protein-rich feeds when compared to their biologically synthesised counterparts. For instance, the oral administration of nanochitosan and zeolite composites has shown a notable increase in enzymatic activity and enhanced growth in *Oncorhynchus mykiss*, commonly known as Rainbow trout, a type of salmonid fish. Some research suggests that nanochitosan may also have immunostimulatory effects on aquatic organisms, bolstering their resilience against diseases and environmental stressors (Ahmed et al., 2021; Sheikhzadeh et al., 2017). For example, in *Oreochromis niloticus*, also known as Nile tilapia, dietary nanochitosan supplementation through feed has been found to enhance growth, improve production performance, and boost immunity (Elabd et al., 2023). This underscores the pivotal role of chitosan in fish feed formulations.

3.3 Disease Control

Nanochitosan exhibits antimicrobial attributes that can prove invaluable in the prevention and management of diseases within the aquaculture sector. Its potential lies in its ability to impede the proliferation of detrimental bacteria and pathogens that pose a threat to fish and shrimp populations. Recent research has even explored the synthesis of nanochitosan through biological means, incorporating selenium nanoparticles and cinnamon extracts. These bioengineered nanochitosan variants have demonstrated enhanced antimicrobial efficacy, particularly against a spectrum of pathogens like *S. aureus*, *Salmonella typhimurium*, and *Listeria monocytogenes* (Chellapandian et al., 2023).

The aquaculture industry faces a substantial risk from disease-causing fish pathogens, particularly with the rise of antibiotic-resistant bacterial strains. Preserving the well-being of farmed fish is of paramount importance, carrying economic and ecological significance. However, the overuse of antimicrobial drugs, such as antibiotics, contributes to the development of antibiotic-resistant pathogens, which is a major concern in aquaculture. Nanotechnology offers a modern and innovative approach to address fish disease diagnostics and therapy by utilizing nano-sized particles. Natural products are seen as sustainable alternatives to synthetic antibiotics for preventing or treating pathogenic attacks in aquaculture (El-Naggar et al., 2022).

In a study conducted by Ahmed et al. (2020), nanochitosan exhibited antimicrobial effectiveness in combating prevalent bacterial and oomycete pathogens in fish. It showed promising antibacterial effectiveness against pathogens like *Pseudomonas fluorescens*, *Aeromonas hydrophila*, and *Yersinia ruckeri*. Higher doses were used to combat *Pseudomonas aeruginosa, Pseudomonas putida, Edwardsiella tarda*, and *Aeromonas salmonicida* subsp. salmonicida. Nanochitosan also interacted with and weakened the viability of pathogens such as *Edwardsiella ictaluri, Francisella noatunensis* subsp. orientalis, *Aeromonas caviae, Aeromonas veronii, Aeromonas invadans,* and *Saprolegnia parasitica.*

In aquaculture, maintaining stable environmental conditions and providing well-balanced diets are essential for promoting fish growth and protecting against stress and diseases. Chitosan and chitosan nanoparticles are used as safe and natural feed additives that enhance fish growth and boost the immune response (Olaniyan et al., 2023). These nanochitosan additives are non-toxic and support improved fish growth performance while inhibiting intestinal microbial pathogens (Dawood et al., 2020; Abd El-Naby et al., 2019). Studies have also explored the use of nanochitosan in combination with vitamins and DNA vaccines to stimulate the fish's immune system. This approach has shown promise in enhancing growth, meat quality, and survival rates of fish challenged with pathogens like *A. hydrophila*. Additionally, nanochitosan has been used as a carrier for oral DNA vaccines against viral and bacterial diseases in fish, including shrimps. The water-solubility, biodegradability, and non-toxic properties of nanochitosan make it effective for these applications (Wu et al., 2020; Rajeshkumar et al., 2009). Furthermore, nanochitosan-based vaccines have demonstrated effectiveness against specific fish diseases, such as infectious salmon anaemia virus (ISAV) and *Vibrio parahemolyticus*. These vaccines can elicit protective immune responses, and in the case of ISAV, they have shown protection rates exceeding 77% (Okeke et al., 2022).

4 Conclusion

Nanochitosan offers promising avenues for disease prevention and therapy in aquaculture. It exhibits antimicrobial properties against various fish pathogens, enhances fish growth and immunity, and serves as a carrier for oral DNA vaccines, providing a sustainable and effective approach to maintaining the health of farmed fish.

References

Abd El-Naby, F. S., Naiel, M. A. E., Al-Sagheer, A. A., & Negm, S. S. (2019). Dietary chitosan nanoparticles enhance the growth, production performance, and immunity in Oreochromis niloticus. *Aquaculture, 501*, 82–89. https://doi.org/10.1016/j.aquaculture.2018.11.014

Ahmed, F., Soliman, F. M., Adly, M. A., Soliman, H. A. M., El-Matbouli, M., & Saleh, M. (2020). In vitro assessment of the antimicrobial efficacy of chitosan nanoparticles against major fish pathogens and their cytotoxicity to fish cell lines. *Journal of Fish Diseases, 43*(9), 1049–1063. https://doi.org/10.1111/jfd.13212

Ahmed, F., Soliman, F. M., Adly, M. A., Soliman, H. A. M., El-Matbouli, M., & Saleh, M. (2021). Dietary chitosan nanoparticles: Potential role in modulation of rainbow trout (Oncorhynchus mykiss) antibacterial defense and intestinal immunity against enteric Redmouth disease. *Marine Drugs, 19*(2), 72. https://doi.org/10.3390/md19020072

Ahuekwe, E. F., Isibor, P. O., Oziegbe, O., Salami, A. O., Akinyosoye, A. D., Akinhanmi, F., Oyewale, J. O., Taiwo, O. S., Akinwunmi, R. A., Ajiboye, I., & Adekeye, B. T. (2023a). Nanochitosan derived from marine bacteria. In *Next generation Nanochitosan* (pp. 147–168).

Ahuekwe, E. F., Isibor, P. O., Akinhanmi, F., Ajiboye, I., Salami, A. O., Adekeye, B. T., Akinpelu, S. O., Kuye, A. D., Bello, A. O., Oyewale, J. O., & George, D. Z. (2023b). Utilization of nanochitosan in the sterilization of ponds and water treatment for aquaculture. In *Next Generation Nanochitosan* (pp. 301–338).

Alajangi, H. K., Kaur, M., Sharma, A., Rana, S., Thakur, S., Chatterjee, M., Singla, N., Jaiswal, P. K., Singh, G., & Barnwal, R. P. (2022). Blood–brain barrier: Emerging trends on transport models and new-age strategies for therapeutics intervention against neurological disorders. *Molecular Brain, 15*(1), 49. https://doi.org/10.1186/s13041-022-00937-4

Aranaz, I., Alcántara, A. R., Civera, M. C., Arias, C., Elorza, B., Heras Caballero, A., & Acosta, N. (2021). Chitosan: An overview of its properties and applications. *Polymers, 13*(19), 3256. https://doi.org/10.3390/polym13193256

Asl, F. D., Mousazadeh, M., Taji, S., Bahmani, A., Khashayar, P., Azimzadeh, M., & Mostafavi, E. (2023). Nano drug-delivery systems for management of AIDS: Liposomes, dendrimers, gold and silver nanoparticles. *Nanomedicine, 18*(3), 279–302. https://doi.org/10.2217/nnm-2022-0248

Chellapandian, H., Jeyachandran, S., Ilangovan, S., & Aseervatham, S. B. (2023). Nanochitosan for the production of more effective fish feed for aquaculture. In Next Generation Nanochitosan (pp. 339–348). Academic Press.

Chen, H., Xing, X., Tan, H., Jia, Y., Zhou, T., Chen, Y., Ling, Z., & Hu, X. (2017). Covalently antibacterial alginate-chitosan hydrogel dressing integrated gelatin microspheres containing tetracycline hydrochloride for wound healing. *Materials Science and Engineering: C., 70*, 287–295. https://doi.org/10.1016/j.msec.2016.08.086

Cota-Arriola, O., Plascencia-Jatomea, M., Lizardi-Mendoza, J., Robles-Sánchez, R. M., Ezquerra-Brauer, J.M., Ruíz-García, J., Vega-Acosta, J. R., & Cortez-Rocha, M. O. (2016). Preparation of chitosan matrices with ferulic acid: Physicochemical characterization and relationship on the growth of *Aspergillus parasiticus*. *CyTA - Journal of Food* 1–10. https://doi.org/10.1080/19476337.2016.1213317

Covarrubias, C., Trepiana, D., & Corral, C. (2018). Synthesis of hybrid copper-chitosan nanoparticles with antibacterial activity against cariogenic *Streptococcus mutans*. *Dental Materials Journal, 37*(3), 379–384. https://doi.org/10.4012/dmj.2017-195

Dawood, M. A. O., Gewaily, M. S., Soliman, A. A., Shukry, M., Amer, A. A., Younis, E. M., Abdel-Warith, A.-W. A., Van Doan, H., Saad, A. H., Aboubakr, M., Abdel-Latif, H. M. R., & Fadl, S. E. (2020). Marine-derived chitosan nanoparticles improved the intestinal histomorphometrical features in association with the health and immune response of grey mullet (Liza ramada). *Marine Drugs, 18*(12), 611. https://doi.org/10.3390/md18120611

De Sousa Victor, R., Da Cunha, M., Santos, A., Viana De Sousa, B., De Araújo Neves, G., De Lima, N., Santana, L., & Rodrigues Menezes, R. (2020). A review on Chitosan's uses as biomaterial: Tissue engineering, drug delivery systems and cancer treatment. *Materials, 13*(21), 4995. https://doi.org/10.3390/ma13214995

Desai, K., McManus, J. M., & Sharifi, N. (2021). Hormonal therapy for prostate cancer. *Endocrine Reviews, 42*(3), 354–373. https://doi.org/10.1210/endrev/bnab002

Dilnawaz, F., Acharya, S., & Kanungo, A. (2023). A clinical perspective of chitosan nanoparticles for infectious disease management. *Polymer Bulletin*. https://doi.org/10.1007/s00289-023-04755-z

Elabd, H., Mahboub, H. H., Salem, S. M. R., Abdelwahab, A. M., Alwutayd, K. M., Shaalan, M., Ismail, S. H., Abdelfattah, A. M., Khalid, A., Mansour, A. T., Hamed, H. S., & Youssuf, H. (2023). Nano-curcumin/chitosan modulates growth, biochemical, immune, and antioxidative profiles, and the expression of related genes in Nile tilapia, Oreochromis niloticus. *Fishes, 8*(7), 333. https://doi.org/10.3390/fishes8070333

El-Naggar, M., et al. (2022). Applications of chitosan and chitosan nanoparticles in fish aquaculture. *Egyptian Journal of Aquatic Biology and Fisheries, 26*(1), 23–43. https://doi.org/10.21608/ejabf.2022.213365

Fajardo, C., Martinez-Rodriguez, G., Blasco, J., Mancera, J. M., Thomas, B., & De Donato, M. (2022). Nanotechnology in aquaculture: Applications, perspectives and regulatory challenges. *Aquaculture and Fisheries, 7*(2), 185–200.

Feng, S.-S., Mei, L., Anitha, P., Gan, C. W., & Zhou, W. (2009). Poly(lactide)–vitamin E derivative/montmorillonite nanoparticle formulations for the oral delivery of docetaxel. *Biomaterials, 30*(19), 3297–3306. https://doi.org/10.1016/j.biomaterials.2009.02.045

Gaglio, S. C., Perduca, M., Zipeto, D., & Bardi, G. (2023). Efficiency of chitosan nanocarriers in vaccinology for mucosal immunization. *Vaccine, 11*(8), 1333. https://doi.org/10.3390/vaccines11081333

Ghattas, M., Dwivedi, G., Lavertu, M., & Alameh, M.-G. (2021). Vaccine technologies and platforms for infectious diseases: Current progress, challenges, and opportunities. *Vaccine, 9*(12), 1490. https://doi.org/10.3390/vaccines9121490

Hernández-Lauzardo, A. N., Bautista-Baños, S., Velázquez-del Valle, M. G., Méndez-Montealvo, M. G., Sánchez-Rivera, M. M., & Bello-Pérez, L. A. (2008). Antifungal effects of chitosan with different molecular weights on in vitro development of Rhizopus stolonifer (Ehrenb.:Fr.) Vuill. *Carbohydrate Polymers, 73*(4), 541–547. https://doi.org/10.1016/j.carbpol.2007.12.020

Ing, L. Y., Zin, N. M., Sarwar, A., & Katas, H. (2012). Antifungal activity of chitosan nanoparticles and correlation with their physical properties. *International Journal of Biomaterials, 2012*, 1–9. https://doi.org/10.1155/2012/632698

Jana, S., & Jana, S. (Eds.). (2019). *Functional chitosan: Drug delivery and biomedical applications*. Springer.

Janakiraman, K., Krishnaswami, V., Rajendran, V., Natesan, S., & Kandasamy, R. (2018). Novel nano therapeutic materials for the effective treatment of rheumatoid arthritis-recent insights. *Materials Today Communications, 17*, 200–213. https://doi.org/10.1016/j.mtcomm.2018.09.011

Kavitha Sankar, P. C., Rajmohan, G., & Rosemary, M. J. (2017). Physico-chemical characterisation and biological evaluation of freeze dried chitosan sponge for wound care. *Materials Letters, 208*, 130–132. https://doi.org/10.1016/j.matlet.2017.05.010

Khairy, A. M., Tohamy, M. R. A., Zayed, M. A., Mahmoud, S. F., El-Tahan, A. M., El-Saadony, M. T., & Mesiha, P. K. (2022). Eco-friendly application of nano-chitosan for controlling potato and tomato bacterial wilt. *Saudi Journal of Biological Sciences, 29*(4), 2199–2209. https://doi.org/10.1016/j.sjbs.2021.11.041

Kheiri, A., Moosawi Jorf, S. A., Malihipour, A., Saremi, H., & Nikkhah, M. (2017). Synthesis and characterization of chitosan nanoparticles and their effect on fusarium head blight and oxidative activity in wheat. *International Journal of Biological Macromolecules, 102*, 526–538. https://doi.org/10.1016/j.ijbiomac.2017.04.034

Lopez-Moya, F., Suarez-Fernandez, M., & Lopez-Llorca, L. (2019). Molecular mechanisms of chitosan interactions with fungi and plants. *International Journal of Molecular Sciences, 20*(2), 332. https://doi.org/10.3390/ijms20020332

Mamo, T., Moseman, E. A., Kolishetti, N., Salvador-Morales, C., Shi, J., Kuritzkes, D. R., Langer, R., Andrian, U. V., & Farokhzad, O. C. (2010). Emerging nanotechnology approaches for HIV/AIDS treatment and prevention. *Nanomedicine, 5*(2), 269–285. https://doi.org/10.2217/nnm.10.1

Meng, Q., Sun, Y., Cong, H., Hu, H., & Xu, F.-J. (2021). An overview of chitosan and its application in infectious diseases. *Drug Delivery and Translational Research, 11*(4), 1340–1351. https://doi.org/10.1007/s13346-021-00913-w

Mikušová, V., & Mikuš, P. (2021). Advances in chitosan-based nanoparticles for drug delivery. *International Journal of Molecular Sciences, 22*(17), 9652. https://doi.org/10.3390/ijms22179652

Mohammed, M., Syeda, J., Wasan, K., & Wasan, E. (2017). An overview of chitosan nanoparticles and its application in non-parenteral drug delivery. *Pharmaceutics, 9*(4), 53. https://doi.org/10.3390/pharmaceutics9040053

Morin-Crini, N., Lichtfouse, E., Torri, G., & Crini, G. (2019). Applications of chitosan in food, pharmaceuticals, medicine, cosmetics, agriculture, textiles, pulp and paper, biotechnology, and environmental chemistry. *Environmental Chemistry Letters, 17*(4), 1667–1692. https://doi.org/10.1007/s10311-019-00904-x

MubarakAli, D., LewisOscar, F., Gopinath, V., Alharbi, N. S., Alharbi, S. A., & Thajuddin, N. (2018). An inhibitory action of chitosan nanoparticles against pathogenic bacteria and fungi and their potential applications as biocompatible antioxidants. *Microbial Pathogenesis, 114*, 323–327. https://doi.org/10.1016/j.micpath.2017.11.043

Nami, S., Aghebati-Maleki, A., & Aghebati-Maleki, L. (2021). Current applications and prospects of nanoparticles for antifungal drug delivery. *EXCLI Journal, 20:Doc562*; ISSN 1611-2156. https://doi.org/10.17179/EXCLI2020-3068

Nasr-Eldahan, S., Nabil-Adam, A., Shreadah, M. A., Maher, A. M., & El-Sayed Ali, T. (2021). A review article on nanotechnology in aquaculture sustainability as a novel tool in fish disease control. *Aquaculture International, 29*, 1459–1480.

Ng, S. W., Selvarajah, G. T., Hussein, M. Z., Yeap, S. K., & Omar, A. R. (2020). *In vitro* evaluation of curcumin-encapsulated chitosan nanoparticles against feline infectious peritonitis virus and pharmacokinetics study in cats. *BioMed Research International, 2020*, 1–18. https://doi.org/10.1155/2020/3012198

Okeke, E. S., Chukwudozie, K. I., Nyaruaba, R., Ita, R. E., Oladipo, A., Ejeromedoghene, O., Atakpa, E. O., Agu, C. V., & Okoye, C. O. (2022). Antibiotic resistance in aquaculture and aquatic organisms: A review of current nanotechnology applications for sustainable management. *Environmental Science and Pollution Research, 29*(46), 69241–69274. https://doi.org/10.1007/s11356-022-22319-y

Olaniyan, O. T., Adetunji, C. O., Dare, A., Ajayi, O. O., & Adeniyi, M. J. (2023). Roles of nanochitosan and polymeric chitosan for the protection of farmed fish against disease-causing pathogens. In *Next generation nanochitosan* (pp. 287–294). Elsevier. https://doi.org/10.1016/B978-0-323-85593-8.00016-3

Onugwu, A. L., Nwagwu, C. S., Onugwu, O. S., Echezona, A. C., Agbo, C. P., Ihim, S. A., Emeh, P., Nnamani, P. O., Attama, A. A., & Khutoryanskiy, V. V. (2023). Nanotechnology based drug delivery systems for the treatment of anterior segment eye diseases. *Journal of Controlled Release, 354*, 465–488. https://doi.org/10.1016/j.jconrel.2023.01.018

Osarenotor, O., & Adetunji, C. O. (2023). Application of nanochitosan in water filtration with high relevance in aquaculture. In *Next generation nanochitosan* (pp. 295–300). Elsevier. Academic Press.

Pilon, L., Spricigo, P. C., Miranda, M., De Moura, M. R., Assis, O. B. G., Mattoso, L. H. C., & Ferreira, M. D. (2015). Chitosan nanoparticle coatings reduce microbial growth on fresh-cut apples while not affecting quality attributes. *International Journal of Food Science & Technology, 50*(2), 440–448. https://doi.org/10.1111/ijfs.12616

Qin, C., Li, H., Xiao, Q., Liu, Y., Zhu, J., & Du, Y. (2006). Water-solubility of chitosan and its antimicrobial activity. *Carbohydrate Polymers, 63*(3), 367–374. https://doi.org/10.1016/j.carbpol.2005.09.023

Rajeshkumar, S., Venkatesan, C., Sarathi, M., Sarathbabu, V., Thomas, J., Anver Basha, K., & Sahul Hameed, A. S. (2009). Oral delivery of DNA construct using chitosan nanoparticles to protect the shrimp from white spot syndrome virus (WSSV). *Fish & Shellfish Immunology, 26*(3), 429–437. https://doi.org/10.1016/j.fsi.2009.01.003

Rhee, J. H. (2020). Current and new approaches for mucosal vaccine delivery. In *Mucosal vaccines* (pp. 325–356). Elsevier. https://doi.org/10.1016/B978-0-12-811924-2.00019-5

Rozman, N. A. S., Tong, W. Y., Leong, C. R., Tan, W. N., Hasanolbasori, M. A., & Abdullah, S. Z. (2019). Potential antimicrobial applications of chitosan nanoparticles (ChNP). *Journal of Microbiology and Biotechnology, 29*(7), 1009–1013. https://doi.org/10.4014/jmb.1904.04065

Sadoughi, F., Mansournia, M. A., & Mirhashemi, S. M. (2020). The potential role of chitosan-based nanoparticles as drug delivery systems in pancreatic cancer. *IUBMB Life, 72*(5), 872–883.

Safer, A.-M., & Leporatti, S. (2021). Chitosan nanoparticles for antiviral drug delivery: A novel route for COVID-19 treatment. *International Journal of Nanomedicine, 16*, 8141–8158. https://doi.org/10.2147/IJN.S332385

Saharan, V., Sharma, G., Yadav, M., Choudhary, M. K., Sharma, S. S., Pal, A., Raliya, R., & Biswas, P. (2015). Synthesis and in vitro antifungal efficacy of Cu–chitosan nanoparticles against pathogenic fungi of tomato. *International Journal of Biological Macromolecules, 75*, 346–353. https://doi.org/10.1016/j.ijbiomac.2015.01.027

Scutera, S., Argenziano, M., Sparti, R., Bessone, F., Bianco, G., Bastiancich, C., Castagnoli, C., Stella, M., Musso, T., & Cavalli, R. (2021). Enhanced antimicrobial and antibiofilm effect of new colistin-loaded human albumin nanoparticles. *Antibiotics, 10*(1), 57. https://doi.org/10.3390/antibiotics10010057

Sheikhzadeh, N., Kouchaki, M., Mehregan, M., Tayefi-Nasrabadi, H., Divband, B., Khataminan, M., Khani Oushani, A., & Shabanzadeh, S. (2017). Influence of nanochitosan/zeolite composite on growth performance, digestive enzymes and serum biochemical parameters in rainbow trout (*Oncorhynchus mykiss*). *Aquaculture Research., 48*(12), 5955–5964. https://doi.org/10.1111/are.13418

Shohani, S., Mondanizadeh, M., Abdoli, A., Khansarinejad, B., Salimi-Asl, M., Ardestani, M., Ghanbari, M., Haj, M., & Zabihollahi, R. (2017). Trimethyl chitosan improves anti-HIV effects of Atripla as a new nanoformulated drug. *Current HIV Research, 15*(1), 56–65. https://doi.org/10.2174/1570162X14666161216142806

Sivanesan, I., Muthu, M., Gopal, J., Hasan, N., Kashif Ali, S., Shin, J., & Oh, J.-W. (2021). Nanochitosan: Commemorating the metamorphosis of an ExoSkeletal waste to a versatile nutraceutical. *Nanomaterials, 11*(3), 821. https://doi.org/10.3390/nano11030821

Soh, S. H., Shim, S., Im, Y. B., Park, H.-T., Cho, C.-S., & Yoo, H. S. (2019). Induction of Th2-related immune responses and production of systemic IgA in mice intranasally immunized with Brucella abortus malate dehydrogenase loaded chitosan nanoparticles. *Vaccine, 37*(12), 1554–1564. https://doi.org/10.1016/j.vaccine.2019.02.005

Sohail, A., Khan, R. U., Khan, M., Khokhar, M., Ullah, S., Ali, A., Bilal, H., Khattak, S., Khan, M., & Ahmad, B. (2021). Comparative efficacy of amphotericin B-loaded chitosan nanoparticles and free amphotericin B drug against Leishmania tropica. *Bulletin of the National Research Centre., 45*(1), 187. https://doi.org/10.1186/s42269-021-00644-5

Venugopal, V., & Sasidharan, A. (2021). Seafood industry effluents: Environmental hazards, treatment and resource recovery. *Journal of Environmental Chemical Engineering., 9*(2), 104758. https://doi.org/10.1016/j.jece.2020.104758

Wu, Y., Rashidpour, A., Almajano, M. P., & Metón, I. (2020). Chitosan-based drug delivery system: Applications in fish biotechnology. *Polymers, 12*(5), 1177. https://doi.org/10.3390/polym12051177

Yilmaz Atay, H. (2019). Antibacterial activity of chitosan-based systems. In S. Jana & S. Jana (Eds.), *Functional chitosan* (pp. 457–489). Springer Singapore. https://doi.org/10.1007/978-981-15-0263-7_15

Zamboulis, A., Nanaki, S., Michailidou, G., Koumentakou, I., Lazaridou, M., Ainali, N. M., Xanthopoulou, E., & Bikiaris, D. N. (2020). Chitosan and its derivatives for ocular delivery formulations: Recent advances and developments. *Polymers, 12*(7), 1519. https://doi.org/10.3390/polym12071519

Zhao, D., Yu, S., Sun, B., Gao, S., Guo, S., & Zhao, K. (2018). Biomedical applications of chitosan and its derivative nanoparticles. *Polymers, 10*(4), 462. https://doi.org/10.3390/polym10040462

Nanochitosan-Based Water-Quality Enhancement

Patrick Omoregie Isibor, David Osagie Agbontaen, and Oyewole Oluwafemi Adebayo

Contents

P. O. Isibor (✉)
Department of Biological Sciences, College of Science and Technology, Covenant University, Ota, Ogun State, Nigeria
e-mail: patrick.isibor@covenantuniversity.edu.ng

D. O. Agbontaen
Department of Public Health, University of South Wales, Pontypridd, UK

O. O. Adebayo
Department of Microbiology, Federal University of Technology, Minna, Nigeria

African Center of Excellence for Mycotoxin and Food Safety, Federal University of Technology, Minna, Nigeria

P. O. Isibor et al. (eds.), *Nanochitosan-Based Enhancement of Fisheries and Aquaculture*, https://doi.org/10.1007/978-3-031-52261-1_6

1 Introduction to Nanochitosan in Water-Quality Enhancement

Nanochitosan is a derivative of chitosan, a biopolymer derived from chitin, found in the exoskeletons of crustaceans, insects, and fungal cell walls. Nanochitosan is characterized by its nanoscale dimensions, typically ranging from 1 to 100 nm.

Nanochitosan's existence in nanoparticle form brings about several distinct advantages owing to its nanoscale dimensions. One of the key benefits is its significantly increased surface area-to-volume ratio compared to larger chitosan particles or bulk material. This elevated ratio arises from the reduction in particle size, resulting in a more extensive surface area per unit volume (Ali et al., 2018).

The wide surface area facilitates a greater exposure of nanochitosan's functional groups, particularly the amino and hydroxyl groups inherent in chitosan. These groups become more accessible and available for interactions with surrounding substances, irrespective of the other molecules, cells, pathogens, or environmental contaminants. The increased reactivity is a direct consequence of this amplified surface area, allowing for enhanced and more efficient chemical interactions (Benettayeb et al., 2023). This unique characteristic also influences its performance in various applications. For instance, in drug delivery systems, the increased surface area of nanochitosan nanoparticles provides more sites for drug loading and release, improving their efficacy. In environmental remediation, the elevated reactivity enhances its capability in binding to and neutralizing pollutants in water or soil. Furthermore, the higher surface area-to-volume ratio amplifies nanochitosan's potential for functionalization or modification. This allows researchers to tailor its properties further, adjusting its reactivity, surface charge, or compatibility for specific applications. Overall, the nanoscale nature of nanochitosan nanoparticles substantially augments their reactivity and interaction potential, making them versatile and valuable across diverse fields and applications (Alonso Fernandez et al., 2016).

Nanochitosan's status as a biocompatible and non-toxic material contributes significantly to its versatility and broad applicability in various domains. Its biocompatibility is characterized by its ability to interact with biological systems without causing adverse reactions or harm. This characteristic makes nanochitosan an ideal candidate for biomedical applications, such as drug delivery, tissue engineering, wound healing, and regenerative medicine. When used in these contexts, nanochitosan demonstrates minimal cytotoxicity and compatibility with biological systems, ensuring its safety and suitability for medical use. Furthermore, its non-toxic nature adds to its appeal for environmental applications. In fields like water treatment, nanochitosan can be employed for remediation purposes without posing risks to ecosystems or living organisms (Ali et al., 2018). Its ability to bind to heavy metals, pesticides, or other pollutants in water while being non-toxic ensures that it can effectively remove contaminants without introducing further harm to the environment.

The combination of biocompatibility and non-toxicity makes nanochitosan an attractive material for applications that necessitate interactions with living organisms, tissues, or the environment. Its safety profile, coupled with its advantageous properties like biodegradability and antimicrobial activity, positions nanochitosan as a promising candidate for a wide array of biomedical and environmental applications. Nanochitosan inherits the biodegradability characteristic from its parent compound, chitosan. This property underscores its environmental friendliness and sustainability, particularly in applications where degradation into harmless byproducts is crucial. Being biodegradable means that nanochitosan can naturally break down into smaller molecules or compounds over time through the actions of enzymes or microorganisms (Ali et al., 2018). This process occurs in various environmental settings, including soil, water, and biological systems. As nanochitosan disintegrates, it yields non-toxic degradation products, predominantly oligosaccharides, which are benign and can be assimilated or metabolized by microorganisms or other biological entities. This innate biodegradability significantly diminishes the environmental impact of nanochitosan. When used in applications such as environmental remediation or biomedical devices, its eventual breakdown into harmless constituents ensures that it does not persist in ecosystems or pose long-term threats to environmental integrity. This aspect aligns with sustainability efforts, contributing to the reduced accumulation of persistent materials in the environment. Nanochitosan's biodegradability is a pivotal feature that adds to its appeal in environmentally conscious applications. Its ability to degrade into non-toxic components after fulfilling its intended purpose not only minimizes its environmental footprint but also supports the broader goal of sustainable material usage and waste reduction.

The conservation of functional groups from chitosan in nanochitosan is a fundamental aspect that defines its chemical behaviour and reactivity. Primarily, nanochitosan preserves the amino ($-NH_2$) and hydroxyl ($-OH$) groups, which are intrinsic to chitosan's molecular structure. These functional groups are crucial contributors to the versatile chemical reactivity exhibited by nanochitosan (Ngah & Fatinathan, 2010).

The amino groups, with their inherent positive charge under certain pH conditions, confer excellent cationic properties on nanochitosan. This positive charge facilitates interactions with negatively charged entities, such as organic molecules, pollutants, or even cell membranes. It enables nanochitosan to act as an effective adsorbent or binder, selectively attracting and trapping various substances through electrostatic interactions. This attribute finds utility in applications like water treatment, where nanochitosan can adsorb heavy metals, dyes, or organic contaminants. Conversely, the hydroxyl groups present in nanochitosan's structure contribute to its hydrophilic nature. This hydrophilicity plays a pivotal role in enhancing its solubility in aqueous environments and influences its interactions with water molecules. These groups also provide sites for chemical modifications or functionalization, enabling the customization of nanochitosan for specific applications (Benettayeb et al., 2021). For instance, these hydroxyl groups can undergo reactions to attach different functional molecules or nanoparticles, expanding its scope for tailored applications in drug delivery or bioengineering.

Overall, the retention of amino and hydroxyl functional groups in nanochitosan confers a broad spectrum of reactivity. This unique chemical makeup allows nanochitosan to participate in various chemical interactions, making it a versatile material with applications across diverse fields, ranging from biomedicine to environmental remediation. Nanochitosan's innate antimicrobial properties are a result of its unique molecular structure and composition. With a high surface area due to its nanoparticle form, nanochitosan offers a platform for enhanced interactions with microorganisms, making it effective in combating waterborne diseases caused by various pathogens. The antimicrobial action of nanochitosan is multifaceted. Its positively charged amino groups interact with the negatively charged components on the surface of microorganisms, disrupting their cellular structures (Ayati et al., 2019; Benettayeb et al., 2022). This disruption can lead to damage or destabilization of the cell membrane, inhibiting essential cellular functions, and ultimately causing microbial death. Moreover, nanochitosan's small size enables efficient penetration into the cell walls of microorganisms, facilitating interactions with intracellular components. This interaction may interfere with vital cellular processes, such as DNA replication or protein synthesis, leading to the inhibition of microbial growth or proliferation. Nanochitosan's antimicrobial activity extends to its ability to generate reactive oxygen species (ROS) upon interaction with microbial cells. ROS, such as free radicals, can induce oxidative stress within the cells, damaging their proteins, DNA, and other biomolecules, ultimately leading to cell death (Choi et al., 2008).

These combined mechanisms make nanochitosan an effective antimicrobial agent against a spectrum of pathogens, including bacteria, fungi, and certain viruses, thereby holding promise for applications in combatting waterborne diseases. Its effectiveness in inhibiting microbial growth and proliferation makes it a potential candidate for various water treatment and purification systems aimed at providing safer drinking water and reducing the spread of waterborne illnesses.

2 Nanochitosan's Role in Water Purification

Nanochitosan's role in water purification is multifaceted and holds promise in addressing various challenges associated with water quality. Its unique properties make it a valuable tool in several water treatment applications (Brião et al., 2020).

Nanochitosan's high surface area and functional groups allow it to adsorb heavy metals like lead, cadmium, and copper from water, reducing their concentrations to safer levels. Nanochitosan is a remarkable material derived from chitosan, a natural biopolymer usually extracted from the shells of crustaceans like shrimp and crabs. When chitosan is broken down into nanoscale particles, it creates nanochitosan, which possesses an incredibly high surface area due to its nanosized structure. This high surface area is a key factor in its ability to adsorb heavy metals effectively. The surface area-to-volume ratio of nanochitosan increases significantly when compared to larger particles of chitosan. This enlarged surface area provides a greater interface for interactions between the nanochitosan and heavy metal ions present in water (Du et al., 2008).

Moreover, nanochitosan contains amino and hydroxyl groups, which are available for chemical interactions. These functional groups have a high affinity for heavy metal ions, allowing nanochitosan to attract, bind, and immobilize metals like lead, cadmium, and copper from water solutions. Through a process known as adsorption, nanochitosan effectively captures and holds these heavy metal ions on its surface or within its porous structure. As a result, the concentrations of these harmful metals in water are significantly reduced, often to levels considered safe for human consumption or environmental standards (Fan et al., 2013). This ability of nanochitosan to adsorb heavy metals makes it a promising material for water purification and remediation processes, offering a sustainable and eco-friendly solution to mitigate the detrimental effects of heavy metal pollution in water bodies. Nanochitosan's versatility extends beyond heavy metals; it is also highly effective in adsorbing organic dyes and various pollutants present in wastewater. Organic dyes, commonly used in industries like textile, leather, and food processing, pose a significant threat to water quality due to their vibrant colours and potential toxicity (Obeid et al., 2013; Elwakeel et al., 2014).

Nanochitosan's porous structure and abundant functional groups make it an excellent adsorbent for these organic compounds. Similar to its interaction with heavy metals, the surface area-to-volume ratio and the available functional groups play a crucial role in its ability to attract and retain organic dyes. The porous nature of nanochitosan provides ample sites for the molecules of organic dyes to adhere to its surface or penetrate its structure. Meanwhile, the functional groups, such as amino and hydroxyl groups, exhibit strong interactions with the chemical structures of organic dyes, leading to adsorption and immobilization (Kadam & Lee, 2015). This adsorption process facilitates the removal of coloured compounds from wastewater, effectively reducing the colour intensity and eliminating organic pollutants (Li et al., 2015). By doing so, nanochitosan aids in improving the overall quality of water, making it safer for discharge or reuse in various industrial or agricultural processes. The environmentally friendly and sustainable nature of nanochitosan's dye removal capabilities makes it an attractive option for wastewater treatment (Jaafari et al., 2020). Its effectiveness in adsorbing both heavy metals (Hritcu et al., 2012) and organic pollutants underscores its potential as a multifunctional material in addressing diverse water-quality issues, contributing significantly to the remediation and purification of contaminated water sources. Nanochitosan's remarkable adsorptive capacity isn't limited to heavy metals and organic dyes, it also extends to pesticides and various organic chemicals commonly found in agricultural runoffs and contaminated water sources. Pesticides, such as herbicides, insecticides, and fungicides, are extensively used in agricultural practices to protect crops from pests and diseases. However, these chemicals can leach into water bodies through runoff, posing significant environmental and health risks. Nanochitosan's adsorption properties make it an effective tool for mitigating the impact of these agricultural pollutants. Its high surface area and abundance of functional groups provide an ideal environment for the interaction and binding of pesticide molecules (Moradi et al., 2014).

Similar to its mechanisms with heavy metals and organic dyes, nanochitosan's porous structure offers sites for the physical adsorption of pesticides, while its functional groups facilitate chemical interactions with these organic compounds. This interaction leads to the adsorption and sequestration of pesticides, effectively reducing their concentrations in water. Moreover, certain organic chemicals, such as hydrocarbons and industrial pollutants, can also be targeted by nanochitosan's adsorptive capabilities. Whether these contaminants originate from industrial discharges, oil spills, or other sources, nanochitosan can play a role in their removal from water sources (Hamza et al., 2019).

By adsorbing pesticides and organic chemicals, nanochitosan contributes to the purification of agricultural runoffs, contaminated water bodies, and industrial effluents. Its ability to mitigate the presence of these harmful substances holds promise for addressing water pollution issues, safeguarding both the environment and human health from the adverse effects of these contaminants. Nanochitosan's versatility in adsorption extends its capabilities to target pesticides and specific organic chemicals, playing a pivotal role in purifying water sources contaminated by agricultural runoff and industrial discharge. In agricultural practices, pesticides are extensively used to protect crops from pests, diseases, and weeds. Aside the surface water bodies, these chemicals can also leach into soil and contaminate groundwater. Nanochitosan's adsorptive properties can also serve as a remediation tool to effectively mitigate the adverse effects of these agricultural pollutants (Moradi et al., 2014).

The unique structure of nanochitosan, characterized by its high surface area and numerous functional groups, provides an ideal platform for interacting with and immobilizing pesticide molecules. The abundance of sites on its surface and within its porous structure allows for the physical adsorption of these chemicals, while the functional groups, such as amino and hydroxyl groups, enable chemical interactions, facilitating the binding and retention of pesticides. Moreover, nanochitosan's adsorptive capacity isn't limited to pesticides alone, it extends to various organic chemicals commonly found in agricultural runoff and industrial discharge. These could include herbicides, insecticides, fertilizers, other agrochemicals, as well as a plethora of organic pollutants. When introduced into contaminated water sources, nanochitosan acts as a powerful adsorbent, attracting and trapping these organic pollutants. By doing so, it effectively reduces their concentrations in water, contributing significantly to the purification of agricultural runoff and contaminated water bodies (Osagie et al., 2021).

The application of nanochitosan in water treatment processes offers a sustainable and eco-friendly approach to remediate water contaminated by agricultural activities and industrial processes. Its ability to target pesticides and organic chemicals underscores its potential as a versatile and effective tool in addressing the diverse challenges associated with water pollution, safeguarding ecosystems and ensuring safer water sources for both human consumption and environmental sustainability (Massoudinejad et al., 2019).

3 Mechanisms of Action in Water-Quality Enhancement

There are several conventional water purification techniques, each method with its specific setback. However, improving water quality involves various mechanisms that work individually or in combination to address specific pollutants or issues. Physical filtration involves passing water through different materials, such as sand, gravel, or membranes, to remove suspended particles, sediments, and larger contaminants. Adding chemicals like chlorine, ozone, or potassium permanganate can disinfect water by killing harmful bacteria, viruses, and pathogens. Coagulants and flocculants are also used to clump together fine particles, aiding in their removal, using living organisms, such as bacteria, algae, or aquatic plants, to break down pollutants in water. For example, in wastewater treatment, bacteria helps break down organic matter through processes like aerobic or anaerobic digestion. Activated carbon or other specialized materials can adsorb contaminants by attracting and binding them to their surface, effectively removing them from the water (Yang et al., 2012). This process involves replacing undesirable ions in the water with more acceptable ions. For instance, water softeners use ion exchange resins to remove calcium and magnesium ions that cause water hardness. This method uses a semi-permeable membrane to remove ions, molecules, and larger particles from water, producing purified water. Increasing the oxygen content in water by exposing it to air helps in reducing odours and improving taste. Aeration also assists in removing volatile organic compounds and gases, allowing suspended particles to settle down by gravity, typically in a settling basin or tank, where they can be removed from the water. These are engineered systems that mimic natural wetlands and use vegetation, soils, and microbes to treat contaminants in water. They are effective in removing pollutants through biological, physical, and chemical processes. Exposing water to ultraviolet (UV) light can deactivate bacteria, viruses, and other pathogens by disrupting their DNA, rendering them unable to reproduce.

Combining several of these mechanisms in water treatment systems with nanochitosan-based enhancement techniques ascertains more optimized results due to the compatibility of nanochitosan with other conventional purification techniques described. This combination may be tailored to specific contaminants or issues and is often necessary to achieve the desired water-quality standards. The selection of methods depends on the nature of pollutants, available technology, and the intended use of the treated water (Olivera et al., 2016).

4 Nanochitosan's High Surface Area and Adsorption Capacity

Adsorption is a fundamental process in water treatment and environmental remediation, where adsorbent materials capture and remove contaminants from aqueous solutions. Nanochitosan, a nanoscale derivative of chitosan, has garnered significant attention for its exceptional adsorption properties. This review elucidates the underlying mechanisms of nanochitosan's high surface area and adsorption capacity, exploring its applications and potential for addressing water pollution challenges (Kuang et al., 2013; Liu et al., 2015).

Adsorption is a crucial method for removing diverse contaminants, including heavy metals, organic pollutants, and dyes, from water sources. Nanochitosan, derived from chitosan, exhibits remarkable adsorption capabilities attributed to its unique structural and physicochemical properties. This chapter delves into the factors contributing to nanochitosan's high surface area and adsorption capacity, shedding light on its potential applications in water treatment and environmental remediation. Nanochitosan is a nanostructured adsorbent. Nanochitosan is characterized by its nanoscale size, high surface area, and surface functionality (Pap et al., 2020).

Nanochitosan's small particle size provides an exceptionally high specific surface area. The increased surface area offers more active sites for interactions with contaminants. The presence of amino ($-NH_2$) and hydroxyl ($-OH$) groups on nanochitosan's surface contributes to its adsorption capacity. These functional groups can form various chemical interactions, such as hydrogen bonding and electrostatic attraction, with target contaminants (Vakili et al., 2014).

The adsorption mechanisms of nanochitosan are multifaceted and encompass several key processes. Van der Waals forces and London dispersion forces contribute to the physical adsorption of contaminants onto nanochitosan's surface. This process is especially effective for nonpolar organic compounds. Chemical adsorption involves the formation of covalent or ionic bonds between nanochitosan's functional groups and polar contaminants, such as heavy metal ions or polar organic compounds. Nanochitosan can exchange ions on its surface with ions in solution, leading to the removal of metal ions and other charged species. The exceptional adsorption capacity of nanochitosan finds applications in various fields. Nanochitosan effectively adsorbs heavy metal ions, such as lead, cadmium, and copper, from contaminated water sources. It efficiently removes organic pollutants, including dyes, pesticides, and pharmaceuticals, contributing to the purification of wastewater. Nanochitosan can remove nutrients like phosphorus and nitrogen, addressing eutrophication issues in aquatic ecosystems. Nanochitosan's applications extend to drinking water treatment, ensuring the removal of potentially harmful contaminants (Ranjbari et al., 2020). Nanochitosan's high surface area and adsorption capacity make it a versatile and effective adsorbent material for the removal of various contaminants from aqueous solutions. Understanding the mechanisms underlying nanochitosan's adsorption properties is critical for optimizing its use in water treatment and environmental remediation applications. As research continues to uncover new insights into the potential of nanochitosan, its role in mitigating water pollution challenges and promoting environmental sustainability is expected to expand, contributing to cleaner and safer water resources.

5 Complexation and Ion Exchange Processes

Complexation and ion exchange are fundamental chemical processes used in various fields, including chemistry, environmental science, and water treatment. These processes involve the interaction between ions or molecules in solution and play crucial roles in the removal and separation of substances (Yang et al., 2012).

Complexation refers to the formation of complexes, which are stable chemical species consisting of a central atom or ion (usually a metal cation) bonded to one or more surrounding molecules or ions, known as ligands. Complexation reactions can occur in aqueous solutions or various other environments. The process of complexation is aided by ligands, which are molecules or ions that donate electron pairs to the central metal ion, forming coordination bonds. Ligands can be neutral molecules or negatively charged ions.

(i) Coordination Number: The coordination number refers to the number of ligands bonded to the central metal ion in a complex. It can vary depending on the metal ion and ligands involved.
(ii) Chelation: Chelation is a special form of complexation in which a ligand forms multiple bonds with the central metal ion, creating a ring-like structure called a chelate. Chelation is often used in the removal of heavy metals from solution.
(iii) Stability Constants: Complexes have associated stability constants (formation constants) that describe the equilibrium constant for the formation of the complex. Higher stability constants indicate more stable complexes.

6 Applications of Nanochitosan in Water Treatment

Nanochitosan's applications in various water systems offer a versatile and effective approach to addressing diverse water-quality challenges. Nanochitosan's adsorption capabilities effectively reduce heavy metal concentrations, ensuring safe drinking water. Its antimicrobial properties combat bacteria, viruses, and fungi, safeguarding against waterborne diseases. Nanochitosan's high surface area adsorbs dyes, chemicals, and organic pollutants, enhancing wastewater quality. It prevents biofilm formation and assists in controlling microbial growth in treatment systems. Nanochitosan helps in adsorbing pesticides and chemicals from agricultural runoff, preventing contamination of water bodies. It aids in removing heavy metals and pollutants from industrial effluents, contributing to cleaner discharge. Nanochitosan's eco-friendly nature ensures minimal environmental impact after use, aligning with sustainability goals.

6.1 Nanochiosan-Based Water Purification Techniques

Purifying drinking water is critical for ensuring safe and healthy consumption. Nanochitosan presents several applications in the context of drinking water purification.

6.1.1 Trace Metal Removal

Nanochitosan exhibits an incredibly high surface area due to the increased surface-to-volume ratio at the nanoscale. This increased surface area is pivotal in the adsorption of heavy metals like lead, arsenic, and cadmium from contaminated water. Adsorption is the process by which molecules or particles adhere to the surface of nanochitosan, which effectively removes these pollutants from the water. Nanochitosan's high surface area provides more active sites for these heavy metal ions to bind, allowing for greater adsorption capacity compared to conventional chitosan or other adsorbents (Abd-Elhakeem et al., 2016).

The chemical structure of nanochitosan, with its amino and hydroxyl functional groups, plays a significant role in attracting heavy metal ions. These functional groups exhibit an affinity for metal ions due to their electrostatic interactions, complexation, and ion exchange capabilities. As a result, when water contaminated with heavy metals comes into contact with nanochitosan, these ions adhere to the surface of the nanoparticles, effectively reducing their concentration in the water (Seyedi et al., 2013; Salehi et al., 2020).

This adsorption process is highly efficient and has been demonstrated in various studies and applications. By effectively reducing the concentrations of lead, arsenic, cadmium, and other heavy metals in drinking water, nanochitosan contributes to making the water safe for consumption within permissible levels set by regulatory standards. Moreover, nanochitosan's eco-friendliness and biodegradability further enhance its appeal as a water purification agent. Its natural origin and biocompatible nature makes it a sustainable solution for addressing water pollution without introducing harmful chemicals into the environment (Saxena et al., 2020).

Overall, nanochitosan's high surface area, coupled with its chemical properties, makes it a promising and effective adsorbent for removing heavy metals from drinking water, thereby significantly contributing to the enhancement of water quality and safeguarding public health.

6.1.2 Pathogen Control

Nanochitosan, derived from chitosan, a natural biopolymer found in the shells of crustaceans like shrimp and crabs, boasts remarkable antimicrobial properties. This substance's ability to inhibit the growth of both bacteria and fungi has garnered significant attention, particularly in applications aimed at improving water quality and safety. Microbial contamination severely compromises the quality of drinking water. The nano-sized particles of nanochitosan possess a larger surface area, allowing for enhanced interactions with pathogens (Sivakami et al., 2013). By interacting with the cell walls of microorganisms, nanochitosan disrupts their structure and function, effectively hindering their growth and proliferation. The effectiveness of nanochitosan in combating pathogens makes it a valuable tool in reducing the risk of waterborne diseases. Bacterial contaminants like E. coli, Salmonella, and various fungi responsible for diseases such as cholera, typhoid, and gastrointestinal infections can be targeted and mitigated by incorporating nanochitosan into water treatment processes. Furthermore, nanochitosan's eco-friendly nature adds to its appeal. Being a derivative of chitosan, which is biodegradable and non-toxic, enhances its suitability for various environmental applications without adverse effects on ecosystems. By leveraging nanochitosan in water treatment, the safety and quality of drinking water can be significantly improved (Benettayeb et al., 2023). This technology holds immense potential in addressing global challenges related to waterborne illnesses, thereby enhancing public health and well-being on a large scale.

6.1.3 Organic Compound Filtration

The adsorption capabilities of nanochitosan extend beyond microbial inhibition, encompassing a wide range of pollutants present in water sources. Its remarkable ability to adsorb various organic compounds, pesticides, and dyes positions it as a valuable tool in the remediation and purification of water. Nanochitosan's structure, characterized by a high surface area due to its nano-sized particles, provides numerous active sites for interactions with diverse contaminants. This feature allows it to effectively attract and bind with organic molecules, pesticides, and dyes present in water (Ranjbari et al., 2019). Organic compounds, including pollutants like industrial chemicals, pharmaceuticals, and hydrocarbons, often pose a significant threat to water quality. Nanochitosan's adsorption properties come into play here, as these compounds can be trapped and removed from the water matrix through the process of adsorption. The interaction between the surface functional groups of nanochitosan and the organic molecules leads to their immobilization, effectively reducing their concentration in the water. Similarly, nanochitosan demonstrates a capacity to adsorb pesticides, which are widely used in agriculture but can leach into water sources, causing environmental concerns. By binding to these pesticide molecules, nanochitosan aids in their removal, thereby mitigating the potential harm they can cause to aquatic ecosystems and human health.

Additionally, the adsorption capabilities of nanochitosan extend to dyes commonly found in industrial wastewater. These dyes, if discharged untreated, can severely impact aquatic life and water quality. Nanochitosan's adsorption properties enable it to effectively capture and remove these dyes from water, contributing to the purification and remediation efforts. The versatility of nanochitosan in adsorbing a broad spectrum of contaminants underscores its potential as an eco-friendly and efficient agent in water treatment technologies. Its application in adsorbing organic pollutants, pesticides, and dyes plays a crucial role in the purification of water sources, ensuring cleaner and safer water for various uses, from drinking to industrial purposes, and preserving environmental integrity (Jeevanandam et al., 2018).

6.1.4 Nanocomposite Filters

Membrane technology has revolutionized water purification processes, and the integration of nanochitosan into membranes or filters has notably enhanced their efficacy in selectively removing contaminants from water sources (Yu et al., 2021).

Nanochitosan's incorporation into membranes or filters capitalizes on its unique properties, such as its high surface area, biocompatibility, and adsorption capabilities. When integrated into membranes, nanochitosan serves as a functional additive that significantly improves the performance of these filtration systems. These membranes or filters, when enhanced with nanochitosan, act as barriers with selective permeability. They effectively trap and remove a wide range of contaminants, including bacteria, viruses, heavy metals, organic compounds, and particulate matter, while allowing safe water molecules to pass through it.

The incorporation of nanochitosan into membranes or filters enhances their adsorption capacity, credited to the increased active sites available for interaction with contaminants. This leads to improved efficiency in removing various pollutants and impurities from water, contributing to higher quality treated water.

Furthermore, nanochitosan's antimicrobial properties play a crucial role in preventing biofouling, a common issue in membrane filtration systems where microorganisms accumulate on the surface, reducing filtration efficiency. By inhibiting microbial growth, nanochitosan helps maintain the longevity and effectiveness of the membranes, ensuring sustained performance over time. Moreover, nanochitosan's biodegradability makes it an environmentally friendly choice for membrane technology. As these membranes reach the end of their lifespan, the biodegradable nature of nanochitosan minimizes environmental impact during disposal (Huang et al., 2009).

Overall, the integration of nanochitosan into membrane technology represents a significant advancement in water purification processes. It not only enhances filtration efficiency by selectively trapping contaminants but also addresses challenges such as biofouling, contributing to the production of cleaner and safer water for various applications, including drinking water supply, industrial processes, and wastewater treatment.

6.2 *Environmental Impact and Safety Considerations*

6.2.1 Ecotoxicity and Biodegradability

Nanochitosan's potential impact on ecosystems and organisms in the environment is a subject of research interest. While chitosan, the precursor of nanochitosan, is generally considered non-toxic and biocompatible, the introduction of nanoparticles might alter its behaviour. Studies are on-going to understand the long-term effects of nanochitosan on aquatic life and soil organisms to ensure its safe use without adverse environmental impacts (Yuwei & Jianlong, 2011).

6.2.2 Health and Safety Concerns

Inhalation or direct contact with nanochitosan during manufacturing or handling procedures raises potential concerns, particularly for workers involved in these processes. Research endeavors are directed towards a comprehensive evaluation of potential health hazards stemming from nanochitosan exposure in diverse forms, including its powdered state, solution, or incorporation into end products. The objective of these studies is to delineate and establish safe exposure levels, providing critical insights for occupational health and safety guidelines. Understanding the potential risks associated with nanochitosan exposure in its various manifestations is imperative for safeguarding the well-being of individuals engaged in its production and application.

In biomedical applications, nanochitosan's biocompatibility is a crucial consideration. While it often shows promise due to its natural origin and low toxicity, thorough testing is necessary to ensure its safety when used in medical implants, drug delivery systems, or other healthcare applications In the domain of biomedical applications, the imperative of evaluating the biocompatibility of nanochitosan emerges as a cardinal consideration. Despite its propitious attributes emanating from its innate origin and a profile marked by low toxicity, the imperative of a comprehensive and meticulous testing regimen is underscored, aiming to unequivocally ascertain its safety and compatibility upon integration into medical implants, drug delivery systems, or other facets of healthcare applications. Nanochitosan, a derivative of chitin, evinces intrinsic biocompatibility attributable to its attributes of biodegradability, low immunogenicity, and minimal cytotoxicity. These attributes render it an enticing candidate for a spectrum of biomedical applications, spanning tissue engineering, wound healing, and drug delivery. Nevertheless, the nuanced interplay between nanochitosan and biological systems necessitates a stringent evaluation to decipher its behavioral intricacies at the cellular and molecular strata. Biocompatibility assessments encompass a multifaceted approach, enshrining both in vitro and in vivo investigations. In vitro experiments discern cell viability, adhesion, and proliferation under nanochitosan exposure, providing nuanced insights into its cellular interactions. Concurrently, in vivo studies scrutinize nanochitosan's in situ performance within living organisms, delineating its impact on tissues, organs, and systemic responses. For medical implants, the integration of nanochitosan with host tissues and the potential instigation of inflammatory responses demand systematic inquiry. Critical facets such as the degradation kinetics of the material, its biostability, and the concomitant release of by-products during decomposition impose pivotal considerations influencing its sustained performance within the biological system. In drug delivery systems, nanochitosan's biocompatibility assumes centrality to preempt the induction of adverse reactions, both locally at the administration site and systemically. Scrutinizing the release kinetics of encapsulated therapeutic agents and ensuring the sustained delivery of drugs with deleterious effects on surrounding tissues mandates a circumspect examination. Furthermore, healthcare applications entailing direct interactions with biological systems, exemplified by wound dressings or implant coatings, necessitate a nuanced comprehension of nanochitosan's influence on tissue regeneration, inflammation dynamics, and the broader landscape of the healing process. In summation, while the promise of nanochitosan in biomedical applications is underscored by its natural origin and minimal toxicity, the translational pathway to efficacious and safe medical solutions mandates exhaustive and systematic biocompatibility assessments. The intricacies characterizing nanochitosan's interactions with biological systems necessitate sustained research efforts to delineate robust safety profiles, ensuring its seamless integration into diverse healthcare applications with the utmost standards of biocompatibility (Zhou et al., 2014).

6.2.3 Regulatory Aspects and Guidelines

Organizations such as the U.S. Food and Drug Administration (FDA), the European Medicines Agency (EMA), and various national regulatory agencies guide the use of nanomaterials in different applications. These guidelines focus on safety assessments, risk management, labelling requirements, and environmental impact evaluations to ensure the responsible use of nanochitosan (Hu et al., 2008).

The judicious harnessing of the advantageous attributes of nanochitosan, while concurrently mitigating potential risks to human health and the environment, hinges on the conduction of comprehensive studies and unwavering adherence to regulatory guidelines. In this intricate paradigm, continuous research endeavors and synergistic collaboration among scientific communities, regulatory bodies, and industries constitute imperative components essential for the establishment of robust safety protocols and guidelines governing the myriad applications of nanochitosan. The multifaceted nature of nanochitosan necessitates an intricate understanding of its physicochemical properties, biocompatibility, and potential toxicity profiles through exhaustive and systematic scientific investigations. Such comprehensive studies form the bedrock upon which informed decisions pertaining to its safe utilization are predicated. The elucidation of intricate interaction mechanisms between nanochitosan and biological entities, coupled with an in-depth exploration of its environmental fate and transport dynamics, provides crucial insights that underpin regulatory considerations. Regulatory guidelines, reflecting a synthesis of scientific insights and risk management principles, play a pivotal role in providing a structured framework for the safe deployment of nanochitosan across diverse applications. These guidelines encapsulate critical facets such as recommended exposure limits, handling procedures, and environmental impact assessments. Rigorous adherence to such regulatory directives ensures the harmonious integration of nanochitosan into industries, healthcare, and other domains without compromising safety standards. Continuous research constitutes a dynamic and iterative process that remains integral to refining our understanding of nanochitosan's behavior in varied contexts. The evolving landscape of scientific knowledge necessitates ongoing investigations to address emerging concerns and optimize safety protocols. Collaborative efforts, involving academia, regulatory agencies, and industry stakeholders, foster a synergetic exchange of expertise, contributing to the development of robust safety frameworks. This collaborative approach aligns with the dynamic nature of nanotechnology, requiring adaptable regulatory responses to evolving scientific insights and technological advancements. Furthermore, the collaborative interplay among scientific communities, regulatory bodies, and industries is vital for the establishment of standardized safety protocols that transcend geographical boundaries. The dissemination of best practices and harmonized safety guidelines ensures a globally consistent approach, thereby facilitating the seamless and secure integration of nanochitosan across diverse applications on an international scale. In conclusion, the prudent utilization of nanochitosan mandates a steadfast commitment to a scientific foundation, regulatory compliance, and collaborative endeavors. The synergy between continuous research, stringent adherence to regulatory guidelines, and collaborative engagement forms the linchpin for maximizing the benefits of nanochitosan while minimizing potential risks to both human health and the environment. This multifaceted approach epitomizes a holistic strategy, reflective of the intricate interplay between scientific inquiry, regulatory governance, and industrial implementation in the realm of nanotechnology.

7 Future Prospects and Challenges

Advancements in nanochitosan for water purification continue to evolve, promising more effective and sustainable solutions. Integration of nanochitosan into "smart" membranes is capable of self-regeneration, adapting to changing water quality, and improving their efficiency over time. Developing nanochitosan-based composite materials with other nanoparticles or polymers will enhance filtration capabilities, selectivity, and durability. Engineering nanochitosan for specific pollutant removal by modifying its surface properties or creating tailored nanoparticles designed to target certain contaminants more effectively. Designing multifunctional nanochitosan-based systems is capable of simultaneously removing various contaminants like heavy metals, microplastics, pharmaceuticals, and pathogens from water sources. Advancing eco-friendly methods to synthesize nanochitosan will reduce energy consumption and waste generation during production. Drawing inspiration from natural systems to create nanochitosan-based materials that mimic biological processes for more efficient water purification (Thirunavukkarasu et al., 2020).

Developing portable and affordable nanochitosan-based filtration devices is suitable for households or communities lacking access to clean water. Implementing nanochitosan in industrial-scale water treatment facilities will address complex contamination issues in municipal water systems, industries, and wastewater treatment plants. Furthermore in-dept knowledge through research on surface modifications or encapsulation techniques is essential to minimize potential ecotoxicity while maintaining the efficacy of nanochitosan in water purification. Rsearch will further enable comprehensive assessments to understand the environmental impact of nanochitosan-based water purification technologies from production to disposal (Sreeram et al., 2017; Jjagwe et al., 2021).

There is need to also establish standardized testing methods and regulations for nanochitosan-based water purification systems to ensure safety, efficacy, and consistent performance. It is essential to provide clear guidelines and protocols for the application, handling, and disposal of nanochitosan-based materials in water treatment processes (Rashtbari et al., 2022).

As research and development in nanochitosan-based water purification advances, these innovations hold the potential to significantly improve water quality, enhance sustainability, and expand access to safe drinking water globally. However, despite its promise, nanochitosan faces several challenges and limitations in water treatment applications. One of such challenges is particle stability and agglomeration. Maintaining the stability of nanochitosan particles in water over extended periods can be challenging, as they might agglomerate or lose their desired properties, impacting their effectiveness in water treatment. Achieving uniform dispersion of nanochitosan in water or within filtration systems without agglomeration or settling poses a challenge, affecting its efficiency in contaminant removal (Saad et al., 2021; Shaumbwa et al., 2021).

Another challenge is scalability and cost-effectiveness; as scaling up nanochitosan production while maintaining consistent quality and properties at an economically feasible level presents a hurdle for large-scale water treatment applications. The cost of producing nanochitosan and its incorporation into water treatment technologies might currently be higher compared to conventional purification methods, limiting its widespread adoption, especially in resource-constrained regions (Rashtbari et al., 2022). Furthermore, while nanochitosan demonstrates adsorption capabilities, ensuring high selectivity for specific contaminants without compromising the removal of other essential elements or compounds remains a challenge. Achieving consistent and optimized performance across a wide range of water qualities, contaminant concentrations, and environmental conditions requires further refinement and customization (Obey et al., 2022).

Larger scale nanochitosan-based water purification may also raise environmental and health concerns, hence understanding the long-term environmental impact of nanochitosan residues or by-products after water treatment and their potential effects on ecosystems is an on-going challenge. Ensuring the safety of workers involved in the production and handling of nanochitosan, along with potential health risks associated with prolonged exposure, requires thorough investigation. Therefore, establishing comprehensive regulatory frameworks and guidelines to assess nanochitosan's safety, efficacy, and environmental impact in water treatment applications is essential but currently lacks standardization. Developing standardized testing methods and protocols to evaluate the performance, durability, and long-term effects of nanochitosan-based water treatment systems remains an on-going challenge (Galhoum et al., 2015).

Integrating nanochitosan-based technologies into existing water treatment infrastructure or developing compatible systems that can effectively incorporate these advancements without major modifications poses a challenge. Ensuring the longevity and reliability of nanochitosan-based filtration systems without compromising their effectiveness or requiring frequent maintenance is critical for sustainable implementation (Galhoum et al., 2017).

Addressing these challenges will require concerted efforts in research, innovation, regulation, and industry collaboration to harness the full potential of nanochitosan in water treatment while overcoming its current limitations.

8 Conclusion

The future of nanochitosan in water treatment holds significant potential, and further research can advance its capabilities in several key areas, which include: (i) Exploring novel surface modifications or functionalization techniques to enhance the adsorption capacity, selectivity, and stability of nanochitosan for targeted contaminant removal. (ii) Investigating methods to control the size, morphology, and properties of nanochitosan particles to optimize their performance in water treatment applications. (iii) Developing next-generation membranes with nanochitosan to achieve higher selectivity, improved fouling resistance, and adaptive properties for various water purification needs. (iv) Integrating nanochitosan with other nanomaterials or polymers to create hybrid filtration systems that synergistically enhance contaminant removal and water purification efficiency.

Researching environmentally friendly and cost-effective methods for synthesizing nanochitosan, such as utilizing renewable resources or innovative manufacturing processes is central to the success of application of nanochitosan in water treatment. It is also imperative to develop strategies to repurpose chitosan waste from the seafood industry into nanochitosan, contributing to sustainability and reducing waste. Expanding research into biomedical uses of nanochitosan, including drug delivery systems, wound healing, and tissue engineering also ascertains safety and efficacy. A crucial objective in that may advance the application of nanochitosan in water treatment is to investigate the potential of nanochitosan in tackling emerging environmental issues, including the remediation of emerging pollutants, management of microplastics, and mitigation of oil spills (Fu & Wang, 2011).

This however requires conducting comprehensive studies on the ecotoxicity, long-term environmental impact, and potential health risks associated with nanochitosan to establish safe exposure limits. Collaborating with regulatory bodies to develop standardized protocols, safety guidelines, and regulatory frameworks for the responsible use of nanochitosan in water treatment. Investigating the integration of nanochitosan-based technologies with sensing and monitoring systems for real-time water-quality assessment and adaptive treatment. Designing compact and portable nanochitosan-based water treatment devices suitable for decentralized applications in remote or disaster-affected areas. Advancements in these directions can significantly improve the efficiency, sustainability, and safety of nanochitosan-based water treatment technologies, paving the way for their broader adoption and impact on global water quality and accessibility.

References

Abd-Elhakeem, M. A., Ramadan, M. M., & Basaad, F. S. (2016). Removing of heavymetals from water by chitosan nanoparticles. *Journal of Advances in Chemistry, 11*, 3765–3771. https://doi.org/10.24297/jac.v11i7.2200

Ali, M. E. A., Aboelfadl, M. M. S., Selim, A. M., Khalil, H. F., & Elkady, G. M. (2018). Chitosan nanoparticles extracted from shrimp shells, application for removal of Fe(II) and Mn(II) from aqueous phases. *Separation Science and Technology, 53*, 2870–2881. https://doi.org/10.1080/01496395.2018.1489845

Alonso Fernandez, M. J., Calvo Salve, P., Remunan Lopes, C., & Vila Jato, J. L. (2016). *B05s Application of nanoparticles based on hydrophilic polymers as pharmaceutical forms*. No. EP0860166A1. Patent Cooperation Treaty. October 22, 2016.

Ayati, A., Ranjbari, S., Tanhaei, B., & Sillanpää, M. (2019). Ionic liquid-modified composites for the adsorptive removal of emerging water contaminants: A review. *Journal of Molecular Liquids, 275*, 71–83. https://doi.org/10.1016/j.molliq.2018.11.016

Benettayeb, A., Morsli, A., Elwakeel, K. Z., Hamza, M. F., & Guibal, E. (2021). Recovery of heavy metal ions using magnetic glycine—Modified chitosan—Application to aqueous solutions and tailing leachate. *Applied Sciences, 11*, 8377. https://doi.org/10.3390/app11188377

Benettayeb, A., Ghosh, S., Usman, M., & Seihoub, F. Z. (2022). Some well-known alginate and chitosan modifications used in adsorption: A review. *Water, 14*, 1353. https://doi.org/10.3390/w14091353

Benettayeb, A., Seihoub, F. Z., Pal, P., Ghosh, S., Usman, M., Chia, C. H., Usman, M., & Sillanpää, M. (2023). Chitosan nanoparticles as potential nano-sorbent for removal of toxic environmental pollutants. *Nanomaterials (Basel), 13*(3), 447. https://doi.org/10.3390/nano13030447. PMID: 36770407; PMCID: PMC9920024.

Brião, G. D. V., de Andrade, J. R., da Silva, M. G. C., & Vieira, M. G. A. (2020). Removal of toxic metals from water using chitosan-based magnetic adsorbents. A review. *Environmental Chemistry Letters, 18*, 1145–1168. https://doi.org/10.1007/s10311-020-01003-y

Choi, O., Deng, K. K., Kim, N. J., Ross, L., Surampalli, R. Y., & Hu, Z. (2008). The inhibitory effects of silver nanoparticles, silver ions, and silver chloride colloids on microbial growth. *Water Research, 42*, 3066–3074. https://doi.org/10.1016/j.watres.2008.02.021

Du, W. L., Xu, Z. R., Han, X. Y., Xu, Y. L., & Miao, Z. G. (2008). Preparation, characterization and adsorption properties of chitosan nanoparticles for eosin Y as a model anionic dye. *Journal of Hazardous Materials, 153*, 152–156. https://doi.org/10.1016/j.jhazmat.2007.08.040

Elwakeel, K. Z., Atia, A. A., & Guibal, E. (2014). Fast removal of uranium from aqueous solutions using tetraethylenepentamine modified magnetic chitosan resin. *Bioresource Technology, 160*, 107–114. https://doi.org/10.1016/j.biortech.2014.01.037

Fan, L., Luo, C., Sun, M., Qiu, H., & Li, X. (2013). Synthesis of magnetic β-cyclodextrin-chitosan/graphene oxide as nanoadsorbent and its application in dye adsorption and removal. *Colloids and Surfaces. B, Biointerfaces, 103*, 601–607. https://doi.org/10.1016/j.colsurfb.2012.11.023

Fu, F., & Wang, Q. (2011). Removal of heavy metal ions from wastewaters: A review. *Journal of Environmental Management, 92*, 407–418. https://doi.org/10.1016/j.jenvman.2010.11.011

Galhoum, A. A., Mahfouz, M. G., Gomaa, N. A. M., Vincent, T., Guibal, E., Atia, A. A., Mahfouz, M. G., Abdel-Rehem, S. T., Gomaa, N. A. M., & Vincent, T. (2015). Amino acid functionalized chitosan magnetic nanobased particles for uranyl sorption. *Industrial and Engineering Chemistry Research, 54*, 12374–12385. https://doi.org/10.1021/acs.iecr.5b03331

Galhoum, A. A., Mahfouz, M. G., Gomaa, N. M., Vincent, T., & Guibal, E. (2017). Chemical modifications of chitosan nano-based magnetic particles for enhanced uranyl sorption. *Hydrometallurgy, 168*, 127–134. https://doi.org/10.1016/j.hydromet.2016.08.011

Hamza, M. F., Wei, Y., Benettayeb, A., Wang, X., & Guibal, E. (2019). Efficient removal of uranium, cadmium and mercury from aqueous solutions using grafted hydrazide-micro-magnetite chitosan derivative. *Journal of Materials Science, 55*, 4193–4212. https://doi.org/10.1007/s10853-019-04235-8

Hritcu, D., Dodi, G., & Popa, M. I. (2012). Heavy metal ions adsorption on chitosan-magnetite microspheres. *International Journal of Chemical Engineering, 4*, 364–368.

Hu, B., Pan, C., Sun, Y., Hou, Z., Ye, H., Hu, B., & Zeng, X. (2008). Optimization of fabrication parameters to produce chitosan-tripolyphosphate nanoparticles for delivery of tea catechins. *Journal of Agricultural and Food Chemistry, 56*, 7451–7458. https://doi.org/10.1021/jf801111c

Huang, G., Yang, C., Zhang, K., & Jeffrey, S. (2009). Adsorptive removal of copper ions from aqueous solution using cross-linked magnetic chitosan beads. *Chinese Journal of Chemical Engineering, 17*, 960–966. https://doi.org/10.1016/S1004-9541(08)60303-1

Jaafari, J., Barzanouni, H., Mazloomi, S., Amir Abadi Farahani, N., Sharafi, K., Soleimani, P., & Haghighat, G. A. (2020). Effective adsorptive removal of reactive dyes by magnetic chitosan nanoparticles: Kinetic, isothermal studies and response surface methodology. *International Journal of Biological Macromolecules, 164*, 344–355. https://doi.org/10.1016/j.ijbiomac.2020.07.042

Jeevanandam, J., Barhoum, A., Chan, Y. S., Dufresne, A., & Danquah, M. K. (2018). Review on nanoparticles and nanostructured materials: History, sources, toxicity and regulations. *Beilstein Journal of Nanotechnology, 9*, 1050–1074. https://doi.org/10.3762/bjnano.9.98

Jjagwe, J., Olupot, P. W., Menya, E., & Kalibbala, H. M. (2021). Synthesis and application of granular activated carbon from biomass waste materials for water treatment: A review. *Journal of Bioresources and Bioproducts, 6*, 292–322. https://doi.org/10.1016/j.jobab.2021.03.003

Kadam, A. A., & Lee, D. S. (2015). Glutaraldehyde cross-linked magnetic chitosan nanocomposites: Reduction precipitation synthesis, characterization, and application for removal of hazardous textile dyes. *Bioresource Technology, 193*, 563–567. https://doi.org/10.1016/j.biortech.2015.06.148

Kuang, S. P., Wang, Z. Z., Liu, J., & Wu, Z. C. (2013). Preparation of triethylene-tetramine grafted magnetic chitosan for adsorption of Pb(II) ion from aqueous solutions. *Journal of Hazardous Materials, 260*, 210–219. https://doi.org/10.1016/j.jhazmat.2013.05.019

Li, Y., Zhou, Y., Nie, W., Song, L., & Chen, P. (2015). Highly efficient methylene blue dyes removal from aqueous systems by chitosan coated magnetic mesoporous silica nanoparticles. *Journal of Porous Materials, 22*, 1383–1392. https://doi.org/10.1007/s10934-015-0017-7

Liu, C., Wang, B., Deng, Y., Cui, B., Wang, J., Chen, W., & He, S. Y. (2015). Performance of a new magnetic chitosan nanoparticle to remove arsenic and its separation from water. *Journal of Nanomaterials, 2015*, 964–967. https://doi.org/10.1155/2015/191829

Massoudinejad, M., Rasoulzadeh, H., & Ghaderpoori, M. (2019). Magnetic chitosan nanocomposite: Fabrication, properties, and optimization for adsorptive removal of crystal violet from aqueous solutions. *Carbohydrate Polymers, 206*, 844–853. https://doi.org/10.1016/j.carbpol.2018.11.048

Moradi, D. S., Rahmanifar, B., Moradi, A. M., & Azar, P. A. (2014). Removal of permethrin pesticide from water by chitosan-zinc oxide nanoparticles composite as an adsorbent. *Journal of Saudi Chemical Society, 18*, 348–355. https://doi.org/10.1016/j.jscs.2014.01.004

Ngah, W. S. W., & Fatinathan, S. (2010). Pb(II) biosorption using chitosan and chitosan derivatives beads: Equilibrium, ion exchange and mechanism studies. *Journal of Environmental Sciences, 22*, 338–346. https://doi.org/10.1016/S1001-0742(09)60113-3

Obeid, L., Bée, A., Talbot, D., Jaafar, S. B., Dupuis, V., Abramson, S., Cabuil, V., & Welschbillig, M. (2013). Chitosan/maghemite composite: A magsorbent for the adsorption of methyl orange. *Journal of Colloid and Interface Science, 410*, 52–58. https://doi.org/10.1016/j.jcis.2013.07.057

Obey, G., Adelaide, M., & Ramaraj, R. (2022). Biochar derived from non-customized matamba fruit shell as an adsorbent for wastewater treatment. *Journal of Bioresources and Bioproducts, 7*, 109–115. https://doi.org/10.1016/j.jobab.2021.12.001

Olivera, S., Muralidhara, H. B., Venkatesh, K., Guna, V. K., Gopalakrishna, K., & Kumar, K. Y. (2016). Potential applications of cellulose and chitosan nanoparticles/composites in wastewater treatment: A review. *Carbohydrate Polymers, 153*, 600–618. https://doi.org/10.1016/j.carbpol.2016.08.017

Osagie, C., Othmani, A., Ghosh, S., Malloum, A., Kashitarash Esfahani, Z., & Ahmadi, S. (2021). Dyes adsorption from aqueous media through the nanotechnology: A review. *Journal of Materials Research and Technology, 14*, 2195–2218. https://doi.org/10.1016/j.jmrt.2021.07.085

Pap, S., Kirk, C., Bremner, B., Turk Sekulic, M., Gibb, S. W., Maletic, S., & Taggart, M. A. (2020). Synthesis optimisation and characterisation of chitosan-calcite adsorbent from fishery-food waste for phosphorus removal. *Environmental Science and Pollution Research, 27*, 9790–9802. https://doi.org/10.1007/s11356-019-07570-0

Ranjbari, S., Tanhaei, B., Ayati, A., & Sillanpää, M. (2019). Novel Aliquat-336 impregnated chitosan beads for the adsorptive removal of anionic azo dyes. *International Journal of Biological Macromolecules, 125*, 989–998. https://doi.org/10.1016/j.ijbiomac.2018.12.139

Ranjbari, S., Tanhaei, B., Ayati, A., Khadempir, S., & Sillanpää, M. (2020). Efficient tetracycline adsorptive removal using tricaprylmethylammonium chloride conjugated chitosan hydrogel beads: Mechanism, kinetic, isotherms and thermodynamic study. *International Journal of Biological Macromolecules, 155*, 421–426. https://doi.org/10.1016/j.ijbiomac.2020.03.188

Rashtbari, Y., Sher, F., Afshin, S., Hamzezadeh Bahrami, A., Ahmadi, S., Azhar, O., Rastegar, A., Ghosh, S., & Poureshgh, Y. (2022). Green synthesis of zero-valent iron nanoparticles and loading effect on activated carbon for furfural adsorption. *Chemosphere, 287*, 132114. https://doi.org/10.1016/j.chemosphere.2021.132114

Saad, E. M., Elshaarawy, R. F., Mahmoud, S. A., & El-Moselhy, K. M. (2021). New ulva lactuca algae based chitosan bio-composites for bioremediation of Cd(II) ions. *Journal of Bioresources and Bioproducts, 6*, 223–242. https://doi.org/10.1016/j.jobab.2021.04.002

Salehi, N., Moghimi, A., & Shahbazi, H. (2020). Preparation of cross-linked magnetic chitosan with methionine-glutaraldehyde for removal of heavy metals from aqueous solutions. *International Journal of Environmental Analytical Chemistry, 102*, 2305–2321. https://doi.org/10.1080/03067319.2020.1753718

Saxena, R., Saxena, M., & Lochab, A. (2020). Recent progress in nanomaterials for adsorptive removal of organic contaminants from wastewater. *ChemistrySelect, 5*, 335–353. https://doi.org/10.1002/slct.201903542

Seyedi, S. M., Anvaripour, B., Motavassel, M., & Jadidi, N. (2013). Comparative cadmium adsorption from water by nanochitosan and chitosan. *International Journal of Engineering and Innovative Technology, 5*, 145–148.

Shaumbwa, V. R., Liu, D., Archer, B., Li, J., & Su, F. (2021). Preparation and application of magnetic chitosan in environmental remediation and other fields: A review. *Journal of Applied Polymer Science, 138*, 1–25. https://doi.org/10.1002/app.51241

Sivakami, M. S., Gomathi, T., Venkatesan, J., Jeong, H. S., Kim, S. K., & Sudha, P. N. (2013). Preparation and characterization of nano chitosan for treatment wastewaters. *International Journal of Biological Macromolecules, 57*, 204–212. https://doi.org/10.1016/j.ijbiomac.2013.03.005

Sreeram, A., Hadi, P., Hui, C. W., Al Ansari, T., & McKay, G. (2017). Optimisation of the removal of arsenate from water using nanochitosan. *Desalination and Water Treatment, 70*, 235–243. https://doi.org/10.5004/dwt.2017.20561

Thirunavukkarasu, A., Nithya, R., & Sivashankar, R. (2020). A review on the role of nanomaterials in the removal of organic pollutants from wastewater. *Reviews in Environmental Science and Biotechnology, 19*, 751–778. https://doi.org/10.1007/s11157-020-09548-8

Vakili, M., Rafatullah, M., Salamatinia, B., Abdullah, A. Z., Ibrahim, M. H., Tan, K. B., Gholami, Z., & Amouzgar, P. (2014). Application of chitosan and its derivatives as adsorbents for dye removal from water and wastewater: A review. *Carbohydrate Polymers, 113*, 115–130. https://doi.org/10.1016/j.carbpol.2014.07.007

Yang, C. H., Wang, C. Y., Huang, K. S., Yeh, C. S., Wang, A. H. J., Wang, W. T., & Lin, M. Y. (2012). Facile synthesis of radial-like macroporous superparamagnetic chitosan spheres with in-situ Co-precipitation and gelation of ferro-gels. *PLoS One, 7*, 49329. https://doi.org/10.1371/journal.pone.0049329

Yu, S., Pang, H., Huang, S., Tang, H., Wang, S., Qiu, M., Chen, Z., Yang, H., Song, G., & Fu, D. (2021). Recent advances in metal-organic framework membranes for water treatment: A review. *Science of the Total Environment, 800*, 149662. https://doi.org/10.1016/j.scitotenv.2021.149662

Yuwei, C., & Jianlong, W. (2011). Preparation and characterization of magnetic chitosan nanoparticles and its application for Cu(II) removal. *Chemical Engineering Journal, 168*, 286–292. https://doi.org/10.1016/j.cej.2011.01.006

Zhou, Z., Lin, S., Yue, T., & Lee, T. C. (2014). Adsorption of food dyes from aqueous solution by glutaraldehyde cross-linked magnetic chitosan nanoparticles. *Journal of Food Engineering, 126*, 133–141. https://doi.org/10.1016/j.jfoodeng.2013.11.014

Nutrient and Drug Delivery Systems

Franklyn Nonso Iheagwam, Adegbolagun Grace Adegboro, and Collins Ojonugwa Mamudu

Contents

F. N. Iheagwam (✉)
Department of Biochemistry, Covenant University, Ota, Nigeria

Covenant University Public Health and Wellness Research Cluster, Ota, Nigeria
e-mail: franklyn.iheagwam@covenantuniversity.edu.ng

A. G. Adegboro
Department of Biochemistry, Covenant University, Ota, Nigeria

Covenant Applied Informatics and Communication Africa Centre of Excellence, Ota, Nigeria

C. O. Mamudu
Biochemistry Program, City University of New York Graduate Center, New York, NY, USA

P. O. Isibor et al. (eds.), *Nanochitosan-Based Enhancement of Fisheries and Aquaculture*, https://doi.org/10.1007/978-3-031-52261-1_7

1 Introduction

Healthy aquatic habitats are imperative to human existence. They dominate about 70% of the Earth's surface, and we depend on them for food, energy and water. Aquatic protein options are considered a preferable source due to their optimal health effects and the presence of important food constituents (Nasr-Eldahan et al., 2021). Fish is regarded as an important component of the human meal in almost all nations of the world (Mohanty, 2015). Fish provides vast amounts of other nutrients including essential amino acids and fatty acids needed by the body. Additionally, the various minerals present in fish are tremendously bioavailable once ingested into the body (Nasr-Eldahan et al., 2021).

Nevertheless, the aquaculture sector is still under variability as its sustainability is controversial due to the deleterious effect of poor management and pollution on productivity and the environment. On this note, nanotechnology is rapidly emerging as a novel approach for science and technology for sustainable food production, addressing disease breakouts, minimal nutrient absorption and environmental safety (Rodrigues et al., 2017). There is an increased interest in chitosan nanoparticles as a result of their inherent features but not limited to biodegradability, biocompatibility, non-toxicity and specificity (Kumaran, 2020).

2 Enhanced Nutrient Absorption Using Nanochitosan-Based Formulations

Fish may be subjected to nutritional inadequacies due to the unevenness, absence or surplus of food constituents. The conventional method of feeding fish is dependent on the provision of food in a pellet form, which is primarily produced with respect to the nutritional needs for essential constituents including fats, minerals, vitamins, carbohydrates, etc. of the fish per day (Gabriel et al., 2022). Nanotechnology is a recently employed field by nutritionists to ensure the availability of diverse delivery media for the promotion of nutrient absorption and bioavailability. These are mainly composed of micronutrients encapsulated within nanoparticles that may be delivery systems constructed from various sources such as lipids, proteins, carbohydrates (as in the case of nanochitosan) and others (Joye et al., 2014). Micronutrients are beneficial in fish and other aquatic creatures as they are essential for health maintenance and stress alleviation. In spite of the minimal requirement of micronutrients, the exorbitant application cost is implicated (Tayel et al., 2019). When encapsulated in a nanoparticle and included in the feeds of aquaculture, these micronutrients have the capacity to permeate cells more effectively, consequently increasing the rate of absorption and managing their application difficulties at a larger scale (Fajardo et al., 2022).

Diets supplemented with nanochitosan-incorporated clinoptilolite were found to increase the total protein level and lysozyme activity in rainbow trout (*Oncorhynchus mykiss*) when compared to the control, potentially enhancing the rate of growth and immune response (Khani Oushani et al., 2020). In another study, nanochitosan was employed in the supply of vitamin C bringing about increased shelf life and release of vitamin C in rainbow trout (Alishahi et al., 2014). Additionally, the same author reported a controlled release of vitamin C in the gut and its protection from enzymatic degradation by the nanochitosan compared to unencapsulated vitamin C. Similarly, nanochitosan was capable of permeating intestinal epithelium and substantially enhancing ascorbic acid (AA) absorption in *Brachionus plicatilis* and zebrafish liver cell line when utilised for AA's delivery (Jiménez-Fernández et al., 2014).

Furthermore, nanochitosan-based feed formulation promoted the tilapia fish's health and production performance following its increased feed utilisation, protein usage, free radicals scavenging enzymes, haematological profile and intestinal make-up (Abd El-Naby et al., 2020). The incorporation of nanochitosan into fish diets brought about a substantial increase in blood indices including red blood cells, haemoglobin, etc., in comparison to the control, hence implicating improved oxygen absorption and transportation (El-Naggar et al., 2021). Following the intestinal morphometric analysis of *Liza ramada* fed with nanochitosan-incorporated diets, significant enhancement in the width and height of the villus as well as increased goblet cells were reported (Dawood et al., 2020). In addition, the author also revealed the presence of a valuable association between the dose of nanochitosan and the enterocyte brush border density, depicting an enhanced absorption.

Enhanced digestion of nutrients, villi healthiness and increased action of intestinal protease in *Labeo rohita* fed with nanochitosan encapsulated trypsin in comparison with fish fed with ordinary trypsin (Kumari et al., 2013). The inhibition of potential infections, promotion of the number of good bacteria and stimulation of the activity of microbial enzymes in the gut of the fish may synergistically enhance the digestibility of feed and nutrient absorption following the incorporation of chitosan nanoparticles (Abdel-Tawwab et al., 2019).

Protein efficiency and increased omega-3 quantities were reported following the supplementation of gluten meals with nanochitosan in the Nile tilapia (El-Naggar et al., 2022). Additionally, in order to overcome the slow metabolism of carbohydrates and limit the utilisation of amino acids for gluconeogenic reasons in the carnivorous fish liver, chitosan-TPP nanoparticles fused with plasmid DNA aimed at altering target genes have been developed (Wu et al., 2020). For instance, a shRNA-overexpressing plasmid fused with a chitosan-TPP nanoparticle, synthesised to suppress the expression of cytosolic alanine transaminase (cALT), was injected into *Sparus aurata*. A substantial reduction in the mRNA levels of cALT1, immunodetectable ALT and ALT activity was noticed in treated *S. aurata* liver. Additionally, cALT expression knockout brought about elevated activity of essential enzymes in glycolysis and protein metabolism in the liver (González et al., 2016). According to Asaikkutti et al. (2023), white-leg shrimp fed with vitamin C-loaded nanochitosan depicted increased bioavailability of the vitamin following the observation of substantial enhancement in immunological markers such as transglutaminase, phenoloxidase, respiratory burst and disease resistance. Additionally, a significant increase in the expression of lysozyme, lectin and cytosolic manganese superoxide dismutase was noticed post-feeding.

3 Controlled Release Systems for Drug Delivery in Aquaculture

The aquatic terrain is regarded as an active channel that is capable of transporting pathogens over a couple of miles (Tayel et al., 2019). Additionally, pathogens have the ability to move within farms due to the organism's significant movement and transporter, which constitute a high compound network for disease transmission (Murray, 2013; Munro & Gregory, 2009). Fish are susceptible to numerous infectious diseases and environmental stressors such as nitrite and ammonia poisoning, oxygen deficiency and thermal and pH stress. To combat infectious diseases and increase productivity, enormous quantities of antibiotics are utilised, which sometimes brings about the emergence of resistant organisms as well as making aquaculture habitats resistant stores that could be transmitted to animal and human pathogens (Tayel et al., 2019). Various mode of treatment administration exists for fish including drugging into the water or feed, injection and skin-surface application in which the first two administrative modes are used for exterior infections, while the latter is for interior infections (Bowker et al., 2016).

Nanotechnology has recently been used for the deterrence of diseases such as in water treatment, habitat disinfection, disease diagnosis and regulation, nutrients and drug effective delivery, and enhancement of fish absorption capacity of these components (Tayel et al., 2019). Nanochitosans have some fascinating attributes to be employed in drug delivery systems including biodegradability, controlled release of substances and muco-adhesiveness (De Oliveira et al., 2021).

3.1 Chitosan Loading Nucleic Acids, Proteins and Inactivated Pathogens

Vaccines are substances that generate immunity towards specified diseases, and effective defence is guaranteed following their administration. The most dependable and efficacious way of vaccination in aquaculture is via oral or injection route. The latter, a conventional additional practice entails the formulation of the vaccines with water/oil and is prone to deleterious effects, which might result in the death of fishes occasionally (Shah & Mraz, 2020). To conquer these difficulties, the scientific body in recent years recommended a nano-delivery system as a possible approach for vaccine release in fish that is deemed safer and promotes effectiveness. The nanoscale delivery system's role in vaccine fabrications cannot be overemphasised, and it is approved that small particle-sized materials create better immune responses

(Selvasudha et al., 2022). In this milieu, to date, various encapsulation systems have been manufactured and tried. Due to the exclusive nature of chitosan, it has been largely used in diverse capacities. DNA nanovaccines have been employed in aquaculture to stimulate immune responses in fish (Baskaran, 2023). On the other hand, nanochitosan has been utilised as a carrier for numerous types of DNA and vaccines in fish via diverse routes of administration such as oral or injection. For example, a moderate defence was reported in *Lates calcarifer* challenged with *Vibrio anguillarum* following oral administration of the nanochitosan DNA vaccine (Rajesh Kumar et al., 2008). Similarly, a nanochitosan-based oral vaccine was used to combat viral infection in *Scophthalmus maximus* (Zheng et al., 2016). Also, a study revealed the improved efficacy of a nanochitosan-coated vaccine against columnaris disease in tilapia (*Oreochromis* sp.) (Kitiyodom et al., 2019). In addition, the oral delivery of nanochitosan packed with a DNA construct made up of the VP28 gene of white spot syndrome virus (WSSV) to *Penaeus monodon* brought about a substantial rate of survival in comparison to complete lethality in the control group (Rajeshkumar et al., 2009). According to Kumari et al. (2013), controlled release of the enzyme was observed following the feeding of the fish, *L. rohita* with nanochitosan encapsulated trypsin, implicating it as a suitable delivery system for proteins or drugs where controlled release is wanted. Additionally, the recombinant outer membrane protein A of *Edwardsiella tarda incorporated into nanochitosan yielded substantially increased levels of free antibody and survivability after oral vaccination of Labeo fimbriatus* (Dubey et al., 2016).

Steady and controlled release of therapeutically active constituents in fish models such as antioxidants, hormones and others was observed following formulation with nanochitosan (Ahmed et al., 2019). Diverse studies have effectively executed encapsulation and delivery in aquaculture by the use of chitosan, viz. treatment mechanisms have been generated against bacterial infections such as *Vibrio parahaemolyticus* in the *Acanthopagrus schlegelii* (Li et al., 2013) and ciliate infections such as *Philasterides dicentrarchi* in turbot (León-Rodríguez et al., 2013). Furthermore, substantial protection, enhanced rate of survival and elevated immune response were reported following the injection of chitosan-coated membrane vesicles (cMVs) into *Danio rerio* against *Piscirickettsia salmonis* infection (Tandberg et al., 2018).

3.2 *Chemical Compounds and Metal Ions Loading*

In addition, numerous studies investigated nanochitosan-loading aromatase inhibitors and eurycomanone compounds that enhance the growth of gonads. The delivery of these compounds by nanochitosan elongated the availability of serum, enhanced testicular growth without toxicity and brought about increased serum levels of reproductive hormones (Bhat et al., 2018, 2019a, b; Wisdom et al., 2018). Similarly, nanochitosan mediates the delivery of metal ions that are antimicrobial and micronutrients (e.g. silver, selenium) to fish (Wu et al., 2020). Disease resistance elevation and immunostimulatory action of chitosan-selenium nanoparticles were reported in *Paramisgurnus dabryanus* and zebrafish following the enhancement of enzymatic activities including acid phosphatase, lysozyme and phagocytic respiratory burst and the reaction of the splenocyte against concanavalin A (Xia et al., 2019; Victor et al., 2019).

3.3 Fish Reproduction

Nanochitosans have been utilised as drug and gene delivery systems in studies with the goal of proper development of gonads in aquaculture (Wu et al., 2020). The conjugation of nanochitosan with aromatase inhibitors brought about a controlled release of the drug and a positive result implicated in enhanced testicular growth which was apparent in hormonal analysis, histology of gonads and gonadosomatic index in *Clarias magur* when compared to the control (Wisdom et al., 2018). Similarly, an aromatase inhibitor, letrozole was conjugated with chitosan nanoparticles to evaluate the effect on masculinisation of *O. mykiss* larvae. The result revealed increased testosterone levels, and oestradiol was undetectable, hence depicting the potent delivery of letrozole by nanochitosan in suppressing oestradiol production, and thus altering the sex ratio of the fish (Alijani et al., 2022). Nanochitosan was employed to deliver salmon luteinising hormone-releasing hormone (sLHRH) in fish, evaluating the effect on reproductive gene expression. Following this study, the elevated expression level of Sox9 and circulating steroid hormones were revealed in the gonads and blood, respectively, in both the control and nanochitosan based. However, the entrapment of sLHRH in nanochitosan brought about regulated and sustained release of the hormones with optimum levels noticed at the end of the experiment (36 h after treatment) in comparison to the control with its optimum hormonal levels noticed at 12 h after treatment (Bhat et al., 2016). Additionally, the intramuscular administration of nanochitosans entrapped with the plasmid construct, steroidogenic acute regulatory protein brought about a sustained stimulatory impact on the expression of major reproductive genes when compared to the unencapsulated plasmid construct (Rathor et al., 2017).

4 Potential for Improving Growth Rates, Feed Efficiency and Health Management

Chitosan nanoparticles have been reported for their better functional properties including antimicrobial, antioxidant and film-formation (Kulawik et al., 2019).

4.1 Seafood Preservation, Edible Coating and Shelf Life

Naturally, seafood is highly prone to deterioration with a short shelf-life ranging from a few days to about 2 weeks. However, the use of nanochitosan in seafood processing can enhance their qualities and lifespan due to the potent antimicrobial properties of this nanoparticle. Nanoencapsulation protects seafood from factors such as stressors, taste and odour as well as brings out the original taste (Fathi et al., 2012; George et al., 2023). Another study reported increased antimicrobial activity and suppression ability of the total volatile nitrogen (TVB-N) level for enhancing the general storage capability following the effect of nanochitosan on silver carp fillet (Ramezani et al., 2015). Coating of *Carangoides coeruleopinnatus* fillets with nanochitosan preserved the sensory properties of the fish, regulated lipid oxidation and decreased microbial count under refrigeration when compared to uncoated samples (Alboghbeish & Khodanazary, 2019).

Furthermore, essential oils (EOs) are complex mixtures of high volatility generated by diverse plant species which are composed of medicinal features such as anti-inflammatory and antioxidant. Additionally, they have biocidal action against numerous pathogens including bacteria and viruses (Calo et al., 2015). *In the aquaculture sector, essential oils have intense prospects for use attributable to their reduced cost, availability, diminished side effects, decreased toxicity and increased biodegradability when compared to antibiotics and the challenge associated with their use* (Tayel et al., 2019). *EOs can also aid in fish feed efficiency because they are capable of enhancing the performance and development of the gastrointestinal tract, mostly at the elementary stages of fish development, avoiding mucoadhesion of pathogens to the intestine, promoting glucose utilisation and enhancing the release of digestive enzymes* (Freccia et al., 2014; Hernández et al., 2016; Zeppenfeld et al., 2016). Due to the toxic and unpleasant effects of anaesthetics in aquaculture, EOs are used as replacements in fish because of their higher biodegradability and decreased toxic effect, compared to chemicals. However, in spite of all of these beneficial properties, EOs have minimal solubility in water making their application in aquatic habitats challenging. They also have increased sensitivity to light, reduced stability and strong sensory properties. Therefore, novel approaches such as nanoencapsulation are now being used to enhance their properties. For example, biosynthesised nanochitosan alongside selenium nanoparticles and cinnamon extracts revealed improved antimicrobial effect against numerous pathogens including *Staphylococcus aureus*, *Salmonella typhimurium* and *Listria monocytogenes*, depicting the key impact of chitosan in the fish feed (Alghuthaymi et al., 2021). The incorporation of nanochitosan and *Ziziphora clinopodioides* essential oil into polylactic acid films brought about an increase in the lifespan of *Oncorhynchus mykiss* fillets when compared with the control and additional treatment options available in the study without unpleasant sensory attributes (Shakour et al., 2021). Consequently, the author recommended polylactic acid films treated with the aforementioned essential oil and nanochitosan as suitable antimicrobial films for seafood products packaging for shelf-life enhancement. In a recent study, the coating of *O. mykiss* with nanochitosan containing essential oil was reported to potentially elevate the shelf life of the fish following both chemical and microbiological analysis (Abdollahzadeh et al., 2023). This was further confirmed during storage time as specimens treated with nanochitosan/nano essential oil had the best sensory properties. This further validates an earlier study where nanochitosan containing *Cumino cyminum* L. essential oil increased antimicrobial, antioxidant and sensory properties, hence implicated as a viable coating to enhance the shelf-life of sardine fillet (Homayonpour et al., 2021). Additionally, the synergistic effect of nanochitosan incorporated edible coatings and clove oil produced a significant increase in the chemical, sensory, physical and microbiological features of Mugil cephalus while under refrigeration for about 3 weeks (Aref et al., 2022). The use of nanochitosan coatings composed of *Zataria multiflora* and *Polylophium involucratum* essential oils resulted in the least microbial count and did not compromise the sensory properties of silver carp fillets for about 12 days under refrigeration in comparison to the control (Mohammadkhan et al., 2022).

4.2 Feed Efficiency

Lately, the farming of various categories of seafood such as fish, crab and shrimp has been flourishing, contributing huge impact on the production of seafood and consequently instrumental in determining the country's economy in this aquaculture sector (Chellapandian et al., 2023). However, diverse environmental factors including pH and salt concentration, the nature of the soil, mineral inadequacies, bad water and disease breakouts can influence the manufacture of seafood, especially fish and shrimp. To conquer these drawbacks, nanotechnology has been utilised in the production of protein-dense feeds, which are noticed to reveal optimal results in comparison to other biologically produced feeds (Chellapandian et al., 2023). In addition, during the direct dispensation of feed to water, feed enrichments can be dispensed to water but can simply degrade upon connection with water. Chitosan nanoparticles can be used as an encapsulating medium. For instance, Abd El-Naby et al. (2020) reported enhanced growth, immunity and increased productivity following the dietary administration of nanochitosan in *Oreochromis niloticus. Similarly, the dietary supplementation of nanochitosan conjugated with vitamin C in O. niloticus, enhanced the growth rate, antioxidant level, immune action, resistance to infection and improved the morphology of the intestine* (Ibrahim et al., 2021). Furthermore, dietary supplementation of vitamin C-loaded nanochitosan in *Litopenaeus vannamei* infected with *Vibrio harveyi* brought about enhanced weight gain, feed effectiveness and increased survival rate (Asaikkutti et al., 2023). The incorporation of nanochitosan into the feed of tilapia was more efficient when compared to the control following a substantial increase in the survival rate (Lembang et al., 2023).

4.3 Growth Rates

Many factors contribute to the development of aquatic creatures including temperature, well-management, vaccination, quality of water and most importantly well-formulated feed. The latter aids feed digestion and facilitates health and growth rates (George et al., 2023). The addition of nanoparticles in feed is the most efficient means to have a feed with an adequate amount of nutrients. Significant increases in all growth-rate parameters including weight gain, final weight and survival rate were reported following the fortification of nanochitosan with vitamin C in *O. niloticus* (Naiel et al., 2020). Similarly, specific growth-rate parameters such as weight gain and specific growth rate were substantially increased following *L. ramada* fed with nanochitosan-incorporated diets (Dawood et al., 2020). Furthermore, a decreased death rate was reported following the supplementation of gluten meals with nanochitosan in the Nile tilapia (El-Naggar et al., 2022). The use of nanochitosan on African catfish enhanced the quality of water, feed usage, survival rate and weight gain as well as body makeup (Udo et al., 2018; Fatahi et al., 2018). Another study reported a controlled release of arginine following its encapsulation with nanochitosan which substantially increased the growth factors of *O. mykiss* juveniles (Table 1). Additionally, the growth rate of Koi fish fed with chitosan nano-emulsion made up of protein and carotenoid was greater when compared to other diets (Sari et al., 2020).

4.4 Health Management

4.4.1 Antibacterial Activity

Nanochitosan is important for the stimulation of the bactericidal activity of both serum and phagocytes, hence facilitating the production of diverse humoral constituents that participate in innate and/or adaptive immune response, consequently

Table 1 A case study of growth performance following administration of nanochitosan

	Dosage	Feeding period	Species	Effect	References
1	5 g		Tilapia	Significant increase in all growth-rate parameters	Naiel et al. (2020)
2	0–2 g/kg	8 weeks	Grey mullet	Increased final weight, weight gain and specific growth rate parameters	Dawood et al. (2020)
3	0.5%	82 days	Nile Tilapia	Substantial increase in growth rate parameters	El-Naggar et al. (2022)
4	5 g	91 days	African catfish	Enhanced daily weight gain, meat quality and survival	Udo et al. (2018)
5	5%	60 days	Rainbow trout	Controlled arginine release with significant increase in the growth factors	Fatahi et al. (2018)

protecting the fish from numerous ailments (George et al., 2023). The mechanism of bacteria inhibition by nanochitosan can take place via diverse likelihoods. First due to the existence of charge differences that interact electrostatically, where chitosan has an antibacterial functional group with a positive charge, whereas the bacteria has a negative charge; hence, an alteration in the bacteria membrane's permeability occurs resulting in an imbalance of the osmotic pressure in the bacteria cells and consequently hampers microbial growth (Rosidah & Mulyani, 2022). Additionally, chitosan inhibits the metabolic activities of microbes by altering the makeup of bacteria such as protein, amino acids and glucose. Hydrolysis could occur in the bacteria's cell wall resulting in the release of cell electrolytes, bringing about mortality. The second likelihood is that chitosan's positive charge interacts with the DNA of bacteria, hence suppressing RNA and protein production. For instance, nanochitosan exhibited both in vivo and in vitro inhibitory activity against *Aeromonas hydrophila* subsp. *Hydrophila* in *O. niloticus* with an inhibition zone of 15 mm diameter (Aly et al., 2023). Similarly, gram-negative bacteria growth was reported to be inhibited by nanochitosan with inhibition zones of 25–48 mm diameter in *O. niloticus* (Abdel-Razek, 2019). According to Abdel-Razek (2019), the incorporation of chitosan nanoparticles in the diet had protective effects on Nile tilapia from diverse bacterial pathogens. Similarly, the effect of nanochitosan on the survival rate of rainbow trout *O. mykiss* was evaluated following continuous feeding for 3 weeks. Thereafter, the survival rate was raised to 80% in comparison to the control group (Saleh et al., 2022). In addition, an increased survival rate against *A. hydrophila* in zebrafish larvae was also reported at 5 days post-fertilisation following exposure to chitosan nanoparticles (Nikapitiya et al., 2018).

4.4.2 Immunostimulatory Activity

Diverse studies have confirmed that nanochitosan can be utilised as an immunostimulant that elevates non-specific body resistance in numerous kinds of fish, which is implicated by an elevation in haematological markers such as the number of lysozymes, erythrocytes, leucocytes and phagocytosis rates (Rosidah & Mulyani, 2022). According to Dawood et al. (2020), *L. ramada*-fed nanochitosan-based diets had the highest lysozyme and phagocytic activity, elevated sodium dismutase (SOD), catalase (CAT) and glutathione peroxidase as well as reduced malondialdehyde levels, consequently, depicting the immune and antioxidant-responsive potential of nanochitosan on the health of the fish. Furthermore, nanochitosan-formulated diets played a key role in the abortion of lethal toxicological effects of pesticides in aquatic habitats (Naiel et al., 2020). Nanochitosans utilised as delivery systems for vaccines or as coating or feed additives to fish demonstrated increased resistance against stressors, enhanced immunity and decreased microbial infection in rainbow trout (Muruganandam et al., 2019). Increased levels of immunological parameters including superoxide dismutase, prophenoloxidase and total haemocyte count were reported in comparison to the control group following dietary supplementation with chitosan nanoparticles against WSSV-infected *Procambarus clarkia* (Sun et al., 2016). Additionally, a substantial increase in survivability was reported.

5 Conclusion

This chapter addresses the application of chitosan nanoparticles for nutrient and drug delivery in the aquaculture field. By encapsulating key elements and nutrients within chitosan nanoparticles, aquatic creatures can be provided with a sustained, constant and controlled dissemination of essential elements. The use of nanochitosan in drug delivery can enhance treatment efficacy and lessen the effect of drug usage on the aquatic environment. Consequently, these approaches hold the potential of eradicating nutrient disproportionality, enhancing growth and immunity, avoiding diseases and ultimately keeping the aquaculture structures evergreen.

References

Abd El-Naby, A. S., Al-Sagheer, A. A., Negm, S. S., & Naiel, M. A. E. (2020). The dietary combination of chitosan nanoparticle and thymol affects feed utilisation, digestive enzymes, antioxidant status, and intestinal morphology of Oreochromis niloticus. *Aquaculture, 515*, 734577. https://doi.org/10.1016/j.aquaculture.2019.734577

Abdel-Razek, N. (2019). Antimicrobial activities of chitosan nanoparticles against pathogenic microorganisms in Nile tilapia, Oreochromis niloticus. *Aquaculture International, 27*(5), 1315–1330. https://doi.org/10.1007/s10499-019-00388-0

Abdel-Tawwab, M., Razek, N. A., & Abdel-Rahman, A. M. (2019). Immunostimulatory effect of dietary chitosan nanoparticles on the performance of Nile tilapia, Oreochromis niloticus (L.). *Fish & Shellfish Immunology, 88*, 254–258. https://doi.org/10.1016/j.fsi.2019.02.063

Abdollahzadeh, M., Elhamirad, A. H., Shariatifar, N., Saeidiasl, M., & Armin, M. (2023). Effects of nano-chitosan coatings incorporating with free/nano-encapsulated essential oil of Golpar (Heracleum persicum L.) on quality characteristics and safety of rainbow trout (Oncorhynchus mykiss). *International Journal of Food Microbiology, 385*, 109996. https://doi.org/10.1016/j.ijfoodmicro.2022.109996

Ahmed, F., Soliman, F. M., Adly, M. A., Soliman, H. A., El-Matbouli, M., & Saleh, M. (2019). Recent progress in biomedical applications of chitosan and its nanocomposites in aquaculture: A review. *Research in Veterinary Science, 126*, 68–82.

Alboghbeish, H., & Khodanazary, A. (2019). The comparison of quality characteristics of refrigerated Carangoides coeruleopinnatus fillets with chitosan and nano-chitosan coatings. *Turkish Journal of Fisheries and Aquatic Sciences, 19*(11). https://doi.org/10.4194/1303-2712-v19_11_07

Alghuthaymi, M. A., Diab, A. M., Elzahy, A. F., Mazrou, K. E., Tayel, A. A., & Moussa, S. H. (2021). Green biosynthesised selenium nanoparticles by cinnamon extract and their antimicrobial activity and application as edible coatings with nano-chitosan. L. Arru (ed.). *Journal of Food Quality, 2021*, 1–10. https://doi.org/10.1155/2021/6670709

Alijani, H., Amini Chermahini, M., Zargham, D., & Mohiseni, M. (2022). The effect of letrozole and letrozole-chitosan nanoparticles on masculinization of Rainbow trout larvae (Oncorhynchus mykiss). *Iranian Journal of Fisheries Sciences, 21*(6), 1383–1396.

Alishahi, A., Proulx, J., & Aider, M. (2014). Chitosan as biobased nanocomposite in seafood industry and aquaculture. In S.-K. Kim (Ed.), *Seafood science: Advances in chemistry, technology and applications* (pp. 211–231). CRC Press.

Aly, S. M., Eissa, A. E., Abdel-Razek, N., & El-Ramlawy, A. O. (2023). Chitosan nanoparticles and green synthesised silver nanoparticles as novel alternatives to antibiotics for preventing *A. hydrophila subsp. hydrophila* infection in Nile tilapia, *Oreochromis niloticus. International Journal of Veterinary Science and Medicine., 11*(1), 38–54. https://doi.org/10.1080/23144599.2023.2205338

Aref, S., Habiba, R., Morsy, N., Abdel-Daim, M., & Zayet, F. (2022). Improvement of the shelf life of grey mullet (Mugil cephalus) fish steaks using edible coatings containing chitosan, nanochitosan, and clove oil during refrigerated storage. *Food Production, Processing and Nutrition, 4*(1), 27. https://doi.org/10.1186/s43014-022-00106-z

Asaikkutti, A., Vimala, K., Jha, N., Bhavan, P. S., & Arul, V. (2023). Effect of dietary supplementation of vitamin C-loaded chitosan nanoparticles on growth, immune-physiological parameters, and resistance of white shrimp Litopenaeus vannamei to Vibrio harveyi challenge. *Animal Feed Science and Technology, 305*, 115764. https://doi.org/10.1016/j.anifeedsci.2023.115764

Baskaran, V. (2023). Nanotechnologies in aquatic disease diagnosis and drug delivery. In A. V. Kirthi, K. Loganathan, & I. Karunasagar (Eds.), *Nanotechnological approaches to the advancement of innovations in aquaculture. Nanotechnology in the life sciences* (pp. 1–21). Springer. https://doi.org/10.1007/978-3-031-15519-2_1

Bhat, I. A., Rather, M. A., Saha, R., Pathakota, G.-B., Pavan-Kumar, A., & Sharma, R. (2016). Expression analysis of Sox9 genes during annual reproductive cycles in gonads and after nanodelivery of LHRH in Clarias batrachus. *Research in Veterinary Science, 106*, 100–106. https://doi.org/10.1016/j.rvsc.2016.03.022

Bhat, I. A., Nazir, M. I., Ahmad, I., Pathakota, G.-B., Chanu, T. I., Goswami, M., Sundaray, J. K., & Sharma, R. (2018). Fabrication and characterization of chitosan conjugated eurycomanone nanoparticles: In vivo evaluation of the biodistribution and toxicity in fish. *International Journal of Biological Macromolecules, 112*, 1093–1103. https://doi.org/10.1016/j.ijbiomac.2018.02.067

Bhat, I. A., Ahmad, I., Mir, I. N., Bhat, R. A. H., Gireesh-Babu, P., Goswami, M., Sundaray, J., & Sharma, R. (2019a). Chitosan-eurycomanone nanoformulation acts on steroidogenesis pathway genes to increase the reproduction rate in fish. *The Journal of Steroid Biochemistry and Molecular Biology, 185*, 237–247. https://doi.org/10.1016/j.jsbmb.2018.09.011

Bhat, I. A., Ahmad, I., Mir, I. N., Yousf, D. J., Ganie, P. A., Bhat, R. A. H., Gireesh-Babu, P., & Sharma, R. (2019b). Evaluation of the in vivo effect of chitosan conjugated eurycomanone nanoparticles on the reproductive response in female fish model. *Aquaculture, 510*, 392–399. https://doi.org/10.1016/j.aquaculture.2019.06.002

Bowker, J., Trushenski, J. T., Gaikowski, M. P., et al. (2016). *Guide to using drugs, biologics, and other chemicals in aquaculture*. American Fisheries Society Fish Culture Section. [Cited 22 September 2023.] Available from http://www.fishculturesection.org/DrugGuide/Files/GUIDE_FEB_2011.pdf

Calo, J. R., Crandall, P. G., O'Bryan, C. A., & Ricke, S. C. (2015). Essential oils as antimicrobials in food systems – A review. *Food Control, 54*, 111–119.

Chellapandian, H., Jeyachandran, S., Rajalakshmi, Ilangovan, S., & Aseervatham, S. B. (2023). Nanochitosan for the production of more effective fish feed for aquaculture. In *Next generation nanochitosan* (pp. 339–348). Elsevier. https://doi.org/10.1016/B978-0-323-85593-8.00001-1

Dawood, M. A. O., Gewaily, M. S., Soliman, A. A., Shukry, M., Amer, A. A., Younis, E. M., Abdel-Warith, A.-W. A., Van Doan, H., Saad, A. H., Aboubakr, M., Abdel-Latif, H. M. R., & Fadl, S. E. (2020). Marine-derived chitosan nanoparticles improved the intestinal histo-morphometrical features in association with the health and immune response of grey mullet (Liza ramada). *Marine Drugs, 18*(12), 611. https://doi.org/10.3390/md18120611

De Oliveira, A. L. B., Cavalcante, F. T. T., Da Silva Moreira, K., Lima, P. J. M., De Castro Monteiro, R. R., Pinheiro, B. B., Dos Santos, K. P., & Dos Santos, J. C. S. (2021). Chitosan nanoparticle: Alternative for sustainable agriculture. In R. F. D. Nascimento, V. D. O. S. Neto, P. B. A. Fechine, & P. D. T. C. Freire (Eds.), *Nanomaterials and nanotechnology. Materials horizons: From nature to nanomaterials* (pp. 95–132). Springer. https://doi.org/10.1007/978-981-33-6056-3_4

Dubey, S., Avadhani, K., Mutalik, S., Sivadasan, S., Maiti, B., Girisha, S., Venugopal, M., Mutoloki, S., Evensen, Ø., Karunasagar, I., & Munang'andu, H. (2016). Edwardsiella tarda OmpA encapsulated in chitosan nanoparticles shows superior protection over inactivated whole cell vaccine in orally vaccinated fringed-lipped peninsula carp (Labeo fimbriatus). *Vaccine, 4*(4), 40. https://doi.org/10.3390/vaccines4040040

El-Naggar, M., Salaah, S., El-Shabaka, H., El-Rahman, F. A., Khalil, M., & Suloma, A. (2021). Efficacy of dietary chitosan and chitosan nanoparticles supplementation on health status of Nile tilapia, Oreochromis niloticus (L.). *Aquaculture Reports, 19*, 100628. https://doi.org/10.1016/j.aqrep.2021.100628

El-Naggar, M. M., et al. (2022). Effects of dietary supplementation of chitosan and chitosan nanoparticles on growth performance, fatty acids profile and liver histology of the Nile tilapia (Oreochromis niloticus) fed fish meal-free diets. *Egyptian Journal of Aquatic Biology and Fisheries, 26*(4), 427–448. https://doi.org/10.21608/ejabf.2022.251297

Fajardo, C., Martinez-Rodriguez, G., Blasco, J., Mancera, J. M., Thomas, B., & De Donato, M. (2022). Nanotechnology in aquaculture: Applications, perspectives and regulatory challenges. *Aquaculture and Fisheries, 7*(2), 185–200. https://doi.org/10.1016/j.aaf.2021.12.006

Fatahi, S., Zakeri, M., Varshosaz, J., & Yavari, V. (2018). Nanoencapsulation of Arginine by chitosan and its effect on amino acid survive in diet and growth Rainbow trout larvae (Onchorhynchus mykiss). *Journal of Aquaculture Development, 12*(3), 105–117.

Fathi, M., Mozafari, M. R., & Mohebbi, M. (2012). Nanoencapsulation of food ingredients using lipid based delivery systems. *Trends in Food Science & Technology, 23*(1), 13–27. https://doi.org/10.1016/j.tifs.2011.08.003

Freccia, A., de Sousa, S. M. N., Meurer, F., Butzge, A. J., Mewes, J. K., & Bombardelli, R. A. (2014). Essential oils in the initial phase of broodstock diets of Nile tilapia. *Revista Brasileira de Zootecnia, 43*, 1–7.

Gabriel, N. N., Habte-Tsion, H.-M., & Haulofu, M. (2022). Perspectives of nanotechnology in aquaculture: Fish nutrition, disease, and water treatment. In A. Krishnan, B. Ravindran, B. Balasubramanian, H. C. Swart, S. J. Panchu, & R. Prasad (Eds.), *Emerging nanomaterials for advanced technologies. Nanotechnology in the life sciences* (pp. 463–485). Springer. https://doi.org/10.1007/978-3-030-80371-1_15

George, D., Lakshmi, S., Sharma, A., Prakash, S., Siddiqui, M., Malavika, B. R., & Elumalai, P. (2023). Nanotechnology: A novel tool for aquaculture feed development. In A. V. Kirthi, K. Loganathan, & I. Karunasagar (Eds.), *Nanotechnological approaches to the advancement of innovations in aquaculture. Nanotechnology in the life sciences* (pp. 67–88). Springer. https://doi.org/10.1007/978-3-031-15519-2_4

González, J. D., Silva-Marrero, J. I., Metón, I., Caballero-Solares, A., Viegas, I., Fernández, F., Miñarro, M., Fàbregas, A., Ticó, J. R., Jones, J. G., & Baanante, I. V. (2016). Chitosan-mediated shRNA knockdown of cytosolic alanine aminotransferase improves hepatic carbohydrate metabolism. *Marine Biotechnology, 18*(1), 85–97. https://doi.org/10.1007/s10126-015-9670-8

Hernández, A., García García, B., Caballero, M. J., & Hernández, M. D. (2016). The inclusion of thyme essential oil in the feed of gilthead seabream (Sparus aurata) promotes changes in the frequency of lymphocyte aggregates in gut-associated lymphoid tissue. *Aquaculture Research, 47*(10), 3341–3345.

Homayonpour, P., Jalali, H., Shariatifar, N., & Amanlou, M. (2021). Effects of nano-chitosan coatings incorporating with free/nano-encapsulated cumin (Cuminum cyminum L.) essential oil on quality characteristics of sardine fillet. *International Journal of Food Microbiology, 341*, 109047. https://doi.org/10.1016/j.ijfoodmicro.2021.109047

Ibrahim, R. E., Amer, S. A., Farroh, K. Y., Al-Gabri, N. A., Ahmed, A. I., El-Araby, D. A., & Ahmed, S. A. A. (2021). The effects of chitosan-vitamin C nanocomposite supplementation on the growth performance, antioxidant status, immune response, and disease resistance of Nile tilapia (Oreochromis niloticus) fingerlings. *Aquaculture, 534*, 736269. https://doi.org/10.1016/j.aquaculture.2020.736269

Jiménez-Fernández, E., Ruyra, A., Roher, N., Zuasti, E., Infante, C., & Fernández-Díaz, C. (2014). Nanoparticles as a novel delivery system for vitamin C administration in aquaculture. *Aquaculture, 432*, 426–433.

Joye, I. J., Davidov-Pardo, G., & McClements, D. J. (2014). Nanotechnology for increased micronutrient bioavailability. *Trends in Food Science & Technology, 40*(2), 168–182.

Khani Oushani, A., Soltani, M., Sheikhzadeh, N., Shamsaie Mehrgan, M., & Rajabi Islami, H. (2020). Effects of dietary chitosan and nano-chitosan loaded clinoptilolite on growth and immune responses of rainbow trout (Oncorhynchus mykiss). *Fish & Shellfish Immunology, 98*, 210–217. https://doi.org/10.1016/j.fsi.2020.01.016

Kitiyodom, S., Yata, T., Yostawornkul, J., Kaewmalun, S., Nittayasut, N., Suktham, K., Surassmo, S., Namdee, K., Rodkhum, C., & Pirarat, N. (2019). Enhanced efficacy of immersion vaccination in tilapia against columnaris disease by chitosan-coated "pathogen-like" mucoadhesive nanovaccines. *Fish & Shellfish Immunology, 95*, 213–219. https://doi.org/10.1016/j.fsi.2019.09.064

Kulawik, P., Jamróz, E., & Özogul, F. (2019). Chitosan for seafood processing and preservation. In G. Crini & E. Lichtfouse (Eds.), *Sustainable agriculture reviews 36. Sustainable agriculture reviews* (pp. 45–79). Springer. https://doi.org/10.1007/978-3-030-16581-9_2

Kumaran, S. (2020). Trends in aquaculture feed development with chitosan nano particles– A review. *Bioscience Biotechnology Research Communications, 13*(1), 73–78. https://doi.org/10.21786/bbrc/13.1/13

Kumari, R., Gupta, S., Singh, A. R., Ferosekhan, S., Kothari, D. C., Pal, A. K., & Jadhao, S. B. (2013). Chitosan nanoencapsulated exogenous trypsin biomimics zymogen-like enzyme in fish gastrointestinal tract. *PLoS One, 8*(9), e74743. https://doi.org/10.1371/journal.pone.0074743

Lembang, M. S., Cahyani, R. T., & Nugraeni, C. D. (2023). Efektivitas Penambahan Nanokitosan dalam Pakan Terhadap Kelangsungan Hidup dan Pertumbuhanan Ikan Nila (Oreochromis niloticus). *Jurnal Sumberdaya Akuatik Indopasifik, 7*(1), 93–102. https://doi.org/10.46252/jsai-fpik-unipa.2023.Vol.7.No.1.265

León-Rodríguez, L., Luzardo-Álvarez, A., Blanco-Méndez, J., Lamas, J., & Leiro, J. (2013). Biodegradable microparticles covalently linked to surface antigens of the scuticociliate parasite P. dicentrarchi promote innate immune responses in vitro. *Fish & Shellfish Immunology, 34*(1), 236–243. https://doi.org/10.1016/j.fsi.2012.10.029

Li, L., Lin, S.-L., Deng, L., & Liu, Z.-G. (2013). Potential use of chitosan nanoparticles for oral delivery of DNA vaccine in black seabream *Acanthopagrus schlegelii* Bleeker to protect from *Vibrio parahaemolyticus*. *Journal of Fish Diseases, 36*(12), 987–995. https://doi.org/10.1111/jfd.12032

Mohammadkhan, F., Noori, N., Akhondzadeh Basti, A., Khanjari, A., Vanaki, E., & Farzaneh, M. (2022). The effects of nanochitosan coating integrated to Zataria multiflora Boiss and Polylophium Involucratum essential oils on the shelf-life extension of silver carp fillets. *Journal of Nutrition, Fasting and Health, 10*(3). https://doi.org/10.22038/jnfh.2022.67037.1399

Mohanty, B. (2015). Nutritional value of food fish. *Conspectus of Inland FisheriesManagement, 4*, 15–21.

Munro, L., & Gregory, A. (2009). Application of network analysis to farmed salmonid movement data from Scotland. *Journal of Fish Diseases, 32*(7), 641–644.

Murray, A. G. (2013). Epidemiology of the spread of viral diseases under aquaculture. *Current opinion in Virology, 3*(1), 74–78.

Muruganandam, M., Chipps, S., & Ojasvi, P. (2019). On the advanced technologies to enhance fisheries production and management. *Acta Scientific Agriculture, 3*(8), 216–222.

Naiel, M. A. E., Ismael, N. E. M., Abd El-hameed, S. A. A., & Amer, M. S. (2020). The antioxidative and immunity roles of chitosan nanoparticle and vitamin C-supplemented diets against imidacloprid toxicity on Oreochromis niloticus. *Aquaculture, 523*, 735219. https://doi.org/10.1016/j.aquaculture.2020.735219

Nasr-Eldahan, S., Nabil-Adam, A., Shreadah, M. A., Maher, A. M., & El-Sayed Ali, T. (2021). A review article on nanotechnology in aquaculture sustainability as a novel tool in fish disease control. *Aquaculture International, 29*(4), 1459–1480. https://doi.org/10.1007/s10499-021-00677-7

Nikapitiya, C., Dananjaya, S. H. S., De Silva, B. C. J., Heo, G.-J., Oh, C., De Zoysa, M., & Lee, J. (2018). Chitosan nanoparticles: A positive immune response modulator as display in zebrafish larvae against Aeromonas hydrophila infection. *Fish & Shellfish Immunology, 76*, 240–246. https://doi.org/10.1016/j.fsi.2018.03.010

Rajesh Kumar, S., Ishaq Ahmed, V. P., Parameswaran, V., Sudhakaran, R., Sarath Babu, V., & Sahul Hameed, A. S. (2008). Potential use of chitosan nanoparticles for oral delivery of DNA vaccine in Asian sea bass (Lates calcarifer) to protect from Vibrio (Listonella) anguillarum. *Fish & Shellfish Immunology, 25*(1–2), 47–56. https://doi.org/10.1016/j.fsi.2007.12.004

Rajeshkumar, S., Venkatesan, C., Sarathi, M., Sarathbabu, V., Thomas, J., Anver Basha, K., & Sahul Hameed, A. S. (2009). Oral delivery of DNA construct using chitosan nanoparticles to protect the shrimp from white spot syndrome virus (WSSV). *Fish & Shellfish Immunology, 26*(3), 429–437. https://doi.org/10.1016/j.fsi.2009.01.003

Ramezani, Z., Zarei, M., & Raminnejad, N. (2015). Comparing the effectiveness of chitosan and nanochitosan coatings on the quality of refrigerated silver carp fillets. *Food Control, 51*, 43–48.

Rathor, P. K., Bhat, I. A., Rather, M. A., Gireesh-Babu, P., Kumar, K., Purayil, S. B. P., & Sharma, R. (2017). Steroidogenic acute regulatory protein (StAR) gene expression construct: Development, nanodelivery and effect on reproduction in air-breathing catfish, Clarias batrachus. *International Journal of Biological Macromolecules, 104*, 1082–1090. https://doi.org/10.1016/j.ijbiomac.2017.06.104

Rodrigues, S. M., Demokritou, P., Dokoozlian, N., Hendren, C. O., Karn, B., Mauter, M. S., Sadik, O. A., Safarpour, M., Unrine, J. M., & Viers, J. (2017). Nanotechnology for sustainable food production: Promising opportunities and scientific challenges. *Environmental Science: Nano, 4*(4), 767–781.

Rosidah, & Mulyani, Y. (2022). A mini-review: The role of chitosan in aquaculture fish health management. *Asian Journal of Fisheries and Aquatic Research*, 24–31. https://doi.org/10.9734/ajfar/2022/v17i330405

Saleh, M., Essawy, E., Shaalan, M., Osman, S., Ahmed, F., & El-Matbouli, M. (2022). Therapeutic intervention with dietary chitosan nanoparticles alleviates fish pathological and molecular systemic inflammatory responses against infections. *Marine Drugs, 20*(7), 425. https://doi.org/10.3390/md20070425

Sari, A. M., Hendrawati, T. Y., Erdawati, & Ramadhan, A. I. (2020). Comparison of effect of chitosan and chitosan nano emulsion on growth performance, water quality and protein retention of Koi fish (Cyprinus Caprio Koi). *Journal of Physics: Conference Series, 1517*(1), 012032. https://doi.org/10.1088/1742-6596/1517/1/012032

Selvasudha, N., Sweety, J. P., Dhanalekshmi, U. M., & Ruckmani, K. (2022). Chitosan-based nanocomposite as targeted drug delivery carrier. In S. Jana & S. Jana (Eds.), *Marine biomaterials* (pp. 97–137). Springer. https://doi.org/10.1007/978-981-16-4787-1_4

Shah, B. R., & Mraz, J. (2020). Advances in nanotechnology for sustainable aquaculture and fisheries. *Reviews in Aquaculture, 12*(2), 925–942. https://doi.org/10.1111/raq.12356

Shakour, N., Khoshkhoo, Z., Basti, A. A., Khanjari, A., & Shotorbani, P. M. (2021). Integration of nanochitosan and Ziziphora clinopodioides essential oil into poly lactic acid films; a new method for extending the shelf life of Oncorhynchus mykiss fillets. *Journal of Food Measurement and Characterization, 15*(4), 2922–2931. https://doi.org/10.1007/s11694-020-00779-5

Sun, B., Quan, H., & Zhu, F. (2016). Dietary chitosan nanoparticles protect crayfish Procambarus clarkii against white spot syndrome virus (WSSV) infection. *Fish & Shellfish Immunology, 54*, 241–246. https://doi.org/10.1016/j.fsi.2016.04.009

Tandberg, J., Lagos, L., Ropstad, E., Smistad, G., Hiorth, M., & Winther-Larsen, H. C. (2018). The use of chitosan-coated membrane vesicles for immunization against salmonid rickettsial septicemia in an adult zebrafish model. *Zebrafish, 15*(4), 372–381. https://doi.org/10.1089/zeb.2017.1556

Tayel, A. A., Elsaied, B. E., & Diab, A. M. (2019). Nanotechnology for aquaculture. In R. N. Pudake, N. Chauhan, & C. Kole (Eds.), *Nanoscience for sustainable agriculture* (pp. 479–544). Springer. https://doi.org/10.1007/978-3-319-97852-9_20

Udo, I. U., Etukudo, U., & Anwana, U.-I. U. (2018). Effects of chitosan and chitosan nanoparticles on water quality, growth performance, survival rate and meat quality of the African catfish, Clarias gariepinus. *Nanoscience, 1*(1), 12–25.

Victor, H., Zhao, B., Mu, Y., Dai, X., Wen, Z., Gao, Y., & Chu, Z. (2019). Effects of Se-chitosan on the growth performance and intestinal health of the loach Paramisgurnus dabryanus (Sauvage). *Aquaculture, 498*, 263–270. https://doi.org/10.1016/j.aquaculture.2018.08.067

Wisdom, K. S., Bhat, I. A., Kumar, P., Pathan, M. K., Chanu, T. I., Walke, P., & Sharma, R. (2018). Fabrication of chitosan nanoparticles loaded with aromatase inhibitors for the advancement of gonadal development in Clarias magur (Hamilton, 1822). *Aquaculture, 497*, 125–133. https://doi.org/10.1016/j.aquaculture.2018.07.049

Wu, Y., Rashidpour, A., Almajano, M. P., & Metón, I. (2020). Chitosan-based drug delivery system: Applications in fish biotechnology. *Polymers, 12*(5), 1177. https://doi.org/10.3390/polym12051177

Xia, I. F., Cheung, J. S., Wu, M., Wong, K.-S., Kong, H.-K., Zheng, X.-T., Wong, K.-H., & Kwok, K. W. (2019). Dietary chitosan-selenium nanoparticle (CTS-SeNP) enhance immunity and disease resistance in zebrafish. *Fish & Shellfish Immunology, 87*, 449–459. https://doi.org/10.1016/j.fsi.2019.01.042

Zeppenfeld, C., Hernández, D., Santinón, J., Heinzmann, B., Da Cunha, M., Schmidt, D., & Baldisserotto, B. (2016). Essential oil of A loysia triphylla as feed additive promotes growth of silver catfish (Rhamdia quelen). *Aquaculture Nutrition, 22*(4), 933–940.

Zheng, F., Liu, H., Sun, X., Zhang, Y., Zhang, B., Teng, Z., Hou, Y., & Wang, B. (2016). Development of oral DNA vaccine based on chitosan nanoparticles for the immunization against reddish body iridovirus in turbots (Scophthalmus maximus). *Aquaculture, 452*, 263–271. https://doi.org/10.1016/j.aquaculture.2015.11.013

Feed Enhancement and Nutrition

Patrick Omoregie Isibor, Onwaeze Ogochukwu Oritseweyinmi, Kayode-Edwards Ifeoluwa Ihotu, and Oyewole Oluwafemi Adebayo

Contents

P. O. Isibor (✉) · O. O. Oritseweyinmi · K.-E. I. Ihotu
Department of Biological Sciences, College of Science and Technology, Covenant University, Ota, Ogun State, Nigeria
e-mail: patrick.isibor@covenantuniversity.edu.ng

O. O. Adebayo
Department of Microbiology, Federal University of Technology, Minna, Nigeria

African Center of Excellence for Mycotoxin and Food Safety, Federal University of Technology, Minna, Nigeria

P. O. Isibor et al. (eds.), *Nanochitosan-Based Enhancement of Fisheries and Aquaculture*, https://doi.org/10.1007/978-3-031-52261-1_8

1 Introduction

The world is on the precipice of a technological revolution, and creating dynamic change, eradicating anthropogenic footprint and achieving sustainable living are more achievable than ever. These goals are embodied by and outlined in the United Nations' 2030 manifesto, the sustainable development goals (Persaud & Dagher, 2021). The objective of goal 14 is aimed at preserving aquatic habitats and sustainably utilising the resources; particularly fish and shellfish. Its pursuit and attainment entail mitigation of pollution, promotion of sustainable fishing practices and safeguarding of aquatic ecosystems. According to Persaud and Dagher (2021), SDG 14 encompasses several significant concerns, such as overfishing, water pollution, the rising ocean pH and the imperative to create safe zones for endangered species. As recently outlined by several reports, these challenges threaten the aquatic population as they strive to balance satisfying increasing economic demands and maintaining the fragile aquatic and global ecosystem (Le Blanc et al., 2017; Ye & Gutierrez, 2017; Widjaja et al., 2020).

Aquaculture is a rapidly advancing practice crucial in addressing the worldwide need for nutrition. Its output has leapt over the years, from approximately 35 million metric tons at the start of the century, which accounted for about 30% of the overall fish output, to over 100 million tonnes by 2014 (Wattanakul et al., 2017). At this time, fish farming comprised approximately 52% of the total fish output, a little under 200 million tonnes (Asche et al., 2018). By 2016, fish production soared to about 171 million tons, more than tripled in the last 6 years (Valério et al., 2023). Based on the findings of Giri (2019), this rise was valued at $362 billion, of which around 47% pertains to aquacultural output. Overfishing is exacerbated by intensive aquatic farming operations, which necessitate even higher amounts of wild fish for the production of fishmeal used in the cultivation of farmed fish. Despite these reports, there is a persistent upward trend in global fish consumption (Szuwalski et al., 2020; Naylor et al., 2021), particularly in underdeveloped nations where fish serves as a crucial component of animal-derived protein (Pradeepkiran, 2019).

Based on the findings reported by Prato et al. (2019), health-conscious consumers perceive seafood to be a nutritionally rich and well-rounded dietary option. Seafood is chiefly valued for its abundance of superior protein and beneficial, healthy dietary fat, which can improve wellness, particularly by mitigating the likelihood of developing heart and coronary ailments (Maesano et al., 2020). Furthermore, it should be noted that marine-derived meals possess a high digestibility rate and serve as a noteworthy reservoir of functional nutrients, such as omega-3 s, vitamins C and D, proteins, algal constituents, selenium, antioxidants and chitin (Aware et al., 2022; Aumeerun et al., 2022). Coupling this attractiveness with the exponentially rising global population, it is reasonable to anticipate a corresponding escalation in consumer demand in the coming years (Schar et al., 2020; Mobsby et al., 2020). Therefore, to ensure the long-term growth of aquaculture, it is imperative to explore novel dietary options for farmed fishes that may effectively enhance their metabolic and physiological activities and, ultimately boost the output (Mobsby et al., 2020). This report explores the potential of a promising nanotech-based approach for tackling malnutrition in aquaculture using nanochitosan-enhanced feed. The role of nanochitosan in enhancing aquaculture feed is crucial for optimised nutrition.

2 Nanochitosan as a Feed Additive for Improved Fish Nutrition

In recent years, there has been a substantial effort to identify cost-effective and eco-friendly methods for promoting the welfare of aquatic animals. Researchers have primarily concentrated on the use of various feed additives, including pro- and prebiotics (Amenyogbe et al., 2020; Mugwanya et al., 2021; del Valle et al., 2023), botanical extracts (Kuebutornye & Abarike, 2020; Tadese et al., 2022), nucleic acids (Flegel, 2019; Charoonnart et al., 2023) and immune-stimulating agents (Flegel, 2019; Vijayaram et al., 2022). These supplements exhibit a wide range of biological features, and their incorporation into feed is usually intended to serve a particular objective. It is well acknowledged that chitin and chitosan, derived from Crustaceans, hold significant promise as effective sources of functional nutritional supplements (Abere et al., 2022; Dashputre et al., 2023). According to some reports, functional feed positively impacts the composition of the bacteria in the animals' gut (Mugwanya et al., 2021), contributing to the enhancement of general wellness. Functional feed additives are widely regarded as potential substitutes for antibacterial and similar medications. According to previous studies (Shekarabi et al., 2022), they enhance development, bolster immunological reaction and improve immunity to infection. Recently, nanochitosan and nanochitosan-composite functional feeds have gained notoriety, as they are repeatedly demonstrated as viable solutions for producing sustainable functional feeds (Yadav et al., 2019; Santos et al., 2020; Pakizeh et al., 2021). The implementation of this technology in the field of fish farming has garnered significant attention due to its limited adverse effects, ability to promote growth and development, and capacity to strengthen innate immunity.

Formulation of optimised feed for enhanced nutrition remains a knowledge gap that hinders breakthroughs in the aquaculture sector, thus mitigating the attainment of SDG 14 (Udo et al., 2018; Asche et al., 2022). These feeds often consist of a diverse range of nutrients designed to meet the dietary demands of the animals, enabling them to carry out their regular activities (Encarnação, 2016; Gamboa-Delgado, 2022). In commercial feeds, it is imperative to preserve robust innate immunity, efficient development and successful breeding (Hu et al., 2021; Chen et al., 2023). To facilitate the ingestion, digestion, absorption and transportation of nutrients, there is a growing utilisation of a wide range of synthetic enhancements in aquatic diets (Nankervis & Jones, 2022). These supplements are functional beyond fulfilling essential dietary needs in the intended organisms, enhancing development and optimising feed utilisation (Asche et al., 2022; Chen et al., 2023). Additionally, these substances can promote fish wellness and bolster their ability to withstand stressors. According to the research conducted by Aheto et al. (2019) and

Khan et al. (2021), the expenses associated with these particular kinds of feeds account for approximately half of all operational costs in fish farming. This financial burden often serves as a significant obstacle for potential investors considering venturing into the industry). This could also deter smaller-scale farmers from meeting the demands of aquaculture nutritional needs (Camp et al., 2020). Consequently, numerous research endeavours have been undertaken to substitute or diminish fishmeal content in dietary compositions by incorporating other forms of protein that are characterised by their cost-effectiveness, renewability, intrinsic value and biological sustainability (Lowry et al., 2019; Rambaran & Schirhagl, 2022; Malik et al., 2023; Chen et al., 2023). Nanotechnology can enhance the absorption capabilities of fish, enabling them to effectively uptake medications such as hormones, vaccinations and nourishment (Bayda et al., 2019; Fajardo et al., 2022). Nanochitosan is a bioengineered instrument that exhibits seamless adaptability across several functional domains (Saeedi et al., 2022). Due to their considerable surface area and compatibility with biological systems, they can serve as biochemical transporters, facilitating the delivery of essential nutrients to cells in need, enhancing the protective capabilities of aquatic organisms and even coordinating intricate chemical processes (Ahmed et al., 2021a, b).

The impact of Nanochitosans on the nutritional quality of aquaculture feeds is extensive. Their surface area enhances surface interaction between nutrients and the gastrointestinal tract (Okeke et al., 2022; Ahmed et al., 2023). This enhanced surface interaction augments nutrient assimilation, promoting improved development and overall well-being of the aquatic organisms (Uyanga et al., 2023). Recent studies have also demonstrated the capacity of nanochitosan particles to function as a safeguarding vehicle for delicate nutrients (Nouri, 2019; Kaboudi et al., 2023). Through encapsulation and stabilisation, nanochitosan complements bioactive components in feed such that they can effectively avoid breaking down through storage (Yadav et al., 2022). This guarantees the preservation of vital nutrients such as proteins and vitamins, enhancing their nutrient uptake in farmed organisms. Nanochitosan also exhibits immune-modulating qualities that facilitate regulating immune responses in aquatic organisms (El-Naggar et al., 2022a, b; Chellapandian et al., 2023). They can augment the body's defences of aquatic species, bolstering their ability to withstand infections and stressful stimuli. This, in turn, can prevent overreliance on antibiotics or alternative therapeutic interventions.

3 Benefits in Enhancing Fish Growth, Immune Response and Stress Tolerance

The incorporation of nanochitosan as part of an aquaculture feed might be seen as a strategic approach in line with the objectives of the sustainable development goals (Isibor et al., 2023). Chitosan, which comes from renewable sources, possesses biodegradable properties, rendering it a sustainable option from an environmental standpoint (Funes et al., 2023). Enhancing the dietary effectiveness of feed has an opportunity to decrease the quantity of feed necessary for achieving peak development, hence mitigating waste generation and minimising ecological impacts (Madhu et al., 2022; Funes et al., 2023). Additionally, the impact of the quality of feed produced is critical in aquaculture, as feed interacts with the digestive system of many animals and leads to direct economic and environmental consequences (Diwan et al., 2022). The status of gut health is strongly linked with a properly regulated microbial community in the intestines, which plays a crucial role in facilitating digestion and absorption while also providing protection from harmful infections for the victim organism (Yukgehnaish et al., 2020). Multiple investigations additionally demonstrate that variations in feed components and alterations in diet can exert an impact on the anatomy of the gastrointestinal tract and the equilibrium of gut microorganisms, hence impacting gastrointestinal and consumption processes (El-Saadony et al., 2022; Diwan et al., 2022; Chen et al., 2022; Sumithra et al., 2022). Nanochitosan and nanochitosan composites present an attractive and commercially viable approach to producing feed (Funes et al., 2023), enriching the digestive system (Ahmed et al., 2021a, b), improving organism growth and health (Abd El-Naby et al., 2019, 2020) and preserving the environment (Isibor et al., 2023).

4 Nanochitosan-Enhanced Feed for Fish Growth

In several extensive empirical investigations, which factor different aquatic conditions (Chen et al., 2014; Zaki et al., 2015), different modes of introduction (El-Naggar et al., 2021), different nutritional constituents and different types of concentration (Ali et al., 2021), the role of nanochitosan in promoting development of aquatic species has been established as a crucial factor in fish growth and development. Based on the findings reported by Hamidian et al. (2018), it has been shown that the primary impact of nanochitosan is the enhancement of the structural composition of the gastrointestinal tract. This improvement has the potential to influence the uptake of nutrients positively and, therefore, enhance the rate of development. According to research findings, using the nanochitosan-based feed in regulated diets has been seen to enhance the length of small intestinal villi while reducing the depth of their crypts (Olaniyan et al., 2023). According to the study conducted by Attaran Dowom et al. (2022), even when present at minimal concentrations, non-protein nitrogen compounds were able to improve nitrogen utilisation and promote the breakdown of amino acids. Nanochitosan is employed to enhance the absorption rate of nutrients (Moges et al., 2020; Ahuekwe et al., 2023). Additionally, they serve as supplements and stabilisers for organic materials and dietary elements that possess restricted bioavailability (Olaniyan et al., 2023).

These changes are consistent in several aquatic organisms. The inclusion of nanochitosan in diets resulted in enhanced development, longer lifespan and superior produce quality in a variety of aquatic organisms. Freshwater fishes that have demonstrated positive growth responses to dietary nanochitosan include Grey mullet: *Liza Ramda* (Dawood et al., 2020); Grey Mullet: *Mugil cephalus* (Akbary &

Younesi, 2017); Kelp grouper: *Epinephelus bruneus* (Harikrishnan et al., 2012); Nile tilapia: *Oreochromis niloticus* (Abd El-Naby et al., 2019, 2020); Rainbow trout: *Oncorhynchus mykiss* (Hamidian et al., 2018); Common carp: *Cyprinus carpio* (Mishra et al., 2023; Rad et al., 2014); African catfish: *Clarias gariepinus* (Udo et al., 2018); Tilapia: *Oreochromis nilotica* (Wang & Li, 2011; Wu, 2020); Pacific white shrimp: *Litopenaeus vannamei* (Chantarasataporn et al., 2013); Koi (Cyprinus carpio koi); Loach fish: *Misgurnus anguillicadatus* (Chen & Chen, 2019); Fish fingerlings: *Labeo rohita* (Ferosekhan et al., 2014); and Silver carp: *Hypophthalmichthys molitrix* (Younus et al., 2020).

Saltwater organisms that have demonstrated positive growth responses to dietary nanochitosan include Whiteleg shrimp: *Litopenaeus vannamei* (Asaikkutti et al., 2023); Turbot: *Scophthalmus maximus*; Sea bass: *Dicentrarchus labrax* olive flounder: *Paralichthys olivaceus* (Chellapandian et al., 2023); Asian seabass: *Lates calcarifer* (El-Bab, 2022); Black tiger shrimp: *Penaeus monodon* (Abdel-Warith et al., 2020); Juvenile tiger shrimp: *Penaeus monodon Fabricius* (Niu et al., 2015; Rochana et al., 2019); White leg shrimp: *Litopenaeus vannamei* (Rad et al., 2014); Shrimp: *Penaeus semisulcatus* (Taher et al., 2017); Juvenile yellow catfish: *Pelteobagrus fulvidraco* (Li et al., 2022); Juvenile gibel carp: *Carassius auratus gibelio* (Chen et al., 2014); Olive flounder (Paralichthys olivaceus) (Chellapandian et al., 2023); and gibel carp (Chen et al., 2014). Interestingly, according to Younus et al. (2020) and Abdel-Ghany and Salem (2020), the efficacy of chitosan and nanochitosan particles in encouraging maximal development is highly species specific.

Conversely, some earlier studies observed that the addition of nanochitosan to diets did not have a significant impact on the development efficiency of certain organisms (Kono et al., 1987*).* According to Victor et al. (2019), including selenium-nanochitosan composites at varying doses through 2 months did not culminate in any discernible alterations in the development of loaches. The utilisation of nanochitosan as a supplementary addition in certain feeds even resulted in a decline in the rate of tilapia growth, as shown by Shiau and Yu (1999). Zaki et al. (2015) provided a possible explanation for this when they described the action of nanochitosan particles in terms of their capacity to enhance intestinal health and facilitate the development of microvilli, stating that it results in an expansion of the absorptive surface area in these areas. A potential correlation exists between a greater quantity of nanochitosan in feed and decreased development efficiency, which may be attributed to bowel obstruction from excessive microvilli expansion. Though overwhelming evidence supports the use of nanochitosan-based feed, it is essential to explore the potential adverse effects associated with nanochitosan use to evaluate its impact fully.

5 Nanochitosan-Enhanced Feed for Immune Response

The utilisation of conventional antibiotics or vaccinations to prevent illness has been found to have adverse effects on aquatic life health and their surroundings (Aly & Albutti, 2014; Assefa & Abunna, 2018). Nanoparticles exhibiting potent antibacterial properties have become recognised as a highly sophisticated and viable alternative to replace antibiotics (Franci et al., 2015; El-Naggar et al., 2021). These nanoparticles have been found to effectively battle several ailments caused by various viruses, fungi and other pathogens (Ahmed et al., 2021a, b; Li et al., 2022). Multiple parameters are employed when assessing the immunological impact of a substance, including cytokine and lysozyme activity, antibody production and leukocyte count, the ratio of protein: albumin: globulin in the blood, tumour regression and immune gene expression (Khieokhajonkhet et al., 2022). Many empirical studies have demonstrated the value of nanochitosan particles on all counts using various controls and in many organisms (Niu et al., 2015; Chellapandian et al., 2023). Various studies also demonstrate increased white blood cells and phagocytic ability in nanochitosan-enhanced feed (Abd El-Naby et al., 2019; Dawood et al., 2020).

Dawood et al. (2020) investigated the impact of enzymes on immunity and infection prevention through the disruption of microbial peptidoglycan structures. The addition of anochitosan particles stimulated the reduction in the average microbial population within the gastrointestinal tract. The reduction was correlated with the quantity of incorporated nanochitosan, as previously documented by Hamidian et al. (2018). Furthermore, the administration of chitosan nanoparticles led to a significant elevation in circulating protein levels, as indicated in the studies conducted by Zaki et al. (2015) and Abd El-Naby et al. (2020). This increase in protein concentration had a substantial impact on the survival rates of various species, ultimately enhancing their lifespans.

Building upon these findings, a study by Xu et al. (2023) demonstrated that the inclusion of chitosan nanoparticles in the diet can enhance immune function through several mechanisms. These mechanisms include an increase in the production of immune cells, the promotion of immune-supporting regions such as the spleen, and an influence on the composition of gut microbiota. Rochana et al. (2019) inferred that nanochitosan exhibits immune-enhancing properties, likely attributed to its ability to stimulate innate immune responses.

The literature thus suggests that nanochitosan and chitosan nanoparticles have promising immunomodulatory effects, making them potential candidates for further exploration in the field of immune system enhancement and infection prevention.

Aquatic organisms that have demonstrated positive immune responses to dietary nanochitosan enhancement include Grey Mullet: *Mugil cephalus* (Akbary & Younesi, 2017); Common carp: *Cyprinuscarpio L.* (Ali et al., 2021); Whiteleg shrimp: *Litopenaeus vannamei* (Asaikkutti et al., 2023); Kelp grouper: *Epinephelus bruneus* (Harikrishnan et al., 2012); Nile tilapia: *Oreochromis niloticus* (Abd El-Naby et al., 2019; Abdel-Tawwab et al., 2019); Rainbow trout: *Oncorhynchus mykiss* (Ahmed et al., 2021a, b), Olive flounder: *Paralichthys olivaceus* (Chellapandian et al., 2023); and Koi: Cyprinus carpio koi.

Some fishes exhibited enhanced protection against viral infections when fed with chitosan nanoparticles, resulting in notably increased survival (Robinson et al., 2014). Some researchers discovered that the inclusion of a Selenium-nanochitosan composite resulted in enhanced immune responses in the kelp grouper (Chellapandian et al., 2023). This enhancement was attributed to a surge in antibody production and activity (Wu et al., 2020). Other organisms that have demonstrated positive immune responses to dietary nanochitosan enhancement include juvenile tiger shrimp: *Penaeus monodon Fabricius* (Niu et al., 2015; Rochana et al., 2019); African catfish: *Clarias gariepinus* (Udo et al., 2018); Shrimp: *Penaeus semisulcatus* (Taher et al., 2017); Tilapia: *Oreochromis nilotica* (Wang & Li, 2011; Wu, 2020); Juvenile yellow catfish: *Pelteobagrus fulvidraco* (Li et al., 2022); and juvenile gibel carp: *Carassius auratus gibelio* (Chen et al., 2014). Low concentrations (less than 1%) of nanochitosan were found to improve the haematological composition and reduce controlled survival caused by imposed external stressors. Syed Raffic Ali et al. (2017) fed *Lates calcarifer* with meals containing varying levels of nanochitosan supplementation for 2 months. The findings revealed that the set of animals with the second-highest concentration of nanochitosan had the most elevated haematological and innate immunological markers on day 45 of the experiment. However, further investigations are necessary to understand the immunological systems associated with nanochitosan particles comprehensively.

6 Nanochitosan-Enhanced Feed for Stress Tolerance

In several extensive empirical investigations, which factor different aquatic conditions, different modes of introduction, different nutritional constituents and different types of concentration, nanochitosan-enhanced feed frequently demonstrates its potential to improve resilience to stress in a variety of aquatic species (El-Naggar et al., 2021; Ibrahim et al., 2021). Stressors may manifest as a result of various causes, including but not limited to alterations in the condition of the water, management procedures, transportation methods and contamination by diseases (Zhang et al., 2021). Managing these conditions is relevant for the long-term preservation of these organisms and the aquaculture practice.

The significant antioxidant effects of nanochitosan can be attributed to its structure and the intrinsic antioxidant properties of chitosan (Liang et al., 2021). This phenomenon can potentially assist aquatic species in mitigating the detrimental effects of oxidative stress induced by many circumstances, such as water contamination or prolonged contact with parasites (Zhang et al., 2021). Nanochitosan can rummage and chelation of free radicals by electron transfer, thereby mitigating cell destruction and enhancing general health. This is ascribed to its chelation function and capacity to scavenge free radicals through the transfer of electrons by its hydroxyl group (El-Naggar et al., 2021). Nanochitosan-based feeds have been shown to increase antioxidant activity in Nile tilapia: *Oreochromis niloticus* (Abd El-Naby et al., 2019, 2020); White leg shrimp: *Litopenaeus vannamei* (Rad et al.,

2014); *Salvia abrotanoide* (Attaran Dowom et al., 2022); Sea bass: *Dicentrarchus labrax* Shrimp (Chellapandian et al., 2023); Juvenile yellow catfish: *Pelteobagrus fulvidraco* (Li et al., 2022); and Grey mullet: *Liza Ramda* (Dawood et al., 2020).

The occurrence of inflammation is a prevalent physiological reaction to stressful circumstances. The anti-inflammatory capabilities of Nanochitosan have the potential to ameliorate this reaction, hence decreasing organ destruction and the accompanying physiological stress. Nanochitosan-based feeds have been shown to reflect anti-inflammatory properties in Tilapia: *Oreochromis nilotica* (Wang & Li, 2011; Wu, 2020); Loach fish: *Misgurnus anguillicadatus* (Chen & Chen, 2019); Nile tilapia: *Oreochromis niloticus* (Abd El-Naby et al., 2019, 2020); Rainbow trout: *Oncorhynchus mykiss* (Hamidian et al., 2018); and African catfish: *Clarias gariepinus* (Udo et al., 2018).

Nanochitosan possesses the ability to effectively uphold consistent parameters about water quality through its flocculation properties and impurity adsorption capabilities (Isibor et al., 2023). Maintaining these conditions is of utmost importance in aquaculture, as abrupt fluctuations in temperature, acidity or nutrient concentrations can have deleterious effects on aquatic species (Wu et al., 2020). According to recent field test reports, the reduced particle size of nanochitosan derived from crustaceans results in a greater number of interactions with contaminants (El-Naggar et al., 2021). This enhanced interaction facilitates a more effective adsorption process and subsequent eradication of these pollutants. Aquaculture directly improves the environmental conditions and lowers stress in farmed organisms (Abd El-Naby et al., 2020).

7 Formulation of Nanochitosan-Incorporated Feeds

The formulation of feeds enhanced with nanochitosan necessitates a thorough analysis of several critical variables to guarantee the feed's efficacy and performance in diverse aquaculture applications.

7.1 Cost-to-Benefit Analysis and Scalability

This is perhaps the most relevant consideration in feed development as it concerns commercial aquaculture practice, as it enables administrators to make well-informed and logical choices while proficiently allocating limited resources (Ranjan et al., 2014). The economic viability and potential environmental advantages of nanochitosan in feed formulation are still not established (Chellapandian et al., 2023). Taking into account the advantages mentioned earlier and acknowledging the greater cost per unit of Nanomaterials compared to conventional feed, it is highly probable that there exists an ideal amount of nanochitosan in any feed that maximises income. For instance, when reporting on nanofertilisers, Xu et al. (2023)

assert that by considering a 20% increase in crop production and the associated cost per unit, its implementation can contribute over $130 per hectare to the income generated from maize. It is imperative for developers to carefully evaluate the financial implications of refining nanochitosan-infused feed, adhering to regulatory requirements, and managing the implementation process. This assessment should be conducted concerning the advantages and prevailing market demand patterns.

7.2 Environmental Impact

Intensive aquaculture operations generate a significant quantity of organic waste, comprising both particulate matter and harmful chemicals (Schumann & Brinker, 2020). These waste products contribute to an elevated BOD and the spread of soluble nutrients. Any surplus nutrients that are not utilised can cause eutrophication when released into the surrounding environment, as they necessitate either assimilation or accumulation (Isibor et al., 2023). The potential for a nutrient to function as a pollutant within an aquatic system is contingent upon several factors, including its status as a limiting nutrient within a specific habitat, as well as the amount present and the overall capacity of the aquatic ecosystem in question (Akoma, 2023).

Research has shown that the qualities of nanochitosan have the potential to decrease the reliance on detrimental additives containing chemicals, such as fertilisers and antibiotics, within the field of aquaculture (Moges et al., 2020; El-Naggar et al., 2021). Nevertheless, its introduction into water systems may adversely affect other organisms or cause disturbances in aquatic ecosystems. Developers must evaluate the environmental ramifications associated with using nanochitosan in feeds. This assessment should examine its potential implications on water safety and the broader aspect of sustainability.

7.3 Source and Quality

Previous research has demonstrated that the purity and size of nanochitosan particles vary depending on factors such as the source of the material and the specific production technique (Isibor et al., 2023). Consequently, these factors are significant considerations and warrant careful study. Crustacean exoskeletons are employed mainly due to their abundant chitin composition (Benettayeb et al., 2023). The selection of the source material can have an impact on the degree of purity and specific attributes of chitin, thereby influencing the overall quality of the resulting nanochitosan (Benettayeb et al., 2023).

Additionally, it is crucial to consider the financial implications and long-term viability of the chitosan origin. The implementation of environmentally friendly and ethical sourcing techniques is vital to both reduce the environmental consequences associated with chitin extraction and guarantee its long-term supply (Chellapandian et al., 2023). Moreover, certain sources may exhibit a higher degree of cost-effectiveness relative to others, rendering them more appropriate for particular applications.

7.4 Target Organism

Organisms have diverse nutritional specifications and digestive capacities. It is crucial to consider the individual nutritional requirements of the target species when customising the diet incorporating nanochitosan (Chellapandian et al., 2023). In the process of feed development, developers must take into account several parameters, including but not limited to intestinal enzymes, stomach acidity and digestive capacity (El-Naggar et al., 2021).

The process of nutrient absorption is of great importance in the functioning of the digestive system in the organism under study (Hamidian et al., 2018). The formulation of nanochitosan should be optimised to enhance its digestibility and absorption efficiency in the target species. Certain species may exhibit a higher degree of selectivity in their feeding behaviours (Moges et al., 2020), emphasising the significance of considering the flavour and consistency of nanochitosan-based substances. Moreover, the growth rates and nutritional preferences of organisms might exhibit variability contingent upon their life stage (Arechavala-Lopez et al., 2022). Nanochitosan-based feed may need to be altered accordingly.

7.5 Nutritional Composition and Digestibility

Formulating feed is multifaceted, and it entails creating a well-balanced combination of different components to fulfil the precise nutritional requirements of the intended animals. Nutrient content is of primary significance in producing feed enhanced with nanochitosan (Wu et al., 2020). A properly balanced diet plays a crucial role in facilitating the developmental process of cultured organisms (Arechavala-Lopez et al., 2022). The formulation of feed containing nanochitosan must be optimised to offer a suitable composition of macro and micronutrients to fulfil the specific dietary needs of the targeted species (Wattanakul et al., 2017).

Developers must conduct a comprehensive evaluation of the composition of the feed and determine if the inclusion of nanochitosan is compatible with the preexisting ingredients. Nanochitosan possesses inherent nutrients, so developers need to consider modifying the nutritional composition of the feed to render it complementary (Avila-Quezada et al., 2022). The efficiency of nutrient absorption and use by organisms is influenced by the bioavailability of nutrients (Wang et al., 2022). Hence, caution should be exercised while introducing nanochitosan into feed to ensure that it does not impede the bioavailability and digestion of other feed constituents. The enhanced feed should be designed in a manner readily assimilated by the intended species, considering their specific digestive physiology (Ranjan et al., 2014).

7.6 Bioavailability

The term 'bioavailability' refers to the extent and rate at which a substance, such as a drug or nutrient, is absorbed and becomes available for the intended organisms (Wang et al., 2022). The optimisation of nanochitosan's bioavailability is crucial to enhance its efficacy in improving the well-being and productivity of organisms (Tran et al., 2023). The nanoscale characteristics of nanochitosan have been demonstrated to improve the bioavailability of nutrients in animal feed significantly (Wang et al., 2022). Multiple studies have indicated that nanochitosan can enhance the assimilation of vital nutrients, including amino acids, vitamins and minerals, within the gastrointestinal system of organisms (El-Naggar et al., 2021). This phenomenon has the potential to result in enhanced rates of growth and general well-being.

Enhancing the bioavailability of essential nutrients refers to the ability of animals to derive increased nutritional content from a given quantity of eaten feed. This practice has the potential to improve feed efficiency, resulting in a reduction in the quantity of feed needed to get desired outputs and thus decreasing production costs (Wang et al., 2022). Developers must take into account the bioavailability of nanochitosan and conduct an assessment of its interactions with other constituents of feed to optimise its bioavailability.

7.7 Required Concentrations and Chemical Stability

Determining the optimal concentration of nanochitosan in the feed is crucial to attaining desired outcomes (Dawoud et al., 2023. Consequently, the concentration of substances may exhibit variability contingent upon the specific objective, such as augmenting growth, bolstering illness resistance or optimising feed use (El-Naggar et al., 2021). Inadequate quantities may fail to yield the expected outcomes, while excessive amounts can result in unforeseen repercussions, resource misallocation and a myriad of other undesired impacts (Nuzaiba et al., 2023).

Using excessive concentration may lead to unnecessary expenditures, but employing a too-low concentration may not be optimal financially. Excessive concentration levels may also potentially have deleterious impacts on animal health (Nuzaiba et al., 2023). Hence, a thorough evaluation of the concentration is essential to ascertain the aquatic species' tolerance towards the feed additive and mitigate potential health hazards.

The accumulation of nanochitosan in the aquatic environment may result from an abnormally high concentration in aquaculture feed (Morales and Moyano, 2010; Su et al., 2022). Regulatory organisations frequently establish recommendations or thresholds regarding the concentration of chemicals in animal feed to safeguard the well-being of animals and customers (Sewell et al., 2022). It is imperative to adhere to these standards to ensure compliance and mitigate the risk of any legal complications.

Furthermore, it is crucial to maintain chemical stability to preserve the integrity and efficacy of nanochitosan, as well as other vital nutrients and bioactive substances included in the feed (Okeke et al., 2022). The maintenance of nutritional integrity is crucial to fulfilling the dietary needs of the intended organisms for the duration of the feed's storage period. The preservation of chemical stability and the extension of shelf life are crucial factors in feed production, as they contribute to a consistent quality (Chellapandian et al., 2023). Aquaculturists can depend on the consistent nutrient composition of the feed, hence facilitating foreseeable fish growth and development.

7.8 Particle Size and Incorporation

The nanochitosan particles exhibit a significantly reduced size at the nanoscale, resulting in an increased surface area compared to per unit volume (Hamidian et al., 2018). The enhanced surface area of aquatic species'gastrointestinal systems promotes more successful interactions with feed and intestinal enzymes (Younus et al., 2020). Consequently, the enhanced availability of nutrients facilitates their absorption, resulting in a better state of bioavailability (Wang et al., 2022). The effective enhancement of vital nutrient absorption, leading to the promotion of health and growth in aquatic species, is achieved through the control of particle size and distribution in Nanochitosan (Hamidian et al., 2018).

8 Conclusion

Recent field observations have indicated that attaining a homogeneous dispersion of nanochitosan particles inside the feed composition is paramount (Morales and Moyano, 2010). The irregular particle sizes or uneven distribution might result in the aggregation or separation of the constituent elements within the feed ingredients (Rauscher et al., 2019). The lack of consistency in nutrient distribution within a given aquatic culture can give rise to discrepancies in nutrient intake across different organisms, ultimately resulting in uneven growth patterns and health differences. The implementation of precise control over particle size and dispersion is crucial in achieving a homogeneous mixture, hence facilitating uniform nutrition delivery to all species.

Field trials and regulations are crucial in evaluating and assessing various products and technologies. These trials are conducted in real-world settings to gather empirical data and determine the effectiveness, safety and compliance of the tested products (Flegel, 2019). Before mass manufacturing nanochitosan-enhanced feed, developers must consider conducting field experiments to assess their efficacy in real-world settings. This entails ongoing monitoring of the fish's health and growth to refine the formulation. To ensure the safety of nanochitosan for ingestion by the species of interest and to mitigate any dangers to human health within the food chain, developers must allocate resources within their budget to include toxicity studies.

To minimise the generation of waste in water systems, responsible field surveys should incorporate efficient water management methods. Additionally, it is crucial to ensure the appropriate quantity of enhanced feed is utilised and to communicate information regarding environmental advantages and risks to interested parties and regulatory bodies.

References

Abd El-Naby, F. S., Naiel, M. A., Al-Sagheer, A. A., & Negm, S. S. (2019). Dietary chitosan nanoparticles enhance the growth, production performance, and immunity in Oreochromis niloticus. *Aquaculture, 501*, 82–89.

Abd El-Naby, A. S., Al-Sagheer, A. A., Negm, S. S., & Naiel, M. A. (2020). Dietary combination of chitosan nanoparticle and thymol affects feed utilization, digestive enzymes, antioxidant status, and intestinal morphology of Oreochromis niloticus. *Aquaculture, 515*, 734577.

Abdel-Ghany, H. M., & Salem, M. E. S. (2020). Effects of dietary chitosan supplementation on farmed fish; a review. *Reviews in Aquaculture, 12*(1), 438–452.

Abdel-Tawwab, M., Razek, N. A., & Abdel-Rahman, A. M. (2019). Immunostimulatory effect of dietary chitosan nanoparticles on the performance of Nile tilapia, Oreochromis niloticus (L.). *Fish & Shellfish Immunology, 88*, 254–258.

Abdel-Warith, A. W. A., El-Bab, A. F. F., Younis, E. S. M., Al-Asgah, N. A., Allam, H. Y., Abd-Elghany, M. F., Shata, Y. H., & Shamlol, F. S. (2020). Using of chitosan nanoparticles (CsNPs), Spirulina as a feed additives under intensive culture system for black tiger shrimp (Penaeus monodon). *Journal of King Saud University-Science, 32*(8), 3359–3363.

Abere, D. V., Ojo, S. A., Paredes-Epinosa, M. B., & Hakami, A. (2022). Derivation of composites of chitosan-nanoparticles from crustaceans source for nanomedicine: A mini review. *Biomedical Engineering Advances, 4*, 100058.

Aheto, D. W., Acheampong, E., & Odoi, J. O. (2019). Are small-scale freshwater aquaculture farms in coastal areas of Ghana economically profitable? *Aquaculture International, 27*, 785–805.

Ahmed, S. B., Mohamed, H. I., Al-Subaie, A. M., Al-Ohali, A. I., & Mahmoud, N. M. (2021a). Investigation of the antimicrobial activity and hematological pattern of nano-chitosan and its nano-copper composite. *Scientific Reports, 11*(1), 9540.

Ahmed, F., Soliman, F. M., Adly, M. A., Soliman, H. A., El-Matbouli, M., & Saleh, M. (2021b). Dietary chitosan nanoparticles: Potential role in modulation of rainbow trout (Oncorhynchus mykiss) antibacterial defense and intestinal immunity against enteric redmouth disease. *Marine Drugs, 19*(2), 72.

Ahmed, J., Vasagam, K. K., & Ramalingam, K. (2023). Nanoencapsulated Aquafeeds and current uses in fisheries/shrimps: A review. *Applied Biochemistry and Biotechnology*, 1–22.

Ahuekwe, E. F., Isibor, P. O., Akinhanmi, F., Ajiboye, I., Salami, A. O., Adekeye, B. T., Akinpelu, S. O., Kuye, A. D., Bello, A. O., Oyewale, J. O., & George, D. Z. (2023). Utilization of nano-chitosan in the sterilization of ponds and water treatment for aquaculture. *Next Generation Nanochitosan*, 301–338.

Akbary, P., & Younesi, A. (2017). Effect of dietary supplementation of chitosan on growth, hematology and innate immunity of grey mullet (Mugil cephalus). *Veterinary Research & Biological Products, 30*(3), 194–203.

Akoma, O. C. (2023). The invisible dancers. *Nature's Life Sustaining Powerhouse in the Aquatic Environment.*

Ali, M. N., Al-Saadi, T. M., & AL-Faragi, J.K. (2021). Effect of chitosan nanoparticles loaded oxytetracycline hydrochloride on health status of common carp (Cyprinuscarpio L.) infected with columnaris disease. *Journal of Physics: Conference Series, 1879*(3), 032075. IOP Publishing.

Aly, S. M., & Albutti, A. (2014). Antimicrobials use in aquaculture and their public health impact. *Journal of Aquaculture Research & Development, 5*(4), 1.

Amenyogbe, E., Chen, G., Wang, Z., Huang, J., Huang, B., & Li, H. (2020). The exploitation of probiotics, prebiotics and synbiotics in aquaculture: Present study, limitations and future directions: A review. *Aquaculture International, 28*, 1017–1041.

Arechavala-Lopez, P., Cabrera-Álvarez, M. J., Maia, C. M., & Saraiva, J. L. (2022). Environmental enrichment in fish aquaculture: A review of fundamental and practical aspects. *Reviews in Aquaculture, 14*(2), 704–728.

Asaikkutti, A., Vimala, K., Jha, N., Bhavan, P. S., & Arul, V. (2023). Effect of dietary supplementation of vitamin C-loaded chitosan nanoparticles on growth, immune-physiological parameters, and resistance of white shrimp Litopenaeus vannamei to Vibrio harveyi challenge. *Animal Feed Science and Technology*, 115764.

Asche, F., Cojocaru, A. L., & Roth, B. (2018). The development of large-scale aquaculture production: A comparison of the supply chains for chicken and salmon. *Aquaculture, 493*, 446–455.

Asche, F., Eggert, H., Oglend, A., Roheim, C. A., & Smith, M. D. (2022). Aquaculture: Externalities and policy options. *Review of Environmental Economics and Policy, 16*(2), 282–305.

Assefa, A., & Abunna, F. (2018). Maintenance of fish health in aquaculture: Review of epidemiological approaches for prevention and control of infectious disease of fish. *Veterinary Medicine International, 2018.*

Attaran Dowom, S., Karimian, Z., Mostafaei Dehnavi, M., & Samiei, L. (2022). Chitosan nanoparticles improve physiological and biochemical responses of Salvia abrotanoides (Kar.) under drought stress. *BMC Plant Biology, 22*(1), 364.

Aumeerun, S., Soulange-Govinden, J., Driver, M. F., Rao, A. R., Ravishankar, G. A., & Neetoo, H. (2022). Novel sources of functional food and feed. In *Handbook of algal technologies and phytochemicals* (p. 207).

Avila-Quezada, G. D., Ingle, A. P., Golińska, P., & Rai, M. (2022). Strategic applications of nanofertilizers for sustainable agriculture: Benefits and bottlenecks. *Nanotechnology Reviews, 11*(1), 2123–2140.

Aware, C., Nalavade, V., Jadhav, R., Bhatia, S. K., Yang, Y. H., Jadhav, J., & Gurav, R. (2022). Algae as food and nutraceuticals. *Algal Biorefineries and the Circular Bioeconomy*, 295–337.

Bayda, S., Adeel, M., Tuccinardi, T., Cordani, M., & Rizzolio, F. (2019). The history of nanoscience and nanotechnology: From chemical–physical applications to nanomedicine. *Molecules, 25*(1), 112.

Benettayeb, A., Seihoub, F. Z., Pal, P., Ghosh, S., Usman, M., Chia, C. H., Usman, M., & Sillanpää, M. (2023). Chitosan nanoparticles as potential nano-sorbent for removal of toxic environmental pollutants. *Nanomaterials, 13*(3), 447.

Camp, E., Garlock, T., & Anderson, J. (2020). Opportunities and obstacles to aquaculture in Florida: FA221/FA221, 5/2020. *EDIS, 2020*(3), 6–6.

Chantarasataporn, P., Yoksan, R., Visessanguan, W., & Chirachanchai, S. (2013). Water-based nano-sized chitin and chitosan as seafood additive through a case study of Pacific white shrimp (Litopenaeus vannamei). *Food Hydrocolloids, 32*(2), 341–348.

Charoonnart, P., Taunt, H. N., Yang, L., Webb, C., Robinson, C., Saksmerprome, V., & Purton, S. (2023). Transgenic microalgae expressing double-stranded RNA as potential feed supplements for controlling white spot syndrome in shrimp aquaculture. *Microorganisms, 11*(8), 1893.

Chellapandian, H., Jeyachandran, S., Ilangovan, S., & Aseervatham, S. B. (2023). Nanochitosan for the production of more effective fish feed for aquaculture. *Next Generation Nanochitosan*, 339–348.

Chen, J., & Chen, L. (2019). Effects of chitosan-supplemented diets on the growth performance, nonspecific immunity and health of loach fish (Misgurnus anguillicadatus). *Carbohydrate Polymers, 225*, 115227.

Chen, Y., Zhu, X., Yang, Y., Han, D., Jin, J., & Xie, S. (2014). Effect of dietary chitosan on growth performance, haematology, immune response, intestine morphology, intestine microbiota and disease resistance in gibel carp (C arassius auratus gibelio). *Aquaculture Nutrition, 20*(5), 532–546.

Chen, C. Z., Li, P., Liu, L., & Li, Z. H. (2022). Exploring the interactions between the gut microbiome and the shifting surrounding aquatic environment in fisheries and aquaculture: A review. *Environmental Research, 214*, 114202.

Chen, Y., Mitra, A., Rahimnejad, S., Chi, S., Kumar, V., Tan, B., Niu, J., & Xie, S. (2023). Retrospect of fish meal substitution in Pacific white shrimp (Litopenaeus vannamei) feed: Alternatives, limitations and prospects. *Reviews in Aquaculture*.

Dashputre, N. L., Sable, R. R., Sawant, M., Khairnar, S. J., Ahire, E. D., Patil, S. B., & Kadam, J. D. (2023). Marine-derived sources of nutritional vitamins. *Vitamins as Nutraceuticals: Recent Advances and Applications*, 129–166.

Dawood, M. A., Gewaily, M. S., Soliman, A. A., Shukry, M., Amer, A. A., Younis, E. M., Abdel-Warith, A. W. A., Van Doan, H., Saad, A. H., Aboubakr, M., & Abdel-Latif, H. M. (2020). Marine-derived chitosan nanoparticles improved the intestinal histo-morphometrical features in association with the health and immune response of grey mullet (Liza ramada). *Marine Drugs, 18*(12), 611.

Dawoud, M. H., Mannaa, I. S., Abdel-Daim, A., & Sweed, N. M. (2023). Integrating artificial intelligence with quality by design in the formulation of lecithin/chitosan nanoparticles of a poorly water-soluble drug. *AAPS PharmSciTech, 24*(6), 169.

del Valle, J. C., Bonadero, M. C., & Gimenez, A. V. F. (2023). Saccharomyces cerevisiae as probiotic, prebiotic, synbiotic, postbiotics and parabiotics in aquaculture: An overview. *Aquaculture*, 739342.

Diwan, A. D., Harke, S. N., Gopalkrishna, & Panche, A. N. (2022). Aquaculture industry prospective from gut microbiome of fish and shellfish: An overview. *Journal of Animal Physiology and Animal Nutrition, 106*(2), 441–469.

El-Bab, F. (2022). The synergy effect of Spirulina platensis and/or chitosan nanoparticles on the growth performance, hematological, biochemical parameters, and body composition of the European seabass (Dicentrarchus labrax). *Egyptian Journal Of Aquatic Biology And Fisheries, 26*(4), 255–271.

El-Naggar, M., Salaah, S., El-Shabaka, H., Abd El-Rahman, F., Khalil, M., & Suloma, A. (2021). Efficacy of dietary chitosan and chitosan nanoparticles supplementation on health status of Nile tilapia, Oreochromis niloticus (L.). *Aquaculture Reports, 19*, 100628.

El-Naggar, M., Medhat, F., & Taha, A. (2022a). Applications of chitosan and chitosan nanoparticles in fish aquaculture. *Egyptian Journal of Aquatic Biology & Fisheries, 26*(1).

El-Naggar, N. E. A., Shiha, A. M., Mahrous, H., & Mohammed, A. A. (2022b). Green synthesis of chitosan nanoparticles, optimization, characterization and antibacterial efficacy against multi drug resistant biofilm-forming Acinetobacter baumannii. *Scientific Reports, 12*(1), 19869.

El-Saadony, M. T., Shehata, A. M., Alagawany, M., Abdel-Moneim, A. M. E., Selim, D. A., Abdo, M., Khafaga, A. F., El-Tarabily, K. A., El-Shall, N. A., & Abd El-Hack, M. E. (2022). A review of shrimp aquaculture and factors affecting the gut microbiome. *Aquaculture International, 30*(6), 2847–2869.

Encarnação, P. (2016). Functional feed additives in aquaculture feeds. *Aquafeed formulation*, 217–237.

Fajardo, C., Martinez-Rodriguez, G., Blasco, J., Mancera, J. M., Thomas, B., & De Donato, M. (2022). Nanotechnology in aquaculture: Applications, perspectives and regulatory challenges. *Aquaculture and Fisheries, 7*(2), 185–200.

Ferosekhan, S., Gupta, S., Singh, R., Rather, A., Kumari, R., Kothari, C. D., Kumar Pal, A., & Balkrishna Jadhao, S. (2014). RNA-loaded chitosan nanoparticles for enhanced growth, immunostimulation and disease resistance in fish. *Current Nanoscience, 10*(3), 453–464.

Flegel, T. W. (2019). A future vision for disease control in shrimp aquaculture. *Journal of the World Aquaculture Society, 50*(2), 249–266.

Franci, G., Falanga, A., Galdiero, S., Palomba, L., Rai, M., Morelli, G., & Galdiero, M. (2015). Silver nanoparticles as potential antibacterial agents. *Molecules, 20*(5), 8856–8874.

Funes, C. F., Bouvier, B., Cézard, C., Fuentealba, C., Jamali, A., Courty, M., Hadad, C., & Van Nhien, A. N. (2023). Theoretical and experimental studies of chitin nanocrystals treated with ionic liquid or deep eutectic solvent to afford nanochitosan sheets. *Journal of Molecular Liquids, 375*, 121350.

Gamboa-Delgado, J. (2022). Isotopic techniques in aquaculture nutrition: State of the art and future perspectives. *Reviews in Aquaculture, 14*(1), 456–476.

Giri, S. S. (2019). Issues, challenges and future needs of commercially important finfish aquaculture in South Asia. *Aquaculture of Commercially Important Finfishes in South Asia*, 1.

Hamidian, G., Zirak, K., Sheikhzadeh, N., Khani Oushani, A., Shabanzadeh, S., & Divband, B. (2018). Intestinal histology and stereology in rainbow trout (Oncorhynchus mykiss) administrated with nanochitosan/zeolite and chitosan/zeolite composites. *Aquaculture Research, 49*(5), 1803–1815.

Harikrishnan, R., Kim, J. S., Balasundaram, C., & Heo, M. S. (2012). Immunomodulatory effects of chitin and chitosan enriched diets in Epinephelus bruneus against Vibrio alginolyticus infection. *Aquaculture, 326*, 46–52.

Hu, F., Zhong, H., Wu, C., Wang, S., Guo, Z., Tao, M., Zhang, C., Gong, D., Gao, X., Tang, C., & Wei, Z. (2021). Development of fisheries in China. *Reproduction and Breeding, 1*(1), 64–79.

Ibrahim, D., Neamat-Allah, A. N., Ibrahim, S. M., Eissa, H. M., Fawzey, M. M., Mostafa, D. I., Abd El-Kader, S. A., Khater, S. I., & Khater, S. I. (2021). Dual effect of selenium loaded chitosan nanoparticles on growth, antioxidant, immune related genes expression, transcriptomics modulation of caspase 1, cytochrome P450 and heat shock protein and Aeromonas hydrophila resistance of Nile Tilapia (Oreochromis niloticus). *Fish & Shellfish Immunology, 110*, 91–99.

Isibor, I. P., Osagie, A. D., & Oritseweyinmi, O. O. (2023). Crustacean Nanochitosan-based bioremediation of nanoplastic-polluted aquatic habitat: A review pursuant to SDG 6. *Scientific African*, e01881.

Kaboudi, Z., Peighambardoust, S. H., Nourbakhsh, H., & Soltanzadeh, M. (2023). Nanoencapsulation of Chavir (Ferulago angulata) essential oil in chitosan carrier: Investigating physicochemical, morphological, thermal, antimicrobial and release profile of obtained nanoparticles. *International Journal of Biological Macromolecules, 237*, 123963.

Khan, M. A., Begum, R., Nielsen, R., & Hoff, A. (2021). Production risk, technical efficiency, and input use nexus: Lessons from Bangladesh aquaculture. *Journal of the World Aquaculture Society, 52*(1), 57–72.

Khieokhajonkhet, A., Aeksiri, N., Ratanasut, K., Kannika, K., Suwannalers, P., Tatsapong, P., Inyawilert, W., & Kaneko, G. (2022). Effects of dietary Hericium erinaceus powder on growth, hematology, disease resistance, and expression of genes related immune response against thermal challenge of Nile tilapia (Oreochromis niloticus). *Animal Feed Science and Technology, 290*, 115342.

Kono, M., Matsui, T., & Shimizu, C. (1987). Effect of chitin, chitosan, and cellulose as diet supplements on the growth of cultured fish [red sea bream, eel, yellowtail]. *Bulletin of the Japanese Society of Scientific Fisheries (Japan).*

Kuebutornye, F. K., & Abarike, E. D. (2020). The contribution of medicinal plants to tilapia aquaculture: A review. *Aquaculture International, 28*(3), 965–983.

Le Blanc, D., Freire, C., & Vierros, M. (2017). *Mapping the linkages between oceans and other Sustainable Development Goals: A preliminary exploration.*

Li, S., Li, C., & Wu, S. (2022). Dietary chitosan modulates the growth performance, body composition and nonspecific immunity of juvenile yellow catfish (Pelteobagrus fulvidraco). *International Journal of Biological Macromolecules, 217*, 188–192.

Liang, Q., Sun, X., Raza, H., Khan, M. A., Ma, H., & Ren, X. (2021). Fabrication and characterization of quercetin loaded casein phosphopeptides-chitosan composite nanoparticles by ultrasound treatment: Factor optimization, formation mechanism, physicochemical stability and antioxidant activity. *Ultrasonics Sonochemistry, 80*, 105830.

Lowry, G. V., Avellan, A., & Gilbertson, L. M. (2019). Opportunities and challenges for nanotechnology in the agri-tech revolution. *Nature Nanotechnology, 14*(6), 517–522.

Madhu, S., Devarajan, Y., Balasubramanian, M., & Raj, M. P. (2022). Synthesis and characterization of nano chitosan obtained using different seafood waste. *Materials Letters, 329*, 133195.

Maesano, G., Di Vita, G., Chinnici, G., Pappalardo, G., & D'Amico, M. (2020). The role of credence attributes in consumer choices of sustainable fish products: A review. *Sustainability, 12*(23), 10008.

Malik, S., Muhammad, K., & Waheed, Y. (2023). Nanotechnology: A revolution in modern industry. *Molecules, 28*(2), 661.

Mishra, A., Shah, B. R., Roy, K., Abdelsalam, E. E. E., Piačková, V., Shaik, H. A., Dvořák, P., Velíšek, J., Agbeko, K. F. K., & Mráz, J. (2023). Andrographolide loaded Pickering emulsion: A bioactive component for improved growth, digestibility, and haematological properties in cultured common carp Cyprinus carpio. *Aquaculture, 562*, 738810.

Mobsby, D., Steven, H. A., & Curtotti, R. (2020). Australian fisheries and aquaculture outlook 2020. *Abares Canberra, 9*.

Moges, F. D., Patel, P., Parashar, S. K. S., & Das, B. (2020). Mechanistic insights into diverse nano-based strategies for aquaculture enhancement: A holistic review. *Aquaculture, 519*, 734770.

Morales, G. A., & Moyano, F. J. (2010). Application of an in vitro gastrointestinal model to evaluate nitrogen and phosphorus bioaccessibility and bioavailability in fish feed ingredients. *Aquaculture, 306*(1–4), 244–251.

Mugwanya, M., Dawood, M. A., Kimera, F., & Sewilam, H. (2021). Updating the role of probiotics, prebiotics, and synbiotics for tilapia aquaculture as leading candidates for food sustainability: A review. *Probiotics and Antimicrobial Proteins*, 1–28.

Nankervis, L., & Jones, C. (2022). Recent advances and future directions in practical diet formulation and adoption in tropical Palinurid lobster aquaculture. *Reviews in Aquaculture, 14*(4), 1830–1842.

Naylor, R. L., Hardy, R. W., Buschmann, A. H., Bush, S. R., Cao, L., Klinger, D. H., Little, D. C., Lubchenco, J., Shumway, S. E., & Troell, M. (2021). A 20-year retrospective review of global aquaculture. *Nature, 591*(7851), 551–563.

Niu, J., Li, C. H., Tian, L. X., Liu, Y. J., Chen, X., Wu, K. C., Jun, W., Huang, Z., Wang, Y., & Lin, H. Z. (2015). Suitable dietary chitosan improves the growth performance, survival and immune function of tiger shrimp, Penaeus monodon. *Aquaculture Research, 46*(7), 1668–1678.

Nouri, A. (2019). Chitosan nano-encapsulation improves the effects of mint, thyme, and cinnamon essential oils in broiler chickens. *British Poultry Science, 60*(5), 530–538.

Nuzaiba, P. M., Gupta, S., Gupta, S., & Jadhao, S. B. (2023). Synthesis of L-methionine-loaded chitosan nanoparticles for controlled release and their in vitro and in vivo evaluation. *Scientific Reports, 13*(1), 7606.

Okeke, E. S., Chukwudozie, K. I., Nyaruaba, R., Ita, R. E., Oladipo, A., Ejeromedoghene, O., Atakpa, E. O., Agu, C. V., & Okoye, C. O. (2022). Antibiotic resistance in aquaculture and aquatic organisms: A review of current nanotechnology applications for sustainable management. *Environmental Science and Pollution Research, 29*(46), 69241–69274.

Olaniyan, O. T., Adetunji, C. O., Dare, A., Ajayi, O. O., & Adeniyi, M. J. (2023). Roles of nanochitosan and polymeric chitosan for the protection of farmed fish against disease-causing pathogens. *Next Generation Nanochitosan*, 287–294.

Pakizeh, M., Moradi, A., & Ghassemi, T. (2021). Chemical extraction and modification of chitin and chitosan from shrimp shells. *European Polymer Journal, 159*, 110709.

Persaud, N., & Dagher, R. (2021). The united nations: 2030 sustainable development goals agenda. In *The role of monitoring and evaluation in the UN 2030 SDGs agenda* (pp. 1–41).

Pradeepkiran, J. A. (2019). Aquaculture role in global food security with nutritional value: A review. *Translational Animal Science, 3*(2), 903–910.

Prato, E., Biandolino, F., Parlapiano, I., Giandomenico, S., Denti, G., Calò, M., Spada, L., & Di Leo, A. (2019). Proximate, fatty acids and metals in edible marine bivalves from Italian market: Beneficial and risk for consumers health. *Science of the Total Environment, 648*, 153–163.
Rad, A. E., Alishahi, M., Ghorbanpour, M., & Zarei, M. (2014). The effects of oral administration of extracted chitosan from white leg shrimp (Litopenaeus vannamei) on hematological and growth indices in common carp (Cyprinus carpio). *Journal of Veterinary Research, 69*(4).
Rambaran, T., & Schirhagl, R. (2022). Nanotechnology from lab to industry – A look at current trends. *Nanoscale Advances, 4*(18), 3664–3675.
Ranjan, R., Prasad, K. P., Vani, T., & Kumar, R. (2014). Effect of dietary chitosan on haematology, innate immunity and disease resistance of Asian seabass Lates calcarifer (Bloch). *Aquaculture Research, 45*(6), 983–993.
Rauscher, H., Mech, A., Gibson, N., Gilliland, D., Held, A., Kestens, V., Koeber, R., Linsinger, T. P. J., & Stefaniak, E. A. (2019). *Identification of nanomaterials through measurements*. Publications Office of the European Union.
Robinson, N. A., Gopikrishna, G., Baranski, M., Katneni, V. K., Shekhar, M. S., Shanmugakarthik, J., Jothivel, S., Gopal, C., Ravichandran, P., Gitterle, T., & Ponniah, A. G. (2014). QTL for white spot syndrome virus resistance and the sex-determining locus in the Indian black tiger shrimp (Penaeus monodon). *BMC Genomics, 15*, 1–21.
Rochana, W., Niroshan, W., Tiruchenduran, S., Sulaiman, M. A., & Mahesh, D. (2019). Effects of chitosan on growth, immune responses and survival of juvenile tiger shrimp (Penaeus monodon Fabricius, 1798). *International Journal of Fisheries and Aquatic Studies, 7*, 129–133.
Saeedi, M., Vahidi, O., Moghbeli, M. R., Ahmadi, S., Asadnia, M., Akhavan, O., Seidi, F., Rabiee, M., Saeb, M. R., Webster, T. J., & Varma, R. S. (2022). Customizing nano-chitosan for sustainable drug delivery. *Journal of Controlled Release, 350*, 175–192.
Santos, V. P., Marques, N. S., Maia, P. C., Lima, M. A. B. D., Franco, L. D. O., & Campos-Takaki, G. M. D. (2020). Seafood waste as attractive source of chitin and chitosan production and their applications. *International Journal of Molecular Sciences, 21*(12), 4290.
Schar, D., Klein, E. Y., Laxminarayan, R., Gilbert, M., & Van Boeckel, T. P. (2020). Global trends in antimicrobial use in aquaculture. *Scientific Reports, 10*(1), 21878.
Schumann, M., & Brinker, A. (2020). Understanding and managing suspended solids in intensive salmonid aquaculture: A review. *Reviews in Aquaculture, 12*(4), 2109–2139.
Sewell, F., Corvaro, M., Andrus, A., Burke, J., Daston, G., Delaney, B., Domoradzki, J., Forlini, C., Green, M. L., Hofmann, T., & Jäckel, S. (2022). Recommendations on dose level selection for repeat dose toxicity studies. *Archives of Toxicology, 96*(7), 1921–1934.
Shekarabi, S. P. H., Javarsiani, L., Mehrgan, M. S., Dawood, M. A., & Adel, M. (2022). Growth performance, blood biochemistry profile, and immune response of rainbow trout (Oncorhynchus mykiss) fed dietary Persian shallot (Allium stipitatum) powder. *Aquaculture, 548*, 737627.
Shiau, S. Y., & Yu, Y. P. (1999). Dietary supplementation of chitin and chitosan depresses growth in tilapia, Oreochromis niloticus× O. aureus. *Aquaculture, 179*(1–4), 439–446.
Su, Y., Zhou, X., Meng, H., Xia, T., Liu, H., Rolshausen, P., Roper, C., McLean, J. E., Zhang, Y., Keller, A. A., & Jassby, D. (2022). Cost–benefit analysis of nanofertilizers and nanopesticides emphasizes the need to improve the efficiency of nanoformulations for widescale adoption. *Nature Food, 3*(12), 1020–1030.
Sumithra, T. G., Sharma, K. S., Gangadharan, S., Suresh, G., Prasad, V., Amala, P. V., Sayooj, P., Gop, A. P., Anil, M. K., Patil, P. K., & Achamveetil, G. (2022). Dysbiosis and restoration dynamics of the gut microbiome following therapeutic exposure to florfenicol in snubnose pompano (Trachinotus blochii) to aid in sustainable aquaculture production strategies. *Frontiers in Microbiology, 13*, 1384.
Syed Raffic Ali, S., Ambasankar, K., Ezhil Praveena, P., Nandakumar, S., & Syamadayal, J. (2017). Effect of dietary fructooligosaccharide supplementation on growth, body composition, hematological and immunological parameters of Asian seabass (Lates calcarifer). *Aquaculture International, 25*, 837–848.
Szuwalski, C., Jin, X., Shan, X., & Clavelle, T. (2020). Marine seafood production via intense exploitation and cultivation in China: Costs, benefits, and risks. *PLoS One, 15*(1), e0227106.

Tadese, D. A., Song, C., Sun, C., Liu, B., Liu, B., Zhou, Q., Xu, P., Ge, X., Liu, M., Xu, X., & Tamiru, M. (2022). The role of currently used medicinal plants in aquaculture and their action mechanisms: A review. *Reviews in Aquaculture, 14*(2), 816–847.

Taher, F. A., Amin, R. M., Abd El-Aziz, A., Abou Senna, F. M., Abou El-Nou, M. F., & El-Mehdawy, A. A. (2017). Characterization and antibacterial effect of chitosan nanoparticles prepared by ball-milling from shell of shrimp, Penaeus semisulcatus (Crustacea: Penaeidae), Al Azhar Bulletin of Science. In *Al Azhar bulletin of science conference*.

Tran, H. Q., Nguyen, T. T., Prokešová, M. D., Matoušek, J., Tomčala, A., Van Doan, H., Kiljunen, M., & Stejskal, V. (2023). Insight into bioavailability of various insect meals for European perch (Perca fluviatilis): A nutritional and stable isotopic evaluation. *Aquaculture, 563*, 738912.

Udo, I. U., Etukudo, U., & Anwana, U. I. U. (2018). Effects of chitosan and chitosan nanoparticles on water quality, growth performance, survival rate and meat quality of the African catfish, Clarias gariepinus. *Nanoscience, 1*(1), 12–25.

Uyanga, V. A., Ejeromedoghene, O., Lambo, M. T., Alowakennu, M., Alli, Y. A., Ere-Richard, A. A., Min, L., Zhao, J., Wang, X., Jiao, H., & Onagbesan, O. M. (2023). Chitosan and chitosan-based composites as beneficial compounds for animal health: Impact on gastrointestinal functions and biocarrier application. *Journal of Functional Foods, 104*, 105520.

Valério, N., Soares, M., Vilarinho, C., Correia, M., & Carvalho, J. (2023). Diving into fish valorisation: Review opportunities and analyzing Azorean fish data. *PRO, 11*(7), 1998.

Victor, H., Zhao, B., Mu, Y., Dai, X., Wen, Z., Gao, Y., & Chu, Z. (2019). Effects of Se-chitosan on the growth performance and intestinal health of the loach Paramisgurnus dabryanus (Sauvage). *Aquaculture, 498*, 263–270.

Vijayaram, S., Sun, Y. Z., Zuorro, A., Ghafarifarsani, H., Van Doan, H., & Hoseinifar, S. H. (2022). Bioactive immunostimulants as health-promoting feed additives in aquaculture: A review. *Fish & Shellfish Immunology*.

Wang, Y., & Li, J. (2011). Effects of chitosan nanoparticles on survival, growth and meat quality of tilapia, Oreochromis nilotica. *Nanotoxicology, 5*(3), 425–431.

Wang, L., Sagada, G., Wang, R., Li, P., Xu, B., Zhang, C., Qiao, J., & Yan, Y. (2022). Different forms of selenium supplementation in fish feed: The bioavailability, nutritional functions, and potential toxicity. *Aquaculture, 549*, 737819.

Wattanakul, W., Thongprajukaew, K., Songnui, A., Satjarak, J., & Kanghae, H. (2017). Presoaking feed pellet significantly improved feed utilization in Asian seabass (Lates calcarifer). *Aquaculture, 471*, 106–112.

Widjaja, S., Long, T., & Wirajuda, H. (2020). *Illegal, unreported and unregulated fishing and associated drivers*.

Wu, S. (2020). The growth performance, body composition and nonspecific immunity of Tilapia (Oreochromis niloticus) affected by chitosan. *International Journal of Biological Macromolecules, 145*, 682–685.

Wu, Y., Rashidpour, A., Almajano, M. P., & Metón, I. (2020). Chitosan-based drug delivery system: Applications in fish biotechnology. *Polymers, 12*(5), 1177.

Xu, H., Wang, X., Liang, Q., Xu, R., Liu, J., & Yu, D. (2023). Dietary chitosan moderates the growth rate, antioxidant activity, immunity, intestinal morphology and resistance against Aeromonas hydrophila of juvenile hybrid sturgeon (Acipenser baerii♀× Acipenser schrenckii♂). *International Journal of Biological Macromolecules, 224*, 1012–1024.

Yadav, M., Goswami, P., Paritosh, K., Kumar, M., Pareek, N., & Vivekanand, V. (2019). Seafood waste: A source for preparation of commercially employable chitin/chitosan materials. *Bioresources and Bioprocessing, 6*(1), 1–20.

Yadav, N., Mudgal, D., Anand, R., Jindal, S., & Mishra, V. (2022). Recent development in nanoencapsulation and delivery of natural bioactives through chitosan scaffolds for various biological applications. *International Journal of Biological Macromolecules*.

Ye, Y., & Gutierrez, N. L. (2017). Ending fishery overexploitation by expanding from local successes to globalized solutions. *Nature Ecology & Evolution, 1*(7), 0179.

Younus, N., Zuberi, A., Mahmoood, T., Akram, W., & Ahmad, M. (2020). Comparative effects of dietary micro-and nano-scale chitosan on the growth performance, non-specific immunity, and resistance of silver carp Hypophthalmichthys molitrix against Staphylococcus aureus infection. *Aquaculture International, 28*, 2363–2378.

Yukgehnaish, K., Kumar, P., Sivachandran, P., Marimuthu, K., Arshad, A., Paray, B. A., & Arockiaraj, J. (2020). Gut microbiota metagenomics in aquaculture: Factors influencing gut microbiome and its physiological role in fish. *Reviews in Aquaculture, 12*(3), 1903–1927.

Zaki, M. A., Shatby, E., & Shatby, E. (2015). Effect of chitosan supplemented diet on survival, growth, feed utilization, body composition & histology of sea bass (Dicentrarchus labrax). *World Journal of Engineering and Technology, 3*(04), 38.

Zhang, M., Hou, G., Hu, P., Feng, D., Wang, J., & Zhu, W. (2021). Nano chitosan–zinc complex improves the growth performance and antioxidant capacity of the small intestine in weaned piglets. *British Journal of Nutrition, 126*(6), 801–812.

Fish Nanotagging and Barcoding

Patrick Omoregie Isibor

Contents

1 Introduction

Tagging programs in aquaculture serve various purposes and objectives, primarily aimed at improving the management, sustainability, and overall success of aquaculture operations. These programs often involve the use of tags, such as physical tags, electronic transponders (e.g. Radio-Frequency Identification [RFID]), or other tracking mechanisms (Macaulay et al., 2021).

Tagging allows for the unique identification of individual aquatic organisms, such as fish or shellfish. This is particularly important for selective breeding programs and tracking the growth and performance of individual animals. Tags help in

P. O. Isibor (✉)
Department of Biological Sciences, College of Science and Technology, Covenant University, Ota, Ogun State, Nigeria
e-mail: patrick.isibor@covenantuniversity.edu.ng

P. O. Isibor et al. (eds.), *Nanochitosan-Based Enhancement of Fisheries and Aquaculture*, https://doi.org/10.1007/978-3-031-52261-1_9

monitoring the health, growth, and overall performance of farmed fish or shellfish populations, contributing to effective stock management (Wilder et al., 2016). They also provide a means to trace the origin of aquaculture products, ensuring transparency and quality control in the supply chain. Furthermore, tags can be used in combination with sensors to monitor the health of aquatic organisms. Any deviations in vital signs or behaviour can signal the early onset of disease, allowing for prompt intervention. In the event of disease outbreaks, tags facilitate the identification and quarantine of affected batches or individuals, preventing the spread of diseases within aquaculture facilities. Tagging programs can be used to assess feeding patterns and efficiency, helping to minimize food wastage and reduce production costs (Dawit et al., 2020).

Tags can provide insights into nutrient uptake by individual organisms, aiding in the formulation of more precise and cost-effective feeding regimes. When equipped with sensors, it can monitor water quality parameters (e.g. temperature, oxygen levels, pH) in real time, which is crucial for maintaining optimal conditions for aquatic organisms (Yoshida et al., 2014). Tags, such as acoustic tags, are used to track the movements and behaviours of aquatic organisms, shedding light on their natural behaviours and habitat preferences. This aids in informed decision making in fisheries and aquaculture. Tagging programs can help identify preferred habitats within aquaculture systems, aiding in the design and management of aquatic environments for optimum productivity. In genetic improvement, individual tagging is essential for monitoring and selecting specific traits in breeding programs aimed at producing healthier, faster-growing, or disease-resistant strains of aquatic organisms (Han et al., 2013; Dewar et al., 2018).

Tagging programs assist in regulatory compliance by providing data on stock management, health monitoring, and traceability, which may be required by government authorities and certification bodies. Tagging also aids in monitoring and treating fish pond effluent before release into the environment. The proper treatment of pond effluent ensures strict adherence to regulatory guidelines for the protection of soil surface and groundwater. The programs generate valuable data for scientific research, innovation, and the development of best aquaculture practices. In technology development, these programs often drive advances in tagging technologies and data analysis methods, contributing to the aquaculture industry's growth (Fajardo et al., 2022).

Tagging programs in aquaculture are thus instrumental in enhancing the management, sustainability, and productivity of aquaculture operations. They enable detailed monitoring of aquatic organisms, data collection for research and innovation, and compliance with regulatory requirements, ultimately leading to more efficient and environmentally responsible aquaculture practices (Postulkova et al., 2016).

2 Types of Tags and Tagging Methods

The use of various types of tags and tagging methods in the realms of fisheries and aquaculture constitutes a fundamental and indispensable component of research and management efforts within these fields. Tags, in their diverse forms, serve as powerful tools that offer profound insights into the intricate behaviours, movements, and population dynamics of aquatic species. These insights, in turn, have far-reaching implications for our understanding of aquatic ecosystems and the sustainable management of their resources.

In both scientific research and resource management, the importance of tagging is indispensable and pivotal to the success of fisheries and aquaculture management. These methods are not merely convenient but are, in fact, an imperative means of gaining knowledge that is otherwise unattainable. The critical value lies in their ability to track and monitor individual fish or aquatic organisms with a precision that traditional observation methods cannot achieve. By affixing tags to these organisms, researchers are empowered to gather a wealth of data that encompasses an array of critical aspects of their lives.

One of the foremost purposes of tagging is to elucidate the complex migratory patterns of aquatic species. This entails tracing their movements across vast expanses of water, offering insights into breeding and feeding grounds, seasonal shifts, and long-distance travel that would otherwise remain shrouded in mystery. Moreover, tagging allows researchers to gain a deeper understanding of habitat use. By tracking the movements of tagged fish, we can discern their preferred environments, the conditions that attract them, and the challenges they encounter as they navigate different habitats.

Another crucial facet of tagging involves monitoring the survival rates of tagged individuals. This information is of paramount importance in both scientific research and the management of fish populations. It enables scientists to gauge the effectiveness of conservation measures, track the success of stocking programs, and evaluate the impact of various environmental factors on the survival of aquatic organisms.

The categories of tags employed in fisheries and aquaculture are diverse and tailored to specific research or management needs. Physical tags, including T-bar anchor tags, Floy tags, and spaghetti tags, are often utilized for their simplicity and cost-effectiveness. These tags can provide valuable information about movement and behaviour when recovered. On the other hand, electronic tags, such as acoustic tags, satellite tags, and Radio-Frequency Identification (RFID) tags, offer a more advanced and high-tech approach. They enable real-time tracking and data collection, allowing for a deeper understanding of aquatic organisms' activities. These electronic tags are particularly valuable for studying fine-scale movements, migration routes, and oceanic behaviours.

The wide-ranging applications of tagging in fisheries and aquaculture encompass a myriad of critical research and management objectives. From deciphering the enigmatic journeys of aquatic species to unravelling the mysteries of their habitat preferences and survival rates, tagging methods are the key to unlocking the secrets of aquatic ecosystems. Understanding the distinctions between various types of tags and their applications is an essential foundation for fish farmers, researchers, and resource managers, as it equips them with the tools needed to make informed decisions for the sustainable utilization and conservation of aquatic resources.

Physical Tags

Physical tags are a category of tagging methods employed in fisheries and wildlife research, as well as aquaculture, for tracking and monitoring individual organisms. These tags are typically tangible, physically attached to the organisms, and are designed to withstand the aquatic environment. Physical tags serve as essential tools for collecting data on the movement, behaviour, and survival of aquatic species. They are particularly useful for studies in which researchers aim to recover the tags at a later time to gather valuable information about the tagged organisms. There are various types of physical tags, each with its design, attachment method, and application (Macaulay et al., 2021).

Common types of physical tags include:

T-Bar Anchor Tags T-bar anchor tags consist of a small, plastic, or metal tag anchored to the body of the fish or aquatic organism using a T-shaped anchor or pin. These tags are often used in fisheries research to mark and track fish populations. T-bar anchor tags are durable and can be easily applied, making them a practical choice for marking larger fish.

Floy Tags Floy tags are brightly coloured plastic tags designed to be attached externally to the dorsal fin or other suitable locations on fish. They are widely used in fisheries research and management, providing a visible marker that allows for easy visual identification. Floy tags are useful for marking fish without the need for recapture and are particularly valuable in catch-and-release studies.

Spaghetti Tags Spaghetti tags are slender, plastic, or metal tags that are typically inserted under the skin of the fish or aquatic organism using a needle or applicator. These tags are often used for marking valuable or commercially significant species in fisheries research. They are less conspicuous than external tags, making them suitable for species where tag visibility may be a concern.

The choice of physical tag depends on the research objectives, the size and species of the organisms being tagged, and the method of recovery. These tags allow researchers to track the movement of tagged individuals, understand migration patterns, assess survival rates, and gain insights into the behaviour of aquatic species. When combined with data retrieval efforts, physical tags provide valuable information that contributes to the management and conservation of fish and aquatic resources.

Electronic Tagging

Electronic tagging, also known as electronic telemetry, is a sophisticated and high-tech method used in fisheries, wildlife research, and aquaculture to track and monitor the movement and behaviour of individual aquatic organisms. Unlike physical tags that are recovered upon recapture, electronic tags are designed to continuously collect and transmit data in real time or at a later point, providing researchers with detailed and precise information about the tagged organisms. Electronic tags offer a deeper understanding of the activities and fine-scale movements of aquatic species, making them valuable tools for a wide range of research and management applications.

Some of the key types of electronic tags used in fisheries and aquaculture include:

Acoustic Tags Acoustic tags are small devices that emit sound signals at specific frequencies. These tags are commonly used in studies involving fish and marine animals. Acoustic receivers placed in the water can detect these signals, enabling researchers to track the movements and behaviour of tagged organisms. Acoustic tagging is particularly effective for studying migration routes, underwater behaviour, and interactions between aquatic species.

Satellite Tags Satellite tags are designed to transmit data to orbiting satellites. They are commonly used in tracking the movements of large, highly mobile aquatic species, such as sharks, sea turtles, and marine mammals. The transmitted data include the location of the tagged organism, allowing researchers to monitor long-distance migrations and oceanic behaviours.

Radio-Frequency Identification (RFID) Tags RFID tags consist of a microchip and antenna that communicate with RFID readers (Keena, 2023). These tags are often used in aquaculture for tracking fish within fish farms or hatcheries. RFID technology allows for the automatic and non-invasive identification of individual fish, making it useful for monitoring growth, feed efficiency, and individual fish health.

Electronic tags offer several advantages over physical tags. They provide continuous and detailed data, allowing researchers to observe fine-scale movements and behaviours over extended periods. This technology is particularly valuable for studying elusive or highly mobile species. However, electronic tagging can be more expensive and may require more advanced equipment and expertise for data collection and analysis.

Overall, electronic tagging enhances our understanding of aquatic organisms, their ecology, and their interactions with their environment. It is an essential tool in fisheries and aquaculture research, contributing to the sustainable management and conservation of aquatic resources (Macaulay et al., 2021).

3 Tagging Techniques

Tagging techniques encompass a range of methods used to attach tags to aquatic organisms for tracking and monitoring purposes in fisheries, wildlife research, and aquaculture. The choice of tagging technique depends on the species being studied, the research objectives, and the ease of tag attachment and recovery. These techniques are crucial for collecting data on the movement, behaviour, and survival of individual organisms. Here are some common tagging techniques:

(i) External Tagging: This technique involves attaching tags to the exterior of an organism's body. Commonly used external tags include Floy tags, external anchor tags, and dart tags. These tags are visible and easily observed without recapturing the animal. External tagging is often employed in fisheries for studies that involve catch-and-release, and it is particularly useful for marking fish with minimal handling and intrusion (Lü et al., 2019).
(ii) Internal Tagging: Internal tagging involves implanting tags within the body of the organism. Spaghetti tags and passive integrated transponder (PIT) tags are examples of internal tags. Internal tagging is a less conspicuous method that reduces tag visibility and potential interference with the organism's natural behaviour. It is commonly used for species where external tags may not be practical (Musselman et al., 2017).
(iii) Fin Clipping: In some cases, researchers use fin clipping as a tagging technique. This involves removing a small portion of a fin (usually the adipose fin) and marking it with specific patterns or codes. Fin clipping is commonly used in fisheries to identify hatchery-reared fish and distinguish them from wild counterparts.
(iv) Injection Tags: Injection tags, such as coded wire tags (CWTs), are tiny wires or pieces of metal implanted into the fish's snout or body. Each tag contains a unique code that is specific to an individual fish. These tags are used in fisheries for marking fish populations and can provide valuable information about migration and stock assessment when retrieved (Lü et al., 2016).
(v) Genetic Tags: Genetic tagging methods involve using genetic markers to identify individuals or populations. DNA profiling, microsatellite analysis, and DNA barcoding are techniques used to genetically tag organisms. Genetic tags provide insights into the relatedness, lineage, and population structure of aquatic species.
(vi) Surgical Implantation: For larger and long-lived species, such as marine mammals or sea turtles, researchers may use surgical implantation to attach satellite tags or other electronic devices. This technique requires surgical procedures to insert the tag within the animal's body cavity, ensuring it remains secure and functional (Meerbeek, 2017).
(vii) Glue-On Tags: In some cases, researchers use non-invasive glue-on tags for smaller aquatic organisms. These tags are attached using adhesive compounds and can be used in studies of invertebrates, juvenile fish, or other species where traditional tagging methods are not practical.

The choice of tagging technique depends on the specific research goals and the biology of the target species. Researchers must consider factors such as the tag's visibility, durability, potential impacts on the organism, and the feasibility of tag recovery. Proper tagging techniques are crucial for the ethical treatment of study subjects and the reliability of data collected in fisheries, aquaculture, and aquatic ecology research.

4 Applications of Tagging in Fisheries

Applications of tagging in fisheries are diverse and vital for understanding and managing aquatic ecosystems and fish populations. Fisheries researchers and managers utilize various tagging techniques to gather data on individual fish and other aquatic organisms, enabling insights into their behaviour, movement, and population dynamics. The applications of tagging in fisheries encompass a wide range of objectives and include the following:

(i) Stock Assessment and Management: Tagging plays a crucial role in estimating fish populations and their trends. By tagging a subset of fish and monitoring recaptures, researchers can estimate population size, growth rates, and mortality rates. This information is used to set fishing quotas, design sustainable management strategies, and assess the health of fish stocks.

(ii) Migration and Movement Studies: Understanding the migratory patterns of fish is essential for effective management. Tagging helps trace fish migrations, revealing critical information about spawning and feeding grounds, seasonal movements, and the connectivity between different regions of aquatic ecosystems.

(iii) Behaviour and Habitat Analysis: Tagging allows researchers to investigate fish behaviour in response to environmental changes, predation, and other factors. It helps in identifying preferred habitats, feeding behaviours, and diurnal or nocturnal activities. This information is valuable for habitat conservation and fisheries management.

(iv) Bycatch Reduction and Gear Modification: Tagging studies assist in developing and refining fishing gear to reduce bycatch, which refers to the unintended capture of non-target species. By understanding the behaviour of both target and non-target species, fisheries can modify gear and practices to minimize bycatch and reduce the environmental impact of fishing operations.

(v) Conservation and Protected Species Management: Tagging is instrumental in tracking and conserving protected or endangered species. It helps assess the effectiveness of conservation measures, monitor the recovery of populations, and ensure compliance with legal protections for endangered or threatened species.

(vi) Invasive Species Control: Invasive species can have detrimental effects on native ecosystems. Tagging can help identify and monitor invasive species, aiding in their control and management efforts.

(vii) Hydrological and Environmental Studies: Tagging is used to study the effects of environmental factors, such as water temperature, flow, and quality, on fish behaviour and distribution. This information is essential for understanding the impact of environmental changes on fish populations and ecosystems.

(viii) Catch and Release Research: Tagging supports catch-and-release practices by providing information on the post-release behaviour and survival of fish. This is valuable for recreational fisheries and ensuring sustainable angling practices.

(ix) Fish Marking for Hatchery Programs: In hatcheries, fish are tagged with identifiers to differentiate hatchery-reared individuals from wild fish. This helps monitor the success of stock enhancement and restoration programs.
(x) Data for Ecological Models: Tagging data is crucial for building ecological models that simulate the dynamics of fish populations and their interactions with the environment. These models guide management decisions and help predict the consequences of various management scenarios.

Tagging in fisheries is a versatile and essential tool that contributes to sustainable fisheries management, environmental conservation, and the protection of aquatic ecosystems. The data collected through tagging studies inform policies and practices that aim to balance the preservation of fish populations with the needs of fisheries and the broader ecosystem. Tagging in aquaculture is an important practice that involves the attachment of various types of tags to cultured aquatic organisms for identification, monitoring, research, and management purposes. Aquaculture, the farming of aquatic species, benefits from tagging in several ways, contributing to efficient operations, product quality, and research advancement. Here are some of the key applications of tagging in aquaculture:

(i) Broodstock Management: Tagging is used to identify and track broodstock, which are the mature fish or aquatic organisms used for breeding in aquaculture facilities. By tagging broodstock, aquaculturists can monitor the reproductive performance and genetic lineage of individuals, ensuring the selection of high-quality parents for the next generation.
(ii) Disease Management and Control: Tags are employed to mark individual fish or aquatic organisms subjected to specific treatments or disease management protocols. This allows aquaculturists to monitor the effectiveness of treatments, identify affected individuals, and separate them from healthy populations to prevent disease transmission.
(iii) Growth and Feed Efficiency Studies: Tagging facilitates the tracking of individual fish's growth rates and feed efficiency. By tagging fish at a known size and monitoring their growth over time, aquaculturists can optimize feeding regimes and assess the performance of different diets or feeding strategies.
(iv) Monitoring Water Quality: In recirculating aquaculture systems (RAS), where water quality is critical, tags can be used to track the movement of individual fish within the system. This helps in evaluating water quality variations and ensuring that fish are exposed to optimal environmental conditions.
(v) Traceability and Product Quality Assurance: In commercial aquaculture, fish are often tagged to ensure traceability and product quality. This is particularly important for meeting consumer demands, verifying the source of seafood, and adhering to labelling and certification requirements.
(vi) Research and Genetic Studies: Tagging is essential for researching the genetics of farmed fish populations. It allows researchers to study heredity, genetic diversity, and the effectiveness of breeding programs. Genetic tags, such as DNA markers, are used for these purposes.

(vii) Management of Multiple Age Classes: In aquaculture systems where multiple age classes of fish coexist, tagging helps differentiate between cohorts and monitor their development. This information aids in planning harvests and managing stocking schedules.
(viii) Individual Fish Identification: Individual fish can be tagged for precise identification in research and aquaculture settings. This is useful for tracking fish growth, behaviour, and health over time.
(ix) Stress and Welfare Assessment: Tags can be used to monitor the stress levels and welfare of farmed fish. Changes in fish behaviour, swimming patterns, or interactions with other fish can provide insights into their well-being and stress responses.
(x) Reproductive Studies: In controlled breeding programs, tags may be used to identify parent fish and track the success of breeding events, including the number of offspring produced.

Tagging in aquaculture is a versatile practice that enhances management, research, and the overall sustainability of aquaculture operations. The choice of tagging method and tag type depends on the species, purpose, and objectives of the aquaculture program. Proper tagging and data collection contribute to the responsible and efficient production of seafood and the advancement of aquaculture science.

5 Challenges and Limitations

Fish tagging is a valuable tool for fisheries research and management, but it comes with several challenges and limitations that researchers and managers must consider when planning and conducting tagging studies. These challenges can affect the accuracy and effectiveness of data collection and interpretation. Here are some of the key challenges and limitations of fish tagging:

(i) Tag Loss and Mortality: One of the primary challenges in fish tagging is tag loss. Tags may become dislodged or damaged, leading to data gaps. Some tagging methods, such as external tags, are more prone to loss.

Tagging itself can cause stress and injury to fish, potentially resulting in increased mortality rates, which can skew survival estimates.

(ii) Incomplete Data: Recapture rates are often low, especially for highly mobile species or those with large home ranges. This can lead to incomplete datasets, limiting the accuracy of population estimates and other analyses.

Uneven spatial and temporal distribution of recaptures can result in biased or incomplete data.

(iii) Tag Effects: Some tagging methods, especially invasive methods like surgery for implanting electronic tags, can have physiological and behavioural effects on fish. These effects may alter their natural behaviours and skew study results. Behavioural changes due to tagging may affect the accuracy of movement and habitat preference studies.

(iv) Ethical and Welfare Concerns: Tagging, especially invasive methods, raises ethical concerns about animal welfare. Researchers must consider the potential harm and stress inflicted on tagged fish and take measures to minimize these effects. Ethical concerns may also arise when tagging species of conservation concern, as the potential negative impacts on these species must be carefully assessed.

(v) Cost and Resource Intensity: Tagging studies can be resource-intensive, requiring funding, equipment, and trained personnel. The costs associated with electronic tags, tracking equipment, and data analysis can be substantial. The need for long-term monitoring and equipment maintenance adds to the resource demands of tagging projects.

(vi) Species-Specific Challenges: The effectiveness of tagging methods can vary among fish species. Some species may be more susceptible to tag loss or tag effects than others, making it essential to adapt tagging methods to the target species. The size of the fish can also influence the choice of tagging method, as smaller fish may require smaller and less invasive tags.

(vii) Environmental Variability: Environmental factors such as water temperature, salinity, and food availability can affect fish behaviour and movement. These factors can introduce variability in tagging data, making it challenging to distinguish natural changes from the effects of tagging.

(viii) Tagging Technology Limitations: Tagging technology, while advanced, has limitations. For example, some electronic tags have limited battery life, limiting the duration of tracking studies. Range limitations of acoustic or satellite tracking equipment can make it challenging to monitor fish in remote or deep-water habitats.

(ix) Data Interpretation Challenges: Interpreting tagging data requires expertise in statistical analysis and ecological understanding. Misinterpretation or incorrect data analysis can lead to erroneous conclusions. Teasing out the causal relationships between tag effects and observed changes in fish behaviour can be complex.

Despite these challenges and limitations, fish tagging remains a valuable tool for fisheries research and management. Researchers work to address these limitations by refining tagging techniques, considering ethical and welfare concerns, and adapting methods to suit specific species and study objectives. Advances in tagging technology and data analysis continue to improve the accuracy and utility of tagging studies in understanding fish populations and ecosystems. Nanochitosan-based fish tagging holds feasible promises in reaching a milestone in fisheries and aquaculture through cutting-edge application techniques.

6 Application of Nanochitosan in Fish Tagging

Nanochitosan, a nanomaterial derived from chitosan, has a variety of potential applications in fish tagging, particularly in improving the effectiveness and safety of tagging methods (Hamed et al., 2016; Khosravi-Katuli et al., 2017; Gabriel et al., 2022). While it may not be a commonly used technique at present, research into nanochitosan-based tagging is ongoing, and it holds promise for several applications:

(i) Biocompatible Tag Materials: Nanochitosan can be used to develop biocompatible tag materials. Traditional tagging materials, such as plastics or metals, may have limited biocompatibility and could potentially affect fish health. Nanochitosan-based tags have the advantage of being more biocompatible, reducing the risk of negative impacts on the tagged fish (Pudake et al., 2019).
(ii) Enhanced Tag Durability: Nanochitosan-based tags may provide enhanced durability. The mechanical properties of nanochitosan, when properly engineered, can result in tags that are resilient in aquatic environments, resisting wear and tear, and minimizing tag loss, which is a common issue in fish tagging (Lee et al., 2010).
(iii) Reduced Tag Effects: Fish tagging can induce stress and potentially harm fish, especially if invasive tagging methods are used. Nanochitosan-based tags may be designed to minimize these tag effects, making them less invasive and more compatible with the welfare of the tagged fish.
(iv) Biodegradability: Nanochitosan is biodegradable, which can be advantageous in applications where long-term monitoring is not required. Biodegradable tags reduce environmental impacts and are suitable for tagging studies that do not need extended tracking (Zhao et al., 2018).
(v) Drug Delivery and Biomedical Applications: While not directly related to tagging, nanochitosan has applications in the development of drug delivery systems. In aquaculture and fisheries, this could be used for controlled delivery of medications or vaccines to tagged fish to improve their health and reduce disease risks (Zaki et al., 2015; Zhao et al., 2018).
(vi) Integration with Existing Tagging Methods: Nanochitosan-based tags could be integrated with existing tagging methods, such as injecting or embedding the nanochitosan tag into a fish's body. This could provide an additional layer of tracking or data collection.

It's important to note that the application of nanochitosan in fish tagging is an area of ongoing research and development. While it holds promise, more studies are needed to assess its effectiveness, long-term effects on fish, and environmental impacts. Researchers and fisheries managers should carefully evaluate and test nanochitosan-based tagging methods before implementing them on a large scale. The development of biocompatible and eco-friendly tagging materials is a significant step towards more sustainable and ethical fish tagging practices (Wang & Li, 2011).

7 Fish Barcoding

Nanochitosan in nano-barcoding of fish refers to the use of nanochitosan-based materials for creating unique and traceable tags, often at the nanoscale, that can be used to identify and track individual fish or fish populations (Pine et al., 2012). This innovative approach has the potential to revolutionize fish identification, monitoring, and research in several ways:

- (i) Nanoscopic Barcodes: Nanochitosan, due to its nanoscale properties, can be engineered to create extremely small, unique, and durable barcodes that can be incorporated into fish tissues or tags. These barcodes can carry identifying information, such as species data, individual IDs, or location-specific markers.
- (ii) Biocompatibility: Nanochitosan-based barcodes are biocompatible, meaning they are less likely to induce stress or harm in tagged fish. This is essential for ethical and welfare considerations when tagging fish, as it minimizes potential adverse effects.
- (iii) Long-Term Tracking: Nanochitosan-based barcodes can provide a means of long-term tracking, allowing researchers to monitor the movements and behaviour of tagged fish over extended periods. This is particularly valuable for studying fish migration, habitat use, and population dynamics.
- (iv) Environmental Friendliness: Nanochitosan is biodegradable and eco-friendly. This means that if fish are tagged with nanochitosan-based barcodes, there is less risk of introducing non-biodegradable materials into aquatic ecosystems, aligning with sustainability and conservation goals (Zaki et al., 2015).
- (v) Customizable Tags: Researchers can design nanochitosan-based tags with specific characteristics, such as size, shape, or composition, to suit the needs of their study. These customizable tags can be tailored for different species, fish sizes, or research objectives.
- (vi) Non-Invasive Methods: Depending on the tagging method used, nanochitosan-based barcodes can be applied using non-invasive techniques. For example, fish could be exposed to nanochitosan-tagged particles in their environment, which are then absorbed or ingested, resulting in internal barcodes.
- (vii) Data Retrieval: Nanochitosan-based barcodes can be designed to be easily read and retrieved. Advanced technology, such as microscopes or spectroscopy, can be used to scan and decode the barcodes, facilitating data collection.
- (viii) Fish Authentication: Nanochitosan-based barcodes can serve as a means of authenticating fish products in the fisheries and aquaculture industry. By scanning the barcode, the species and origin of the fish can be verified, which is crucial for preventing fraud.

While the concept of nanochitosan in the nano-barcoding of fish holds significant promise, it is an emerging field, and research is ongoing to refine the technology, validate its effectiveness, and address potential challenges. Ethical considerations and the well-being of tagged fish must also be a focus in developing and implementing these innovative tagging methods. As technology and research in this area progress, it may offer new insights and opportunities for the sustainable management of fish populations and the responsible utilization of aquatic resources (Macaulay et al., 2021).

8 Future Directions and Innovations

8.1 Comparison of Nanotags with Other Fish Tagging Techniques

Emerging technologies and advancements in tagging have the potential to revolutionize the way researchers and scientists track and monitor various species, including fish, wildlife, and even microorganisms (Jepsen et al., 2015). These innovations are expanding our understanding of animal behaviour, movement, and ecology. Some of the notable emerging technologies and advancements in tagging include:

(i) Bio-Logging and Biotelemetry: Bio-logging involves the use of sensors and data loggers attached to animals, allowing for the collection of a wide range of data, including GPS coordinates, temperature, depth, and more (Piramuthu & Zhou, 2016; Shen et al., 2020). Biotelemetry systems transmit data to satellites or receivers in real time, providing immediate access to information about animal movements (Fig. 1).

(ii) Environmental DNA (eDNA): eDNA involves the detection and analysis of DNA shed by organisms into the environment, such as water or soil. This non-invasive technique can be used to monitor the presence and distribution of species, including those that are difficult to observe directly.

(iii) Acoustic Telemetry and Passive Integrated Transponders (PIT): Acoustic telemetry involves attaching acoustic transmitters to animals (Wilder et al., 2016). Receivers placed underwater can track the movements and behaviour of tagged organisms. PIT tags are small, passive electronic tags used for tracking fish in aquaculture and fisheries (Wilder et al., 2016; Musselman et al., 2017).

(iv) Satellite and GPS Tracking: Advances in satellite and GPS technology have made it possible to track animals over large geographic areas. These tracking methods provide crucial insights into the migration and navigation of animals, including birds and marine species.

(v) Remote Sensing and Drones: Remote sensing technologies, including satellite imagery and aerial drones, offer a bird's-eye view of ecosystems and animal habitats. This information aids in the monitoring of animal populations and environmental changes.

(vi) Biologging with Miniaturized Devices: Miniaturized tracking devices, such as accelerometers and gyroscopes, can be attached to animals to record their

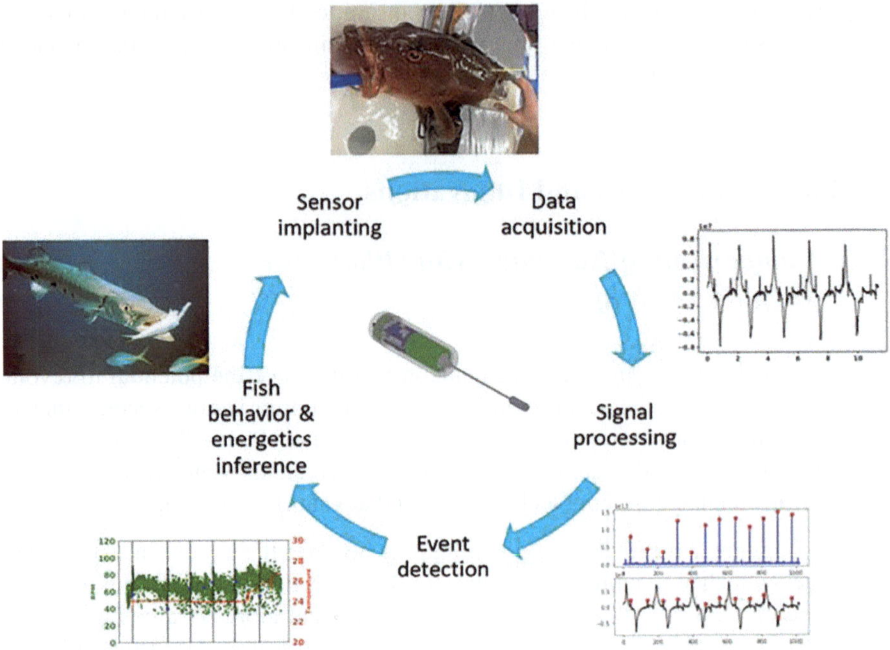

Fig. 1 Application of implantable biosensor for monitoring fish energetics. (Source: Shen et al., 2020)

movements, behaviour, and activity patterns. These devices are particularly useful for studying animal behaviour and physiology.

(vii) Camera Traps and Image Recognition: Camera traps equipped with motion sensors and image recognition technology are used for monitoring terrestrial wildlife. These devices capture images of animals, allowing researchers to assess population density and species diversity.

(viii) RFID and NFC Technology: Radio-frequency identification (RFID) and near-field communication (NFC) technology are used to track and monitor animals, especially in captivity or controlled environments (Fig. 2). These technologies are employed in studies of fish, birds, and other species (Piramuthu & Zhou, 2016; Keena, 2023).

(ix) Machine Learning and Artificial Intelligence: Machine learning and artificial intelligence are used to process and analyze large volumes of tracking data. These technologies help researchers uncover complex patterns in animal behaviour and predict future movements.

(x) Autonomous Underwater Vehicles (AUVs) and Gliders: AUVs and gliders equipped with various sensors are used to study marine life and collect data on water conditions and the behaviour of marine organisms. These autonomous vehicles can operate for extended periods, collecting valuable data in remote ocean regions.

(xi) Nanochitosan tags/Nanotags: Nanotechnology has enabled the development of nanoscale tags and sensors that can be attached to tiny or sensitive organisms. These nanotags are often biocompatible, allowing for the tracking of even smaller species or organisms without causing harm.

These emerging technologies and advancements are enhancing our ability to monitor and understand the natural world. They are contributing to advancements in conservation, ecology, and wildlife management, as well as expanding our knowledge of ecosystems and the behaviour of species. However, ethical considerations, data privacy, and the well-being of tagged or tracked animals remain important aspects of research in this field.

Sustainable and responsible fish tagging practices are therefore essential to ensure that fisheries research and management efforts do not harm the well-being of the fish being studied and maintain the integrity of aquatic ecosystems. When planning and conducting fish tagging studies, researchers and fisheries managers should adhere to several key principles to promote ethical and environmentally conscious practices (Wang et al., 2018). It is thus crucial to consider the welfare of the fish and prioritize their well-being throughout the tagging process. It is also important to minimize stress, injury, and any potential negative impacts on the health and behaviour of tagged fish. Aquaculture centres are therefore encouraged to use tagging methods that are suitable for the specific fish species under study. Not all tagging methods are appropriate for all species, and consideration should be given to the size, physiology, and behaviour of the fish (Lü et al., 2019).

Non-invasive tagging methods that do not require physical intrusion or surgery such as external tags, fin clips, and tissue sampling are highly recommended. The use of biocompatible tagging materials that are less likely to cause harm or irritation to fish is highly recommended. Nanochitosan is a material that is non-toxic and does not adversely affect the fish's health (Meerbeek, 2017).

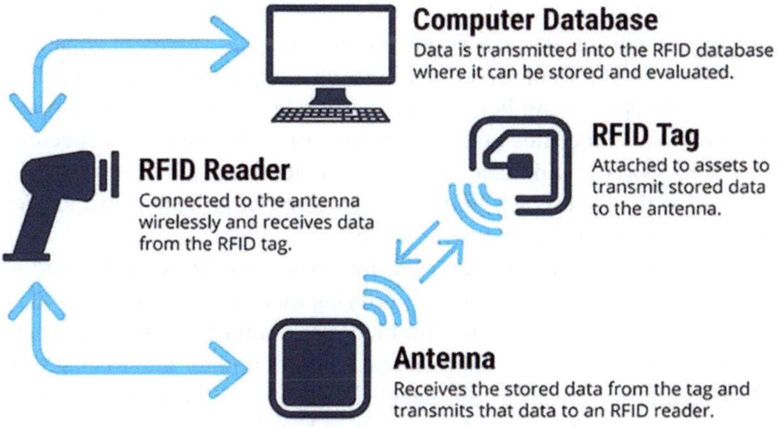

Fig. 2 Components of RFID system. (Source: Keena, 2023)

Furthermore, the size and weight of tags must be appropriate for the size and swimming capabilities of the fish. Oversized or heavy tags can impede fish movement and behaviour. Among the various options, nanotags stand out as a minute material with insignificant weight, yet a novel biopolymer for fish tagging. Careful attachment and deployment of tags are required to minimize the risk of tag loss or displacement. Secure attachment methods are recommended to minimize discomfort (Pine et al., 2012; Postulkova et al., 2016).

Designing tagging programs with mechanisms to retrieve tags or recapture tagged fish at the end of the study is required for non-biodegradable tagging materials so that tags do not persist in the environment after the study is completed (Viacava et al., 2017). Again, nanochitosan being a biocompatible and biodegradable material poses no such challenge of environmental persistence; hence it is an eco-friendly and sustainable tagging technique. Based on these advantages conducting thorough environmental impact assessments before initiating tagging programs is not a critical requirement for nanochitosan-based fish tagging, as nanotags aid seamless adherence to local, national, and international regulations and ethical guidelines related to fish tagging (Macaulay et al., 2021).

8.2 Aquacultural Fish Barcoding

Fish barcoding is a molecular identification technique used in fisheries and aquaculture to accurately and rapidly identify fish species and monitor seafood products throughout the supply chain. It involves the use of DNA barcodes, which are short, standardized DNA sequences, to identify fish species based on their genetic profiles (Viacava et al., 2017).

Fish barcoding provides a highly accurate method for identifying fish species, even in cases where traditional morphological methods may be challenging due to factors like incomplete or processed specimens. It helps confirm the species identity of fish and seafood products, reducing the risk of mislabeling, fraud, or substitution. Fish barcoding can be used to assess and monitor the biodiversity of aquatic ecosystems, which is crucial for conservation efforts and understanding the impact of fishing practices. Barcoding can help ensure traceability in the seafood supply chain, helping regulatory authorities and consumers verify the origin and species of fish products. It supports compliance with legal regulations and international agreements related to the fishing and trade of endangered or protected species (Wang & Li, 2011; Wang et al., 2018; Pudake et al., 2019).

Barcoding can be used to monitor the freshness and safety of fish and seafood products by identifying the species and verifying their quality and origin. It helps in the detection of adulteration or substitution of premium or endangered species with cheaper alternatives. Fish barcoding can be used to monitor and manage stocks in aquaculture facilities, ensuring the genetic integrity of the farmed fish. It helps detect and prevent unintended hybridization in breeding programs. Barcoding can assist in identifying disease-causing pathogens in fish populations, contributing to effective disease management and biosecurity measures (Han et al., 2013; Wang & Li, 2011; Yoshida et al., 2014). Fish barcoding supports research on the evolutionary relationships among fish species, helping scientists understand their genetic diversity and origins. It aids in studying the genetic structure and gene flow in fish populations, as well as detecting invasive fish species early, helping in the management and eradication of these species to protect native ecosystems.

Barcoding can facilitate international trade by providing a standardized method for species identification and meeting the import/export requirements of different countries. It enhances consumer confidence by ensuring the authenticity and safety of fish products. Furthermore, fish barcoding can be used to identify and protect endangered fish species, contributing to their conservation and management (Lee et al., 2010).

9 Conclusion

Responsible and sustainable fish tagging practices are crucial for the ethical and effective study of aquatic ecosystems and the conservation of fish populations. By harnessing the numerous benefits of nanochitosan, consideration of the well-being of fish and the environmental impact of tagging activities are ascertained, hence researchers and fisheries managers can contribute to the responsible and sustainable utilization of aquatic resources.

Fish barcoding is a valuable tool in fisheries and aquaculture, offering precise species identification, ensuring the authenticity of seafood products, supporting biodiversity conservation, and facilitating compliance with regulations and market demands. It is increasingly becoming an integral part of sustainable and responsible fishing and aquaculture practices.

Fish nanotags and barcoding are complementary technologies that provide critical tools for fisheries management, ecological research, and the conservation of fish species. Nanotags offer real-time tracking and behavioural insights, while barcoding ensures accurate and standardized species identification. Together, these technologies play a vital role in promoting sustainable fishing practices, preserving aquatic ecosystems, and safeguarding the seafood supply chain's integrity.

References

Dawit, M. F., Patel, P., Parashar, S. K. S., & Das, B. (2020). Mechanistic insights into diverse nano-based strategies for aquaculture enhancement: A holistic review. *Aquaculture, 519*, 734770. https://doi.org/10.1016/j.aquaculture.2019.734770

Dewar, H., Wilson, S. G., Hyde, J. R., Snodgrass, O. E., Leising, A., & Lam, C. H. (2018). Basking shark (*Cetorhinus maximus*) movements in the eastern North Pacific were determined using satellite telemetry. *Frontiers in Marine Science, 5*, 163.

Fajardo, C., Martinez-Rodriguez, G., Blasco, J., Mancera, J. M., Thomas, B., & de Donato, M. (2022). Nanotechnology in aquaculture: Applications, perspectives and regulatory challenges. *Aquaculture and Fisheries, 7*(2), 185–200. https://doi.org/10.1016/J.AAF.2021.12.006

Gabriel, N. N., Habte-Tsion, H. M., & Haulofu, M. (2022). *Perspectives of nanotechnology in aquaculture: Fish nutrition, disease, and water treatment. Emerging nanomaterials for advanced technologies* (pp. 463–485). Springer. https://doi.org/10.1007/9783-030-80371-1_15

Hamed, I., Özogul, F., & Regenstein, J. M. (2016). Industrial applications of crustacean by-products (chitin, chitosan, and chitooligosaccharides): A review. *Trends in Food Science and Technology, 48*, 40–50.

Han, D., Hong, J., Kim, H. C., Sung, J. H., & Lee, J. B. (2013). Multiplexing enhancement for the detection of multiple pathogen DNA. *Journal of Nanoscience and Nanotechnology, 13*(11), 7295–7299.

Jepsen, N., Thorstad, E. B., & Havn, T. (2015). The use of external electronic tags on fish: An evaluation of tag retention and tagging effects. *Animal Biotelemetry, 3*, 49–54.

Keena, M. (2023). *What is RFID technology and how does it work? A guide for retail (2024).* https://www.shopify.com/ng/retail/rfid-technology. Accessed 4 Nov 2023.

Khosravi-Katuli, K., Prato, E., Lofrano, G., Guida, M., Vale, G., & Libralato, G. (2017). Effects of nanoparticles in species of aquaculture interest. *Environmental Science and Pollution Research, 24*, 17326–17346. https://doi.org/10.1007/s11356-017-9360-3

Lee, J. B., Campolongo, M. J., Kahn, J. S., Roh, Y. H., Hartman, M. R., & Luo, D. (2010). DNA based nanostructures for molecular sensing. *Nanoscale, 2*(2), 188–197.

Lü, H. J., Fu, M., Dai, S., Xi, D., & Zhang, Z. X. (2016). Experimental evaluation of calcein and alizarin red S for immersion marking of silver carp *Hypophthalmichthys molitrix* (Valenciennes, 1844). *Journal of Applied Ichthyology, 32*(1), 83–91.

Lü, H., Fu, M., Zhang, Z., Su, S., & Yao, W. (2019). Marking fish with fluorochrome dyes marking fish with fluorochrome dyes. *Reviews in Fisheries Science, 28*(1), 117–135.

Macaulay, G., Warren-Myers, F., Barrett, L. T., Oppedal, F., Føre, M., & Dempster, T. (2021). Tag used to monitor fish behavior in aquaculture: A review of benefits, problems and solutions. *Reviews in Aquaculture, 13*(2), 1565–1582.

Meerbeek, J. (2017). Retention of visible implant alphanumeric tags in adult Walleye. *Journal of Freshwater Ecology, 32*, 721–725.

Musselman, W. C., Worthington, T. A., Mouser, J., & Williams, D. M. (2017). Passive integrated transponder tags: Review of studies on warm water fishes with notes on additional species. *Journal of Fish and Wildlife Management, 8*(2), 353–364.

Pine, W. E., Hightower, J. E., Coggins, L. G., Lauretta, M. V., & Pollack, K. H. (2012). Design and analysis of tagging studies. In A. V. Zale, D. L. Parrish, & T. M. Sutton (Eds.), *Fisheries techniques* (3rd ed., pp. 521–530). American Fisheries Society.

Piramuthu, S., & Zhou, W. (2016). *RFID and sensor network automation in the food industry: Ensuring quality and safetythrough supply chainvisibility* (pp. 1–34). Wiley.

Postulkova, E., Mares, J., Halacka, K., & Kopp, R. (2016). Toxic effect of fluorescence pigment on zebra fish (*Danio rerio*). *Mendel Net, 23*, 347–351.

Pudake, R., Chauhan, N., & Kole, C. (2019). *Nanoscience for sustainable agriculture*. Springer. https://doi.org/10.1007/978-3-319-97852-9

Shen, Y., Arablouei, R., Hoog, F. D., Malan, J., Sharp, J., Shouri, S., Clark, T. D., Lefevre, C., Kroon, F., Severati, A., and Kusy, B. (2020). Estimating heart rate and detecting feeding events of fish using an implantable biologger. 12

Viacava, G. E., Vázquez, F. J., Ayala-Zavala, J. F., & Ansorena, M. R. (2017). Sustainability challenges involved in use of nanotechnology in the agrofood sector. In *Sustainability challenges in the agrofood sector* (pp. 343–368). Wiley.

Wang, Y., & Li, J. (2011). Effects of chitosan nanoparticles on survival, growth and meat quality of tilapia, *Oreochromis nilotica. Nanotoxicology, 5*(3), 425–431. https://doi.org/10.3109/1743539 0.2010.530354

Wang, Y. C., Mohan, C. O., Guan, J., Ravishankar, C. N., & Gunasekaran, S. (2018). Chitosan and gold nanoparticles-based thermal history indicators and frozen indicators for perishable and temperature-sensitive products. *Food Control, 85*, 186–193. https://doi.org/10.1016/J.FOODCONT.2017.09.031

Wilder, R. M., Hassrick, J. L., Grimaldo, L. F., Greenwood, M. F. D., Acuna, S., & Burns, J. M. (2016). Feasibility of passive integrated transponder and acoustic tagging for endangered adult delta smelt. *North American Journal of Fisheries Management, 36*, 1167–1177.

Yoshida, C. M. P., Maciel, V. B. V., Mendonça, M. E. D., & Franco, T. T. (2014). Chitosan biobased and intelligent films: Monitoring pH variations. *Food Science and Technology, 55*(1), 83–89. https://doi.org/10.1016/J.LWT.2013.09.015

Zaki, M., Salem, M., Gaber, M., & Nour, A. (2015). Effect of chitosan supplemented diet on survival, growth, feed utilization, body composition and histology of Sea Bass (*Dicentrarchus labrax*). *World Journal of Engineering and Technology, 3*, 38–47. https://doi.org/10.4236/wjet.2015.34C005

Zhao, D., Yu, S., Sun, B., Gao, S., Guo, S., & Zhao, K. (2018). Biomedical applications of chitosan and its derivative nanoparticles. *Polymers, 10*(4), 462. https://doi.org/10.3390/polym10040462

Nanochitosan-Based Enhancement of Fish Breeding Programs

Patrick Omoregie Isibor

Contents

1 Introduction

Nanochitosan, a derivative of chitosan, has swiftly emerged as a transformative element in fish breeding programs, significantly influencing aquaculture practices. Its growing prominence stems from distinctive attributes promising diverse

P. O. Isibor (✉)
Department of Biological Sciences, College of Science and Technology, Covenant University, Ota, Ogun State, Nigeria
e-mail: patrick.isibor@covenantuniversity.edu.ng

P. O. Isibor et al. (eds.), *Nanochitosan-Based Enhancement of Fisheries and Aquaculture*, https://doi.org/10.1007/978-3-031-52261-1_10

advantages, not only in enhancing fish health but also in revolutionizing the broader spectrum of aquaculture methodologies.

Distinguished by its minute nanoscale dimensions and exceptional biocompatibility, nanochitosan presents a multifaceted appeal within aquaculture. Its potential to augment fish reproduction is particularly noteworthy. These nanoparticles possess inherent capabilities to enhance the absorption of essential nutrients vital for fish health (Ahmed et al., 2019). This attribute significantly supports the growth and development of reproductive organs in fish, thereby contributing to an observable enhancement in the reproductive cycle, leading to increased spawning frequency. Furthermore, nanochitosan intervention positively influences the quality of fish eggs, resulting in superior egg quality, thereby amplifying the survival rates of fish larvae (Barchanski et al., 2015). This domino effect, originating from nutrient uptake, cascades into improved reproductive success and subsequently augments the sustainability of fish populations. This advancement is pivotal not only for individual fish but also for broader aquatic ecosystems. The amplified spawning frequency and improved egg quality directly impact biodiversity preservation and equilibrium within aquatic environments. By fortifying the reproductive mechanisms of fish, nanochitosan indirectly contributes to the sustainability of fish stocks, pivotal in maintaining the delicate balance of aquatic ecosystems. Additionally, the increased survival rates of fish larvae lead to more robust, resilient populations, alleviating pressures on wild fish resources and contributing to species conservation efforts (Ahmed et al., 2019).

In essence, the integration of nanochitosan into fish breeding programs represents a paradigm shift in aquaculture methodologies. Its capacity to optimize reproductive processes, elevate egg quality, and subsequently bolster fish larvae survival rates not only holds potential in meeting global seafood demands but also underscores its pivotal role in supporting ecological equilibrium essential for sustaining aquatic ecosystems. As ongoing research continues to uncover its potential, nanochitosan emerges as a symbol of innovation in aquaculture, paving the way for more efficient, sustainable, and environmentally conscious approaches to fish breeding (Barchanski et al., 2015).

2 Nanochitosan Applications in Fish Breeding

Nanochitosan, derived from chitosan, has emerged as a promising tool in enhancing fish breeding programs, offering multifaceted applications within aquaculture practices. Its unique properties and biocompatibility render it invaluable in improving various aspects of fish reproduction and health. Nanochitosan plays a pivotal role in bolstering reproductive health among fish. Its nano-sized particles facilitate the efficient uptake of vital nutrients essential for reproductive organ development, consequently heightening spawning frequency. Intervention with nanochitosan yields eggs of superior quality, impacting the survival rates of fish larvae. This enhancement in egg quality significantly contributes to the resilience of fish populations (Bhat et al., 2016).

Nanochitosan's biocompatibility and non-toxic characteristics ensure its suitability in fish breeding without adverse environmental impacts, making it an eco-conscious choice for aquaculture. By elevating spawning frequency and enhancing egg quality, nanochitosan indirectly influences biodiversity maintenance and equilibrium within aquatic environments. It aids in sustaining fish stocks and reduces pressures on wild fish resources, contributing to species conservation (Bhat et al., 2018). Integration of nanochitosan into fish breeding programs signifies a step towards more efficient and sustainable aquaculture practices. Its capacity to optimize reproductive processes holds promise in meeting global seafood demands while preserving ecological balance.

Nanochitosan's applications in fish breeding demonstrate its potential to revolutionize aquaculture practices, promising efficient, eco-conscious, and sustainable solutions for the challenges faced in fish reproduction and conservation (Bhat et al., 2019).

2.1 Nanochitosan-Based Improvement of Reproductive Health

Enhancing gonadal development through nanochitosan application represents a cutting-edge approach in aquaculture, offering the remarkable potential to optimize reproductive processes in fish. The application of nanochitosan, owing to its unique properties, has shown promise in augmenting gonadal development, leading to improved reproductive outcomes (Bhat et al., 2021).

Nanochitosan's nano-sized particles possess exceptional properties that facilitate enhanced nutrient absorption. When administered as a supplement in fish diets, nanochitosan optimizes the uptake of vital nutrients essential for gonadal development. This includes micronutrients critical for reproductive health, such as vitamins, minerals, and amino acids. The improved absorption aids in fortifying the gonadal tissues, promoting their growth and maturity. Nanochitosan's biocompatible nature allows for interaction at the cellular level. It can potentially modulate hormonal pathways involved in reproductive processes. By influencing the release and regulation of reproductive hormones, nanochitosan plays a role in stimulating gonadal development, thereby enhancing the maturation of reproductive organs (Bhat et al., 2021).

Stress can adversely impact gonadal development in fish. Nanochitosan's ability to act as an anti-stress agent due to its immunomodulatory and antioxidant properties can create a conducive environment for reproductive health. Reducing stress levels in fish contributes positively to gonadal development and the overall reproductive cycle. The application of nanochitosan has been associated with improvements in egg quality. Fish fed with diets supplemented with nanochitosan tend to produce eggs of superior quality, ensuring better chances of successful fertilization and hatching. Furthermore, the optimization of gonadal development often leads to increased spawning frequency, enhancing the overall reproductive success of fish populations (Bhat, 2023). Enhancing gonadal development through nanochitosan supplementation offers far-reaching implications for aquaculture sustainability. Improved reproductive outcomes can lead to increased production and yield, addressing the global demand for seafood. Moreover, by boosting the reproductive success of cultured fish, reliance on wild fish stocks can be reduced, contributing to conservation efforts and maintaining ecological balance (Durfey et al., 2019).

While the potential of nanochitosan in enhancing gonadal development is promising, further research is needed to elucidate optimal dosages, application methods, and species-specific responses. Its application in aquaculture holds immense promise in revolutionizing reproductive processes, ensuring sustainable fish breeding practices, and meeting the growing demands for seafood globally. Promoting gamete quality and viability using nanochitosan introduces a novel avenue in aquaculture, offering potential advancements in optimizing reproductive success in fish. Nanochitosan, with its unique properties, has shown promise in enhancing the quality and viability of gametes, significantly impacting reproductive outcomes (Hajiyeva et al., 2022).

Nanochitosan's nanoscale structure facilitates improved nutrient absorption in fish. When integrated into fish diets or as a supplement, nanochitosan enhances the absorption and utilization of essential nutrients vital for gamete development. This optimized nutrient intake supports metabolic processes critical for the production of healthy and high-quality gametes. The inherent antioxidant and anti-inflammatory properties of nanochitosan contribute to reducing oxidative stress and inflammation in fish. By mitigating cellular damage caused by free radicals, nanochitosan helps maintain the integrity of gametes, enhancing their quality and viability. Nanochitosan's biocompatibility allows it to interact at the cellular level, potentially influencing the regulation of reproductive hormones (Hassanein et al., 2021). It can modulate hormone pathways involved in gamete production, maturation, and release. This regulation supports the synchronization and optimization of gamete quality.

Environmental stressors, such as pollutants and fluctuations in water parameters, can impact gamete quality. Nanochitosan acts as a protective shield by adsorbing and neutralizing harmful substances, creating a more conducive environment for gamete development and viability. The application of nanochitosan has shown promise in enhancing fertilization rates. Fish exposed to nanochitosan-supplemented environments exhibit increased rates of successful fertilization due to the improved quality and viability of gametes (Durfey et al., 2019; Jahaabad et al., 2020). By bolstering gamete quality and viability, nanochitosan contributes to the sustainable production of healthy fish populations. The resulting increase in successful fertilization, embryo development, and survival rates positively impacts aquaculture sustainability and reduces reliance on wild fish stocks (Jahaabad et al., 2020).

While the potential of nanochitosan in promoting gamete quality and viability is promising, further research is essential to optimize its application methods, dosages, and species-specific responses. Its integration into aquaculture practices holds substantial promise in elevating reproductive success, ensuring healthier fish populations, and meeting global demands for sustainable seafood production.

2.2 Assisted Reproductive Techniques

Artificial spawning induction, a crucial aspect of aquaculture, relies on efficient techniques to stimulate and regulate the reproductive cycle of fish. Nanochitosan, a nanomaterial derived from chitosan, has emerged as a promising agent in this field, offering innovative solutions to enhance the efficiency and success of artificial spawning induction processes.

Nanochitosan's biocompatibility and nano-sized particles have demonstrated the potential to influence the hormonal pathways involved in fish reproduction. By interacting with cellular mechanisms, nanochitosan can modulate the production, release, and regulation of reproductive hormones. This modulation plays a significant role in stimulating the onset of spawning, synchronizing reproductive cycles, and enhancing the efficiency of artificial spawning induction. In optimization of gonadal development, nanochitosan's unique properties facilitate better nutrient absorption and utilization in fish. When incorporated into fish diets or administered as a supplement, nanochitosan optimizes the uptake of essential nutrients crucial for gonadal development. This enhancement contributes to the maturation and growth of reproductive organs, promoting efficient spawning in induced breeding programs (Joshi et al., 2015; Jehanabad et al., 2019).

In the enhancement of egg quality and quantity, the application of nanochitosan has shown promising results in improving both the quality and quantity of eggs produced during artificial spawning. Fish exposed to nanochitosan-supplemented environments tend to produce eggs of superior quality, with higher fertilization rates. Additionally, nanochitosan's role in stimulating reproductive hormones often leads to increased egg production, further enhancing the success of artificial spawning induction.

Stress and environmental factors can hinder the success of artificial spawning induction. Nanochitosan's antioxidant and immunomodulatory properties help alleviate stress and neutralize pollutants, creating a favourable environment for fish reproduction (Joshi et al., 2019). This mitigation of stressors contributes to enhanced spawning efficiency and higher survival rates of spawned eggs and larvae. The integration of nanochitosan into artificial spawning induction aligns with sustainable aquaculture practices. Its ability to improve spawning efficiency, egg quality, and larval survival rates supports the sustainable production of fish stocks, reducing reliance on wild populations and promoting ecological balance. Nanochitosan holds immense promise in revolutionizing artificial spawning induction in aquaculture. Its multifaceted role in stimulating reproductive hormones, optimizing gonadal development, improving egg quality, and mitigating environmental stressors underscores its potential to elevate the efficiency and sustainability of induced breeding programs (Kookaram et al., 2021). As research continues to unveil its applications, nanochitosan stands as a pivotal innovation, offering pathways to more effective and sustainable fish reproduction in aquaculture.

Artificial spawning induction, a crucial aspect of aquaculture, relies on efficient techniques to stimulate and regulate the reproductive cycle of fish. Nanochitosan, a nanomaterial derived from chitosan, has emerged as a promising agent in this field, offering innovative solutions to enhance the efficiency and success of artificial spawning induction processes. For stimulation of reproductive hormones, nanochitosan's biocompatibility and nano-sized particles have demonstrated the potential to influence the hormonal pathways involved in fish reproduction. By interacting with cellular mechanisms, nanochitosan can modulate the production, release, and regulation of reproductive hormones. This modulation plays a significant role in stimulating the onset of spawning, synchronizing reproductive cycles, and enhancing the efficiency of artificial spawning induction.

2.3 Enhancing Larval Development

Improving fish larval health in aquaculture comes with its share of challenges, impacting successful rearing and the development of robust fish populations (Fatma et al., 2019). Formulating nutritionally balanced feeds for diverse larval species with specific dietary requirements is challenging. Meeting the nutritional needs of various developmental stages and species remains a complex task.

Larvae are highly sensitive to environmental changes. Maintaining optimal water quality parameters consistently throughout their development is challenging, especially when dealing with temperature fluctuations, water chemistry variations, or sudden changes in habitat conditions. Larvae are particularly susceptible to diseases due to their underdeveloped immune systems. Preventing and managing diseases without compromising larval health remains a challenge, especially in high-density rearing environments. Encouraging larval acceptance of artificial feeds can be difficult. Larvae might exhibit finicky feeding behaviour, requiring specialized feeding techniques to ensure adequate nutrient intake. Larvae are sensitive to stressors, including handling, transportation, or changes in environmental conditions. Mitigating stress and maintaining optimal conditions are essential but challenging, especially during critical developmental stages. Maintaining strict biosecurity measures to prevent the introduction of pathogens into rearing systems is challenging. The risk of pathogen transmission from external sources remains a constant concern. Limited research and technological advancements in larval-rearing methods and disease management strategies hinder the development of more effective and innovative solutions for optimizing larval health.

Implementing optimal larval health strategies can be cost-intensive. High-quality feeds, disease prevention measures, and constant monitoring can add to production expenses. Skilled personnel with expertise in larval rearing techniques and disease management may be limited, impacting the implementation of effective strategies. Different fish species have unique requirements and challenges in their larval rearing. Developing species-specific protocols and diets can be demanding and time-consuming.

Addressing these challenges requires continuous research and innovation on nanochitosan augmentation to develop more efficient and sustainable larval-rearing practices in aquaculture. Finding solutions to these challenges is crucial for ensuring healthy and successful larval rearing, contributing to sustainable fish production. Nanochitosan's application in larval feeding and growth represents a promising avenue in aquaculture, offering novel strategies to enhance the health, development, and survival of fish larvae. This nanomaterial, derived from chitosan, possesses unique properties that can significantly impact larval feeding practices and foster optimal growth. Nanochitosan's nano-sized particles facilitate improved nutrient absorption in fish larvae. When incorporated into larval feed formulations, nanochitosan optimizes the assimilation of essential nutrients critical for growth and development. This enhanced nutrient uptake contributes to better overall health and accelerated growth rates in larvae. Furthermore, for improved digestibility and feed conversion, the inclusion of nanochitosan in larval diets has demonstrated improved digestibility and feed conversion efficiency. Its nano-scale structure aids in breaking down feed components, making nutrients more accessible for absorption. This enhancement in digestibility ensures that larvae can efficiently utilize nutrients, promoting robust growth.

Nanochitosan's immunomodulatory properties contribute to bolstering the immune systems of fish larvae. By stimulating immune responses, nanochitosan helps larvae combat potential pathogens, reducing susceptibility to diseases. This reinforcement of immune defences contributes to higher survival rates and healthier larval populations. Nanochitosan's antioxidant properties play a crucial role in reducing oxidative stress in fish larvae. Environmental stressors can adversely impact larval growth. Nanochitosan acts as a shield against oxidative damage, minimizing stress and promoting a favourable environment for optimal growth and development.

The application of nanochitosan in larval feeding aligns with sustainable aquaculture practices. Its biodegradability and eco-friendly nature make it an attractive alternative to synthetic compounds. Additionally, by improving larval health and growth, nanochitosan contributes to the sustainable production of fish stocks, reducing pressure on wild fish populations. Nanochitosan's intervention in larval feeding practices often results in increased survival rates and improved quality of larvae. Enhanced growth, coupled with reinforced immune systems, leads to stronger, more resilient larvae, ensuring higher survival rates and better overall quality.

Nanochitosan's integration into larval feeding regimes holds significant promise in aquaculture. Its ability to enhance nutrient absorption, improve feed conversion, strengthen immune responses, mitigate stress, and promote sustainability underscores its potential to revolutionize larval-rearing practices. As research continues to unveil its applications, nanochitosan stands as a pivotal innovation, offering pathways to more efficient, sustainable, and healthier larval growth in aquaculture.

2.4 Nanochitosan Preparation Techniques for Fish Breeding

The preparation of nanochitosan for fish breeding involves a range of techniques aimed at optimizing its utilization in aquaculture. These methods enable the transformation of chitosan, derived from crustacean shells, into nanoscale particles with properties conducive to enhancing fish health, reproduction, and overall aquaculture sustainability (Kookaram et al., 2021).

Acid hydrolysis involves subjecting chitosan to acidic conditions, breaking down its larger molecular chains into nanoscale particles. By using acids like hydrochloric acid or acetic acid, controlled particle size and properties can be achieved. Adjusting acid concentration, temperature, and reaction time allows fine-tuning of nanochitosan characteristics (Levitin et al., 2014; Wijesena et al., 2015).

Ionic gelation is a technique that involves cross-linking chitosan molecules using ionic interactions, often employing agents like tripolyphosphate (TPP) to form nanoparticles. Controllable nanoparticle size is achievable by adjusting the chitosan-to-TPP ratio and manipulating reaction conditions, offering versatility in particle size modulation (Shard et al., 2014).

Nanoprecipitation entails the rapid mixing of a chitosan solution with a non-solvent such as ethanol or acetone, inducing precipitation to form nanoparticles. Control over nanoparticle size is achieved by adjusting solution concentrations and mixing conditions, allowing precise customization of particle dimensions (Kaya et al., 2013; Elsawy et al., 2016).

Coacervation is a method that induces the phase separation of chitosan from its solution by altering environmental conditions like pH or temperature, resulting in the formation of a coacervate. This coacervate can be further processed into nanoparticles. Coacervation is a versatile approach enabling the encapsulation of various substances within nanoparticles (Dubey et al., 2016).

Emulsion cross-linking involves emulsifying a chitosan solution in an oil phase and then employing agents like glutaraldehyde or genipin to cross-link and form nanoparticles. This technique is particularly useful for encapsulating hydrophobic substances within chitosan nanoparticles (Riegger et al., 2018).

Supercritical fluid technology is a process that dissolves chitosan in a supercritical fluid (e.g. supercritical carbon dioxide) and rapidly depressurizes to induce nanoparticle formation. This method provides control over nanoparticle size and morphology, offering versatility in nanoparticle engineering (Cardoso et al., 2022).

Enzymatic hydrolysis employs enzymes like lysozyme to break down chitosan into smaller particles, allowing controlled hydrolysis. This relatively gentle technique enables the production of nanochitosan with specific properties suited for various aquaculture applications (Fonseca et al., 2020).

These diverse methods offer tailored approaches to preparing nanochitosan optimized for fish breeding programs. The resulting nanoparticles possess specific characteristics tailored to enhance fish health, reproduction, and overall sustainability in aquaculture. By customizing nanoparticle properties, these techniques contribute to advancing aquaculture practices and supporting the long-term health and viability of fish populations.

2.5 Administration and Benefits of Nanochitosan in Fish Breeding

Nanochitosan's biocompatibility and non-toxic nature are key factors that contribute to its attractiveness and broad applicability across various fields, including medicine, food technology, and especially in the realm of environmental sciences and aquaculture.

Nanochitosan exhibits a remarkable affinity with biological tissues. Its nano-sized particles and chemical properties allow for seamless integration with living organisms without eliciting adverse reactions or immune responses. Due to its nanoscale structure, nanochitosan can interact effectively at the cellular level. This interaction facilitates various applications, including drug delivery systems and tissue engineering, as it doesn't interfere with cellular functions. Its biocompatibility minimizes inflammatory responses and the likelihood of rejection by the immune system when used in medical or biological applications, making it an ideal candidate for biomedical purposes. Nanochitosan exhibits a low toxicity profile, making it safer for use in various applications. Its natural origin from chitosan, a derivative of chitin found in crustacean shells, contributes to its non-toxic nature (Kou et al., 2018).

Nanochitosan's ability to break down into harmless byproducts upon degradation ensures minimal environmental impact. Its biodegradability makes it an eco-friendly alternative to synthetic compounds. In aquaculture applications, such as improving water quality or enhancing fish health, nanochitosan's non-toxic nature ensures minimal harm to aquatic ecosystems, fish, or other organisms in the environment. Nanochitosan's non-toxicity makes it suitable for use in food technology and packaging. It can enhance food preservation and safety without posing health risks to consumers. Nanochitosan's biocompatibility and non-toxic attributes make it a versatile and safe material for various applications. Its ability to interact harmoniously with biological systems, coupled with its minimal environmental impact, positions nanochitosan as a promising solution in advancing technologies and practices across multiple domains, including medicine, food science, and environmental management (Wu et al., 2020).

Regulating fish hormonal pathways for reproductive enhancement is a fundamental aspect of aquaculture, enabling controlled spawning, increased fertility, and improved reproductive success. Manipulating hormonal pathways involves influencing the endocrine system to induce or optimize reproductive processes. Hormonal treatments involve administering synthetic or natural hormones to fish to induce spawning or synchronize reproductive cycles. Common hormones include gonadotropin-releasing hormone analogs (GnRHa), luteinizing hormone-releasing hormone analogs (LHRHa), and synthetic mimics of naturally occurring fish hormones, aided by the infusion of nanochitosan for optimum delivery (Mahadevaswamy et al., 2023). Furthermore, manipulating light and temperature conditions mimics natural environmental cues, influencing fish reproductive behaviour. Controlling photoperiods and temperature regimes can further enhance spawning or synchronize reproductive cycles in some species. Fish ovaries release several hormones, primarily involved in the regulation of reproductive processes (Fig. 1). Fish ovaries produce estrogens, which are involved in the regulation of the oestrous or menstrual cycle, vitellogenesis (the process of yolk formation in eggs), and secondary sexual characteristics in females. Progestins are the hormones involved in ovulation and the preparation of the reproductive tract for fertilization and gestation. Gonadotropins are regulated by follicle-stimulating hormone (FSH), and luteinizing hormone (LH) is released from the pituitary gland to regulate the ovarian cycle, including follicular development, ovulation, and corpus luteum formation, while inhibin is the hormone that regulates the secretion of FSH from the pituitary gland, helping to modulate the reproductive cycle by providing negative feedback (Mylonas & Zohar, 2001; Mylonas et al., 2010). These hormones work in concert to regulate the reproductive processes in fish, including oogenesis (egg development), ovulation, and spawning, which may all be enhanced by the infusion of nanochitosan as illustrated in Fig. 1.

Nutritional intervention aided by nanochitosan inclusion in diet regimes plays a role in regulating fish hormones. Specialized diets containing specific nutrients, vitamins, or precursors to hormone synthesis can influence reproductive processes

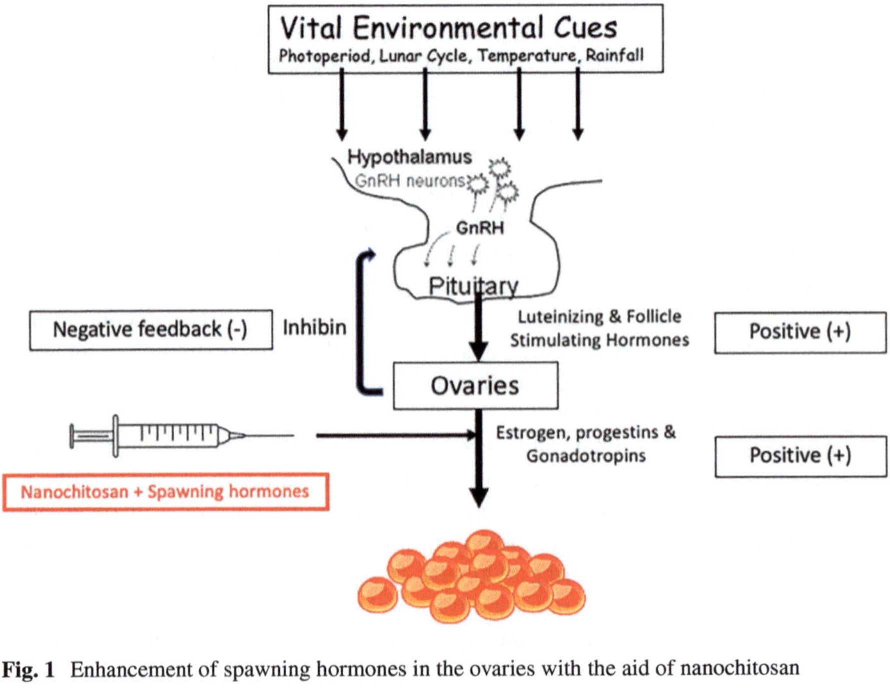

Fig. 1 Enhancement of spawning hormones in the ovaries with the aid of nanochitosan

and gonadal development (Palma et al., 2019). Stressors such as overcrowding, inadequate water quality, or abrupt environmental changes can disrupt hormonal balance. Maintaining optimal environmental conditions reduces stress, allowing fish to maintain healthy hormonal levels necessary for reproduction. Utilizing pheromones or chemical signals released by conspecifics can trigger reproductive behaviours or synchronization of spawning events. Pheromonal cues play a role in communication among fish, influencing their reproductive physiology. Advancements in biotechnology enable genetic manipulation to enhance reproductive traits in fish. Gene editing techniques aim to modify specific genes related to reproductive functions, potentially improving fertility or spawning success. Ongoing research aims to optimize hormonal treatments and strategies for different fish species. Tailoring hormonal treatments to specific species, considering sex, maturity stage, and environmental factors, enhances their efficacy and minimizes adverse effects (Rather et al., 2013).

Regulating fish hormonal pathways requires a comprehensive understanding of species-specific reproductive biology, endocrinology, and environmental factors. Ethical considerations, proper handling, and compliance with regulatory standards are crucial when using hormonal manipulation in aquaculture. Balancing effective reproductive enhancement with minimal stress and long-term health impacts on fish populations remains a focal point in aquaculture practices (Rather et al., 2016).

2.6 Dosage and Administration Strategies

2.6.1 Dosage

The application of nanochitosan in fish spawning involves a nuanced approach to ensure its effectiveness without causing adverse effects on the fish or the aquatic environment.

Initial Trials Commencing with lower concentrations of nanochitosan during the initial trials serves several crucial purposes. First, it allows for the evaluation of fish response to the compound. Observing behavioural changes, reproductive activity, and any signs of stress or discomfort becomes easier when using lower concentrations. Moreover, this approach helps in detecting potential adverse effects or toxicity that might arise from the use of nanochitosan (Rathor et al., 2017).

The concentrations used in these preliminary trials typically range from parts per million (ppm) to several ppm. Researchers often select concentrations based on previous studies, scientific literature, or initial pilot experiments. These concentrations act as a starting point, providing a baseline to assess the compound's impact on fish spawning.

Gradual Increase Following positive outcomes from initial trials without any adverse effects, a gradual increase in nanochitosan concentration can be contemplated. This incremental adjustment allows researchers or aquaculturists to gauge the threshold at which the compound delivers the desired outcomes without causing harm.

Throughout this incremental increase, close monitoring of fish behaviour, health, and reproductive performance becomes paramount. Any signs of stress, altered behaviour, or negative physiological responses should prompt a reassessment of the dosage. By incrementally raising the dosage, it's possible to identify an optimal concentration that enhances spawning success while maintaining fish welfare (Saha et al., 2018).

Species-Specific Consideration Different fish species exhibit distinct physiological and biochemical characteristics. These differences often result in varying tolerances and responses to external substances like nanochitosan. Therefore, tailoring the dosage based on species-specific considerations is crucial. For instance, species with different reproductive behaviours, metabolic rates, or body sizes might respond differently to the same concentration of nanochitosan. Understanding these variations enables aquaculturists or researchers to adjust the dosage accordingly, ensuring that the compound effectively stimulates spawning without causing harm. Considering these nuanced factors in dosage determination and administration strategies is essential for the successful application of nanochitosan in fish spawning. It highlights the importance of a systematic approach, constant monitoring, and the customization of dosage based on the specific needs and characteristics of the fish species involved in aquaculture practices (Shah & Mraz, 2020).

2.6.2 Administration Strategies

Incorporation in Feeding

Incorporating nanochitosan into fish feed represents a practical and non-invasive method to administer this compound in aquaculture settings. This approach capitalizes on the natural feeding behaviour of fish and simplifies the process of delivering nanochitosan without the need for individual handling of fish. Nanochitosan, being a versatile compound, can be effectively blended or coated onto the pellets or feed that fish regularly consume. This process ensures that the nanochitosan becomes an integral part of their diet. The compound may be absorbed through the gastrointestinal tract, allowing it to interact with the fish's internal systems. Fish are accustomed to feeding on pellets or other forms of feed provided to them. Integrating nanochitosan into their feed aligns seamlessly with their natural feeding habits. As fish consume their regular diet, they unknowingly ingest the nanochitosan along with the feed, making it a convenient and non-disruptive way to introduce the compound into their system.

Unlike some other administration methods, such as immersion or injection, mixing nanochitosan with fish feed streamlines the administration process. It eliminates the need for individual handling or direct contact with the fish. This reduces stress on the fish and minimizes potential disturbances to their environment, making it a more practical and efficient technique for large-scale applications in aquaculture. The process of incorporating nanochitosan into fish feed allows for a relatively controlled dosage. Aquaculturists can precisely determine the amount of nanochitosan added to the feed, ensuring a consistent and measured intake by the fish. This control over dosage is crucial in avoiding potential overdosage or under-dosage scenarios, optimizing the compound's effectiveness. Since fish feed is typically provided regularly, integrating nanochitosan into their diet facilitates continuous exposure to the compound over time. This sustained exposure could potentially enhance its effects on fish reproductive processes or other targeted outcomes. In essence, mixing nanochitosan with fish feed offers a practical, non-invasive, and controlled method for delivering the compound to fish in aquaculture settings. By leveraging the fish's natural feeding behaviour, this approach simplifies administration while ensuring consistent exposure, thereby potentially influencing reproductive processes or other desired outcomes in a controlled and manageable manner (Saha et al., 2018).

Immersion or Bath Treatment

In aquaculture, immersion or bath treatment involving nanochitosan entails exposing fish to a solution containing this compound for a designated period. This method aims to enable nanochitosan to penetrate the fish's skin and interact with their internal systems. Fish are placed in a solution that contains a specific concentration of nanochitosan. This solution may be formulated to facilitate the absorption of nanochitosan through the fish's skin, allowing the compound to enter their bloodstream and interact with their internal organs. Nanochitosan, due to its nano-sized particles, may have enhanced permeation capabilities, enabling it to cross biological barriers more effectively. This allows the compound to interact with various physiological systems within the fish, potentially influencing reproductive processes or other targeted physiological responses. The duration of exposure and the concentration of nanochitosan in the solution are critical factors. These parameters are carefully determined based on research or experimental data, aiming to achieve the desired effect without causing harm to the fish.

Optimizing the duration and concentration helps in maximizing the compound's efficacy while minimizing adverse effects. During and after the immersion treatment, continuous monitoring of the fish is essential. Observation of their behaviour, vital signs, and any immediate reactions is crucial. Monitoring allows for the prompt identification of any adverse effects, changes in behaviour, or signs of stress that might arise due to the treatment. Immediate reactions or behavioural changes observed during or after the immersion treatment may indicate potential adverse effects of nanochitosan on the fish. This vigilant monitoring enables aquaculturists or researchers to promptly identify and address any negative impacts, ensuring the well-being of the fish. After the immersion treatment, continued observation is essential to assess the fish's recovery and overall health. Post-treatment monitoring ensures that any lingering effects or delayed responses are recognized and managed appropriately (Shen & Wang, 2018).

Immersion or bath treatment involving nanochitosan offers a method to deliver the compound to fish by allowing it to permeate their skin and interact with internal systems. Careful monitoring throughout the process, from exposure to post-treatment observation, is crucial to ensure the effectiveness of the treatment while safeguarding the health and well-being of the fish subjected to this method.

Injection

Injecting nanochitosan solutions directly into fish is a precise and targeted method of administering this compound in aquaculture. This approach involves the direct introduction of nanochitosan into the fish's body, offering controlled delivery and potentially influencing specific physiological processes. Injecting nanochitosan directly into fish allows for a controlled and precise delivery of the compound. This method bypasses external barriers, ensuring direct access to the fish's bloodstream or targeted tissues. As a result, it may provide a more concentrated and immediate effect compared to other administration methods. Due to its invasive nature, this method demands expertise and careful handling to minimize stress and potential injury to the fish. Skilled personnel, such as experienced aquaculturists or veterinarians, are required to perform the injections accurately, ensuring the welfare of the fish and minimizing any discomfort or adverse effects caused by the procedure.

The process of injecting nanochitosan solutions into fish requires precision to minimize stress and potential injury. Proper techniques, including selecting appropriate injection sites, using correct needle sizes, and employing gentle handling, are crucial to mitigate any negative impact on the fish's health and behaviour. This method offers a targeted approach, allowing nanochitosan to directly enter specific tissues or organs of interest. Depending on the intended outcome, such as influencing reproductive processes or immune responses, targeted injections can potentially achieve localized effects. Due to the specialized nature of this method, expertise and caution are paramount. Understanding fish anatomy, injection techniques, and the physiological response to nanochitosan is essential to ensure the effectiveness of the treatment while safeguarding the fish's well-being. After the injection, continuous monitoring of the fish is crucial to observe any immediate reactions, changes in behaviour, or signs of stress. Post-injection observation helps in identifying and addressing any adverse effects that might arise from the procedure. Injecting nanochitosan solutions directly into fish offers a targeted and controlled approach to administering the compound. While it allows for precise delivery, it requires expertise, careful handling, and vigilant monitoring to minimize stress, potential injury, and adverse effects on the fish, ensuring their well-being throughout the process (Wisdom et al., 2018).

In all of the described techniques, constant vigilance and observation of fish behaviour, health, and reproductive responses are essential aspects of assessing the efficacy and safety of nanochitosan. Any alterations in behaviour, health issues, or unexpected reactions need immediate attention and analysis to determine the cause and potential adjustments needed in dosage or administration methods. Furthermore, understanding the potential environmental impact of nanochitosan is crucial for sustainable aquaculture practices. This includes assessing its degradation, possible effects on water quality, and any implications for other aquatic organisms. Research into the compound's fate in the environment helps in minimizing ecological risks associated with its use. Adherence to local regulations and guidelines concerning the use of additives in aquaculture is critical. Compliance ensures that the application of nanochitosan aligns with legal and ethical standards, safeguarding both the aquatic environment and consumer health.

The various administration methods of nanochitosan in fish spawning each come with specific considerations and benefits. Monitoring fish health and behaviour, assessing environmental impact, and complying with regulations are integral components to ensure the safe and effective application of nanochitosan in aquaculture practices.

2.7 *Monitoring and Assessment of Nanochitosan Effects on Breeding*

Monitoring and assessing the effects of nanochitosan on breeding in fish involves a comprehensive approach that evaluates various parameters before, during, and after its application. It is imperative to start by establishing baseline reproductive parameters such as spawning frequency, egg quality, hatch rates, and larval development in the fish population under normal conditions. Furthermore, an assessment of the overall health status of the fish, including growth rates, behaviour, and general physiological conditions, is essential (Wisdom et al., 2018).

Continuously monitoring fish behaviour during exposure to nanochitosan is required, with close observations of any changes in swimming patterns, feeding habits, or reproductive behaviours. The reproductive performance must be monitored by tracking spawning behaviour, egg production, fertilization rates, and the development of embryos/larvae during the application period. To meet the requirement of post-treatment assessment, reproductive outcomes must be monitored through the evaluation of changes in reproductive parameters post-application. Factors to consider include spawning success, egg quality, hatch rates, and larval survival compared to baseline data. Monitoring fish health and behaviour after exposure to nanochitosan is vital to determine any lingering effects or signs of stress that persist after treatment.

Monitoring and assessment of nanochitosan effects on breeding may also require laboratory analyses such as histological and biochemical examinations. Histological analysis is conducted on fish tissue to assess any structural changes in reproductive organs caused by nanochitosan exposure. Biochemical and hormonal assays are required to measure hormone levels or biochemical markers associated with reproductive function to understand how nanochitosan may have affected hormonal regulation (Wisdom et al., 2022).

To determine environmental impact assessment, water quality monitoring and ecological observations are required. Water quality monitoring is determined to assess the impact of nanochitosan on water quality parameters such as pH, dissolved oxygen, and ammonia levels, while ecological observations consider the effects of nanochitosan on non-target organisms in the aquatic ecosystem. Furthermore, statistical methods are then employed to compare data sets, determining significant changes in reproductive parameters, fish health, or environmental conditions.

Dose–response relationship patterns are required to analyze the relationship between nanochitosan dosage and observed effects. It is thus essential to monitor the persistence of effects and any potential long-term impacts on fish reproduction and ecosystem health. Adaptive management is also vital to consider ongoing monitoring data to adapt and refine application strategies for improved effectiveness and reduced environmental impact.

Regular and comprehensive monitoring throughout the entire process is critical for understanding the effects of nanochitosan on fish breeding. This approach ensures the accurate assessment of its efficacy, safety, and potential impact on both the target species and the aquatic environment.

3 Case Studies and Experimental Findings

Nanochitosan's application in spawning across different fish species has shown promise in various aspects of reproductive enhancement.

Research exploring the application of nanochitosan in stimulating spawning in Salmonids, such as Salmon and Trout, has unveiled its potential to significantly impact various stages of the reproductive process, leading to improved reproductive success. Nanochitosan has been investigated for its ability to stimulate spawning in Salmonids. By promoting the release of mature eggs, it contributes to increasing the availability of eggs for fertilization. Nanochitosan treatments have shown promising results in enhancing egg quality among Salmonids. This includes improvements in the structural integrity of eggs, potentially reducing deformities and increasing their viability. Enhanced egg quality due to nanochitosan treatment correlates with increased fertilization rates. The strengthened egg membrane and reduced deformities contribute to a higher probability of successful fertilization (Ziaei-nejad et al., 2020). Nanochitosan's influence extends to the embryonic stage by promoting healthier and more robust embryonic development. This can lead to better survival rates and increased chances of successful hatchings. The positive impact of nanochitosan on egg quality and embryonic development translates to improved larval survival and growth. Healthier initial development may contribute to stronger, more resilient larvae.

Collectively, nanochitosan's effects on egg quality, fertilization rates, embryonic development, and subsequent larval stages contribute to enhancing the overall reproductive success of Salmonids. Improved reproductive success can lead to increased broodstock productivity, better hatchery outcomes, and potentially higher survival rates in aquaculture settings. Nanochitosan's potential to positively influence various facets of the reproductive process in Salmonids holds promise for aquaculture. However, further studies and refinement of application methods are necessary to fully realize its benefits and optimize its use in improving spawning success and overall reproductive performance in these important fish species. Nanochitosan application in freshwater Teleosts, specifically Tilapia and Catfish, has exhibited promising potential in stimulating crucial aspects of reproductive physiology, leading to enhanced spawning frequency and improved egg quality. Nanochitosan treatments have shown the ability to stimulate gonadal development in Tilapia and Catfish. This process involves enhancing the growth and maturation of reproductive organs, leading to increased gamete production. The application of nanochitosan is associated with promoting the maturation of gonads in these freshwater Teleosts. This advancement in maturity can result in more developed and functional reproductive organs, contributing to improved spawning potential (Palma et al., 2019).

Through its influence on gonadal development and maturation, nanochitosan may induce a higher frequency of spawning events in Tilapia and Catfish. This stimulation can lead to more regular and frequent reproduction cycles. Nanochitosan-treated fish have displayed an improvement in the quality of eggs produced. Enhanced gonadal development often leads to eggs with better structural integrity, increased viability, and potentially improved hatching success. The stimulation of gonadal development and increased spawning frequency due to nanochitosan application can offer significant benefits in aquaculture settings. It can contribute to better reproductive management, increased efficiency in egg production, and potentially higher yields. Further research is essential to delve deeper into the specific mechanisms by which nanochitosan influences gonadal development and spawning frequency in Tilapia and Catfish. Understanding these mechanisms can aid in refining application methods and optimizing their use in aquaculture for consistent and advantageous reproductive outcomes. Nanochitosan's potential in stimulating gonadal development, inducing increased spawning frequency, and enhancing egg quality holds promise for the management of freshwater Teleosts like Tilapia and Catfish in aquaculture. However, ongoing research is crucial to comprehensively understand its mechanisms and optimize its application for achieving consistent and beneficial results in enhancing reproductive success (Kookaram et al., 2021).

Nanochitosan's application in spawning enhancement for Zebrafish and Medaka has been a subject of exploration, primarily focusing on improving egg production, fertilization rates, and subsequent larval rearing in these model fish species. Nanochitosan interventions have shown potential in boosting egg production and fertilization rates in Zebrafish and Medaka. By enhancing the reproductive processes, nanochitosan may positively influence the quantity and quality of eggs produced, consequently increasing successful fertilization rates. Nanochitosan's effects extend beyond spawning, potentially aiding in larval rearing, which is critical for the survival and development of offspring. Studies suggest that nanochitosan treatments may contribute to higher survival rates among larvae and improve the overall quality of fry in Zebrafish and Medaka populations (Hoseinifar et al., 2019; Chou et al., 2020). The mechanisms underlying these effects likely involve nanochitosan's multifaceted properties, including its biocompatibility and potential to influence reproductive physiology. Nanochitosan's interactions with the reproductive systems of these fish species may enhance various stages of reproduction, from egg production to larval development. The utilization of nanochitosan in Zebrafish and Medaka holds promise for advancing research in reproductive biology and aquaculture. Further exploration of nanochitosan's mechanisms and its precise impact on different stages of fish reproduction could offer valuable insights into improving breeding strategies and sustaining these model fish populations. Nanochitosan's exploration of these model fish species presents an exciting avenue for enhancing spawning, larval rearing, and overall fish population management. Continued research in this area could potentially contribute to advancements in aquaculture practices and reproductive biology studies.

Nanochitosan has emerged as a potential agent for enhancing spawning performance, egg quality, and larval survival in shrimp species. Its application is being researched to improve reproductive outcomes and increase the efficiency of larval rearing in shrimp aquaculture. Nanochitosan shows promise in positively influencing spawning performance among shrimp species. Studies indicate that its application may enhance the reproductive capacity of female shrimp, potentially leading to increased spawning rates. The utilization of nanochitosan has demonstrated the potential to enhance the quality of shrimp eggs (Udo et al., 2018; Choi et al., 2020). By influencing reproductive physiology, nanochitosan treatments might result in the production of higher-quality eggs, which could positively impact fertilization rates and subsequent larval development. Research suggests that nanochitosan applications could contribute to higher larval survival rates in shrimp. The treatment may aid in improving the overall health and survival of shrimp larvae, potentially enhancing the efficiency of larval-rearing processes in aquaculture settings. Nanochitosan's effects on shrimp reproductive biology involve interactions with reproductive systems and cellular processes. The potential application of nanochitosan in shrimp aquaculture offers promising prospects for improving reproductive outcomes and enhancing larval rearing efficiency. Continued research into its effects and optimal application methods could lead to advancements in shrimp breeding practices and sustainable aquaculture. Nanochitosan's exploration of shrimp species represents an innovative approach to improving spawning, egg quality, and larval survival, holding potential for advancements in shrimp aquaculture and reproductive biology studies. Continued research endeavours may unveil its efficacy and pave the way for enhanced practices in shrimp breeding and larval rearing.

Nanochitosan has been explored for its potential to enhance spawning, elevate egg quality, improve hatch rates, and support larval health in commercially valuable marine fish species such as Sea Bass and Sea Bream. Studies indicate that nanochitosan applications could lead to an improvement in the quality of eggs produced by marine fish like Sea Bass (Kumar et al., 2008) and Sea Bream (Mechlaoui et al., 2019). This enhancement in egg quality may subsequently result in higher hatch rates, contributing to increased breeding success. Nanochitosan's application shows promise in bolstering the health and vitality of larvae in these marine fish species. It may contribute to enhancing larval robustness, potentially leading to better survival rates and improved overall health during the crucial larval developmental stages. The mechanisms underlying nanochitosan's influence on egg quality, hatch rates, and larval health in Sea Bass and Sea Bream are under investigation. Its biocompatibility and potential interactions with the reproductive systems of these species are being explored for potential applications in aquaculture. The successful utilization of nanochitosan in Sea Bass and Sea Bream could have significant implications for improving reproductive success and larval rearing in aquaculture. It might lead to advancements in breeding strategies and contribute to the sustainable management of these commercially important marine fish species. The investigation of nanochitosan in marine fish species such as Sea Bass and Sea Bream signifies a promising avenue for enhancing reproductive success and larval health. Its potential to improve egg quality, hatch rates, and larval development holds significance for aquaculture practices, aiming to sustainably manage these commercially valuable marine fish species. Continued research can unravel its efficacy and optimize its application for the benefit of marine fish aquaculture.

Nanochitosan has been a subject of study for its potential to augment spawning frequency, elevate egg quality, and bolster fry survival in ornamental fish species like Guppies and Tetras. Its application is geared towards enhancing both the quality and quantity of offspring in captive breeding programs for these ornamental fish. Research suggests that nanochitosan applications may contribute to increasing the spawning frequency among ornamental fish species. This intervention aims to encourage more frequent reproductive cycles in captive environments. The utilization of nanochitosan has demonstrated the potential to enhance the quality of eggs produced by ornamental fish like Guppies and Tetras. By influencing reproductive physiology, nanochitosan treatments might lead to the production of higher-quality eggs, thereby improving hatching success rates (Udo et al., 2018). Nanochitosan's application might positively impact the survival rates of fry in ornamental fish species. Boosting the health and resilience of fry could lead to increased survival rates during the critical early stages of development. The aim of employing nanochitosan in ornamental fish breeding programs is to improve reproductive outcomes, increase the quantity of offspring, and enhance their overall quality in captive settings. Nanochitosan's exploration of ornamental fish species like Guppies and Tetras presents an intriguing avenue for enhancing spawning, egg quality, and fry survival in captive breeding programs. Its potential to positively influence reproductive outcomes and increase the quality and quantity of offspring underscores its role in advancing ornamental fish breeding practices. Continued research endeavours can further unveil its efficacy and optimize its application for the benefit of ornamental fish breeding programs.

Nanochitosan has been investigated for its potential to enhance spawning success and improve fry quality in fish species like Common Carp and Pangasius by leveraging its antimicrobial properties to combat diseases that could impact reproductive health. Studies have also explored nanochitosan's antimicrobial attributes in Common Carp and Pangasius. Its application aims to mitigate diseases that might affect the reproductive health of these fish species, potentially leading to improved spawning success and the overall health of the fry. By targeting microbial threats that could compromise reproductive health, nanochitosan might indirectly enhance spawning success rates in Common Carp and Pangasius. Reducing the incidence of diseases during reproductive cycles may positively influence the reproductive health of these fish species. The utilization of nanochitosan to mitigate diseases could contribute to better fry quality. Potentially minimizing the impact of diseases on larval health may lead to stronger and healthier fry, enhancing their overall quality. Nanochitosan's exploration of Common Carp and Pangasius for disease mitigation during reproductive phases holds significance for aquaculture. Improving reproductive health and fry quality could contribute to better breeding outcomes and bolster the overall productivity of aquaculture practices. Continued research endeavours can unlock its potential applications for disease management in aquaculture and reproductive biology studies. The examination of nanochitosan's antimicrobial properties in fish species like Common Carp and Pangasius demonstrates its potential to mitigate diseases that might impact reproductive health and fry quality. Its application holds promise for advancing disease management strategies in aquaculture and improving breeding outcomes in these fish species. Continued research efforts are crucial to unravel its efficacy and optimize its application for disease mitigation and reproductive health enhancement in aquaculture settings.

The application of nanochitosan in fish spawning aims to enhance various aspects of reproduction, including egg quality, fertilization rates, larval survival, and overall reproductive success. However, the effectiveness of nanochitosan can vary among species due to differences in physiology, reproductive biology, and environmental factors. Continued research efforts are crucial to further elucidate the precise mechanisms through which nanochitosan influences spawning frequency, egg quality, and fry survival in fish and shellfish. This exploration holds promise for advancing captive breeding practices and sustaining fish populations.

Research continues to explore the potential of nanochitosan in optimizing spawning performance across diverse fish species. Understanding species-specific responses and fine-tuning application methods and dosages remains crucial for its successful use in improving spawning outcomes in aquaculture.

4 Challenges and Future Directions

Addressing environmental and regulatory concerns in the context of nanochitosan-based stimulation of spawning in fish is crucial for ensuring responsible and sustainable application within aquaculture practices. Ecotoxicological Effects: Thorough assessments are necessary to evaluate the potential ecotoxicological impacts of nanochitosan on aquatic ecosystems, including its effects on non-target organisms and the overall ecosystem health.

Understanding the bioaccumulation potential and degradation pathways of nanochitosan in aquatic environments is essential to predict its persistence and potential impacts on food webs. Complying with regulatory standards and obtaining necessary approvals for the use of nanochitosan in aquaculture is imperative. This involves adhering to guidelines set by relevant authorities for novel substances used in aquatic environments. Developing clear labelling and usage guidelines for nanochitosan-based products in fish breeding programs is essential to ensure proper application and minimize environmental risks. Implementing precise dosage guidelines and application protocols can minimize unintended environmental exposure while ensuring effective stimulation of spawning in fish. Establishing monitoring mechanisms to track the environmental impact post-application of nanochitosan and reporting any observed effects is crucial for proactive risk management. Involving stakeholders, including regulatory bodies, aquaculture industry players, environmental agencies, and local communities, fosters collaboration and facilitates transparent decision making (Shah & Mraz, 2020).

Educating the public about the benefits, risks, and responsible use of nanochitosan-based technologies in fish breeding promotes informed decision making and fosters public trust. Continued research into alternative eco-friendly compounds or technologies for spawning enhancement in fish can provide safer alternatives with potentially lower environmental impact. Further research into the environmental fate, behaviour, and long-term impacts of nanochitosan in aquatic ecosystems is necessary to inform risk assessment and management strategies.

Addressing environmental and regulatory concerns surrounding nanochitosan-based spawning stimulation in fish requires a holistic approach encompassing rigorous assessments, compliance with regulatory frameworks, risk mitigation strategies, stakeholder engagement, and continued research for sustainable aquaculture practices. Collaborative efforts and a proactive approach are pivotal for ensuring the safe and responsible use of nanochitosan in fish breeding programs while safeguarding aquatic ecosystems.

The utilization of nanochitosan in stimulating spawning in fish, while promising, presents several potential limitations and risks that warrant consideration. Nanochitosan's potential effects on non-target organisms and aquatic ecosystems require a comprehensive assessment to ensure minimal ecological disruption. Assessing potential health risks to workers handling nanochitosan and consumers exposed indirectly through fish consumption is critical. Investigating any allergic reactions or adverse effects that might arise from exposure to nanochitosan-based products. The effectiveness of nanochitosan-based spawning enhancement might vary among different fish species or environmental conditions, necessitating species-specific optimization. Establishing precise dosage guidelines and application protocols for consistent and effective results without causing harm to fish or the environment can be challenging.

Obtaining regulatory approval for nanochitosan-based products might involve lengthy processes and compliance with evolving standards. Ethical concerns related to the use of nanochitosan in fish breeding, including animal welfare considerations and public acceptance, need careful consideration. The cost of nanochitosan production and its application methods might influence its feasibility and adoption within aquaculture. The availability of appropriate technology and infrastructure for large-scale implementation in aquaculture settings may pose limitations. Prolonged use of nanochitosan might lead to the development of resistance in pathogens, potentially reducing its effectiveness over time. There's a possibility of pathogens adapting to nanochitosan exposure, necessitating continuous innovation, and adaptation in spawning enhancement strategies.

Inadequate knowledge about the long-term environmental fate, safety, and ecological impact of nanochitosan in aquatic systems. The absence of standardized protocols for nanochitosan application in different fish species limits its widespread adoption and comparison of results. Addressing these potential limitations and risks associated with nanochitosan-based spawning stimulation in fish requires rigorous research, risk assessment, stakeholder engagement, and adherence to regulatory standards. Collaborative efforts aimed at understanding, mitigating, and managing these challenges are essential for ensuring the safe and effective utilization of nanochitosan in fish breeding programs (Mahadevaswamy et al., 2023).

Advancing research to enhance the applications of nanochitosan-based stimulation of spawning in fish involves several key avenues for exploration: fFurther research into nanochitosan formulations to improve stability, bioavailability, and effectiveness in different aquatic environments; exploring innovative delivery systems or carriers to enhance the targeted and controlled release of nanochitosan for optimal spawning stimulation, conducting comprehensive studies to understand the species-specific responses and requirements for nanochitosan-based spawning stimulation across various fish species, And investigating the influence of environmental variables (pH, temperature, and salinity) on the efficacy of nanochitosan in different fish species. In-depth knowledge of the antimicrobial mechanisms of nanochitosan and its interaction with fish reproductive systems at the molecular level is paramount. Enhancing understanding of the impact of nanochitosan on fish reproductive physiology, hormonal regulation, and gamete quality. Conducting comprehensive studies to assess the environmental fate, persistence, and potential ecological impacts of nanochitosan in aquatic ecosystems. Evaluating the broader ecological effects of nanochitosan exposure on non-target organisms and ecosystem functions. Investigating strategies to mitigate the development of resistance in pathogens exposed to nanochitosan. Conducting thorough assessments to address potential human health risks associated with the handling and consumption of nanochitosan-treated fish. Collaborating with regulatory agencies to establish standardized protocols, safety guidelines, and approval processes for nanochitosan use in aquaculture. Developing ethical guidelines addressing animal welfare, public acceptance, and social implications of nanochitosan-based interventions. Encouraging collaboration among researchers, industry stakeholders, and regulatory bodies to share data, best practices, and advancements. Promoting awareness and education programs to engage stakeholders and the public on the benefits, risks, and responsible use of nanochitosan in fish breeding (Bhat, 2023).

Advancing research in these areas will facilitate the development of safe, effective, and environmentally responsible applications of nanochitosan for enhancing spawning in fish. This multifaceted approach is crucial for harnessing the full potential of nanochitosan in aquaculture while mitigating associated risks and ensuring sustainable practices.

5 Conclusion

The exploration of nanochitosan's role in stimulating spawning in fish represents a compelling avenue in contemporary aquaculture research. Its multifaceted properties, including antimicrobial action, biocompatibility, and potential for improving reproductive outcomes, hold significant promise for revolutionizing fish breeding practices. However, this innovative approach is not without challenges and considerations.

Addressing environmental impact, regulatory compliance, efficacy optimization, and risk mitigation remains imperative in realizing the full potential of nanochitosan-based spawning enhancement. Robust research, encompassing species-specific studies, mechanistic understandings, ecotoxicological assessments, and regulatory frameworks, forms the cornerstone of responsible application. Collaboration among stakeholders, knowledge dissemination, and continual innovation are crucial for advancing this technology in a sustainable, ethical, and effective manner.

In navigating these pathways, harnessing the benefits of nanochitosan while addressing limitations and risks underscores the need for a balanced and comprehensive approach. With ongoing advancements and a commitment to responsible implementation, nanochitosan-based spawning stimulation stands poised to significantly contribute to the evolution of aquaculture practices, fostering sustainable fish breeding, and ensuring the integrity of aquatic ecosystems.

References

Ahmed, F., Soliman, F. M., Adly, M. A., Soliman, H. A. M., El-Matbouli, M., & Saleh, M. (2019). Recent progress in biomedical applications of chitosan and its nanocomposites in aquaculture: A review. *Research in Veterinary Science, 126*, 68–82.

Barchanski, A., Taylor, U., & Sajti, C. L. (2015). Bioconjugated gold nanoparticles penetrate spermatozoa depending on plasma membrane status. *Journal of Biomedical Nanotechnology, 11*(9), 1597–1607.

Bhat, I. A. (2023). Nanotechnology in reproduction, breeding and conservation of fish biodiversity: Current status and future potential. *Reviews in Aquaculture, 15*(2), 557–567.

Bhat, I. A., Rather, M. A., Saha, R., Pathakota, G., Pavan-Kumar, A., & Sharma, R. (2016). Expression analysis of Sox9 genes during annual reproductive cycles in gonads and after nanodelivery of LHRH in *Clarias batrachus*. *Research in Veterinary Science, 106*, 100–106. https://doi.org/10.1016/j.rvsc.2016.03.022. ISSN 0034-5288.

Bhat, I. A., Nazir, M. I., Ahmad, I., Pathakota, G., Chanu, T. I., Goswami, M., Sundaray, J. K., & Sharma, R. (2018). Fabrication and characterization of chitosan conjugated eurycomanone nanoparticles: In vivo evaluation of the biodistribution and toxicity in fish. *International Journal of Biological Macromolecules, 112*, 1093–1103. ISSN 0141-8130.

Bhat, I. A., Ahmad, I., Mir, I. N., Yousf, D. J., Ganie, P. A., Bhat, R. A. H., Gireesh-Babu, P., & Sharma, R. (2019). Evaluation of the in vivo effect of chitosan conjugated eurycomanone nanoparticles on the reproductive response in female fish model. *Aquaculture, 510*, 392–399. ISSN 0044-8486.

Bhat, I. A., Dar, J. Y., & Ahmad. (2021). Testicular development and spermatogenesis in fish: Insights into molecular aspects and regulation of gene expression by different exogenous factors. *Reviews in Aquaculture, 13*(4), 2142–2168.

Cardoso, V. M., Ferreira, L. M. B., Comparetti, E. J., Sampaio, I., Ferreira, N. N., Miranda, R. R., & Zucolotto, V. (2022). Chapter 4 – Stimuli-responsive polymeric nanoparticles as controlled drug delivery systems. In V. Gajbhiye, K. R. Gajbhiye, & S. Hong (Eds.), *Stimuli-responsive nanocarriers* (pp. 87–117). Academic. ISBN 9780128244562.

Choi, Y. S., An, S. K., Jang, S., Jang, M. S., Koo, J. M., & Hwang, S. Y. (2020). Effect of dietary supplementation of nanochitosan on growth performance, feed utilization, and digestive enzyme activity in the Pacific white shrimp, *Litopenaeus vannamei. Aquaculture Reports, 18*, 100504.

Chou, C., Mi, F., Horng, J., Lin, L., Tsai, M., Liu, C., Lu, K., Chu, C., Chen, Y., Lee, Y. A., & Cheng, C.-H. (2020). Characterization and toxicology evaluation of low molecular weight chitosan on zebrafish. *Carbohydrate Polymers, 240*, 116164. https://doi.org/10.1016/j.carbpol.2020.116164. ISSN 0144-8617.

Dubey, S., Mody, N., Sharma, R., Agrawal, U., & Vyas, S. P. (2016). Chapter 4 – Nanobiomaterials: Novel nanoplatforms for protein and peptide delivery. In A. M. Grumezescu (Ed.), *Nanobiomaterials in drug delivery* (pp. 111–146). William Andrew Publishing. ISBN 9780323428668.

Durfey, C. L., Swistek, S. E., & Liao, S. F. (2019). Nanotechnology-based approach for safer enrichment of semen with best spermatozoa. *Journal of Animal Science and Biotechnology, 10*(1), 1–2.

Elsawy, M. A., Saad, G. A., & Sayed, A. M. (2016). Mechanical, thermal, and dielectric properties of poly(lactic acid)/chitosan nanocomposites. *Polymer Engineering and Science, 56*(2016), 987–994. https://doi.org/10.1002/pen.24328

Fatma, S. A. E., Naiel, M. A. E., Al-Sagheer, A. A., & Negm, S. S. (2019). Dietary chitosan nanoparticles enhance the growth, production performance, and immunity in *Oreochromis niloticus*. *Aquaculture, 501*, 82–89. https://doi.org/10.1016/j.aquaculture.2018.11.014. ISSN 0044-8486.

Fonseca, L. R., Santos, T. P., Czaikoski, A., & Cunha, R. L. (2020). Modulating properties of polysaccharides nanocomplexes from enzymatic hydrolysis of chitosan. *Food Research International, 137*, 109642. ISSN 0963-9969.

Hajiyeva, A. D., Mamedov, C. A., & Khalilov, R. I. (2022). Influence of nanoparticles on the yield of embryos of the roe common carp (*Cyprinus carpio* Linnaeus, 1758) in the fermentation process. *Advances in Biology & Earth Sciences, 7*(1), 5–12.

Hassanein, E. M., Hashem, N. M., El-Azrak, K. E., González-Bulnes, A., Hassan, G. A., & Salem, M. H. (2021). Efficiency of GnRH–loaded chitosan nanoparticles for inducing LH secretion and fertile ovulations in protocols for artificial insemination in rabbit does. *Animals: An Open Access Journal from MDPI, 11*(2), 440.

Hoseinifar, S. H., Raeisi, M., & Salimi, M. (2019). Effects of dietary chitosan nanoparticles on growth performance, survival, body composition, and intestinal morphology of the rainbow trout, *Oncorhynchus mykiss*. *Aquaculture, 500*, 1–8.

Jahaabad, M. J., Ziaenejad, S., Ghayedi, A., Moradian, H., & Seyyedi, A. M. (2020). The effects of selenium nanoparticles enriched food on sperm quality and fertilization of rainbow trout (*Oncorhynchus mykiss*) male breeders. *Iranian Veterinary Journal, 16*(3), 106–115.

Jehanabad, M. J., Rastiannasab, A., Ghaedi, A., Mahmodi, R., & Ardakani, S. M. M. (2019). Effect of different levels of selenium nanoparticles on some reproductive indices in rainbow trout (*Oncorhynchus mykiss*). *Aquatic Physiology and Biotechnology, 7*(3), 113–132.

Joshi, H. D., Ghode, G. S., & Gore, S. B. (2015). Efficiency of letrozole loaded PLGA nanoparticles on sex reversal of *Poecilia reticulata* (Peters, 1859). *Journal of Applied and Natural Science, 7*(1), 394–399.

Joshi, H. D., Tiwari, V. K., Gupta, S., Sharma, R., Lakra, W. S., & Sahoo, U. (2019). Application of nanotechnology for the production of masculinized tilapia, *Oreochromis niloticus* (Linnaeus, 1758). *Aquaculture, 511*, 734206.

Kaya, M., Sargin, I., Tozak, K. O., Baran, T., Erdogan, S., & Sezen, G. (2013). Chitin extraction and characterization from *Daphnia magna* resting eggs. *International Journal of Biological Macromolecules, 61*, 459–464.

Kookaram, K., Mojazi, A. B., Dorkoosh, F. A., Nematollahi, M. A., Mortazavian, E., & Abed, E. A. (2021). Effect of oral administration of GnRHa+ nanoparticles of chitosan in oogenesis acceleration of goldfish *Carassius auratus*. *Fish Physiology and Biochemistry, 47*(2), 477–486.

Kou, L., Bhutia, Y. D., Yao, Q., He, Z., Sun, J., & Ganapathy, V. (2018). Transporter-guided delivery of nanoparticles to improve drug permeation across cellular barriers and drug exposure to selective cell types. *Frontiers in Pharmacology, 9*, 27.

Kumar, S. R., Ahmed, V. P. I., Parameswaran, V., Sudhakaran, R., Babu, V. S., & Hameed, A. S. S. (2008). Potential use of chitosan nanoparticles for oral delivery of DNA vaccine in Asian sea bass (Lates calcarifer) to protect from Vibrio (Listonella) anguillarum. *Fish & Shellfish Immunology, 25*(1–2), 47–56.

Levitin, S. V., Gal'braikh, L. S., Grunin, Y. B., & Masas, D. S. (2014). Supramolecular structure of chitosan acid-hydrolysis products. *Fibre Chemistry, 46*(3), 147–150. https://doi.org/10.1007/s10692-014-9579-4

Mahadevaswamy, C. G., Chuphal, N., & Dar, S. A. (2023). Chitosan-hypothalamic hormonal analogue nanoconjugates enhanced the reproductive performance in Indian Major carp, *Labeo rohita*. *Frontiers in Marine Science. Aquatic Physiology, 10*. https://doi.org/10.3389/fmars.2023.1311158

Mechlaoui, M., Dominguez, D., & Robaina, L. (2019). Effects of different dietary selenium sources on growth performance, liver and muscle composition, antioxidant status, stress response and expression of related genes in gilthead seabream (*Sparus aurata*). *Aquaculture, 507*, 251–259.

Mylonas, C. C., & Zohar, Y. (2001). Endocrine regulation and artificial induction of oocyte maturation and spermiation in basses of the genus Morone. *Aquaculture, 202*(3–4), 205–220.

Mylonas, C. C., Fostier, A., & Zanuy, S. (2010). Broodstock management and hormonal manipulations of fish reproduction. *General and Comparative Endocrinology, 165*(3), 516–534.

Palma, P., Nocillado, J., & Superio. (2019). Induction of gonadal development in protogynous grouper with orally delivered FSH DNA. *Marine Biotechnology, 21*(5), 697–706.

Rather, M. A., Sharma, R., Gupta, S., Ferosekhan, S., Ramya, V. L., & Jadhao, S. B. (2013). Chitosan-nanoconjugated hormone nanoparticles for sustained surge of gonadotropins and enhanced reproductive output in female fish. *PLoS One, 8*(2), e57094. https://doi.org/10.1371/journal.pone.0057094. Epub 2013 Feb 27. PMID: 23468918; PMCID: PMC3584147.

Rather, M. A., Bhat, I. A., Gireesh-Babu, P., Chaudhari, A., Sundaray, J. K., & Sharma, R. (2016). Molecular characterization of kisspeptin gene and effect of nano–encapsulted kisspeptin-10 on reproductive maturation in *Catla catla*. *Domestic Animal Endocrinology, 56*, 36–47. https://doi.org/10.1016/j.domaniend.2016.01.005. ISSN 0739-7240.

Rathor, P. K., Bhat, I. A., & Rather, M. A. (2017). Steroidogenic acute regulatory protein (StAR) gene expression construct: Development, nanodelivery and effect on reproduction in air-breathing catfish, *Clarias batrachus*. *International Journal of Biological Macromolecules, 104*, 1082–1090.

Riegger, B. R., Baurer, B., Mirzayeva, A., Tovar, G. E. M., & Bach, M. (2018). Systematic approach for preparation of chitosan nanoparticles via emulsion crosslinking as potential adsorbent in waste-water treatment. *Carbohydrate Polymers, 180*, 46–54.

Saha, R., Bhat, I. A., & Charan, R. (2018). Ameliorative effect of chitosan-conjugated 17α-methyltestosterone on testicular development in *Clarias batrachus*. *Animal Reproduction Science, 193*, 245–254.

Shah, B. R., & Mraz, J. (2020). Advances in nanotechnology for sustainable aquaculture and fisheries. *Reviews in Aquaculture, 12*(2), 925–942.

Shard, P., Bhatia, A., & Sharma, D. (2014). Optimization and physicochemical parameters on synthesis of chitosan nanoparticles by ionic gelation technique. *International Journal of Drug Delivery, 6*, 58–63.

Shen, Z. G., & Wang, H. P. (2018). Environmental sex determination and sex differentiation in teleosts–how sex is established. In *Sex control in aquaculture* (pp. 85–115). Liss Inc.

Udo, I. U., Etukudo, U., & Anwana, U. U. (2018). Effects of chitosan and chitosan nanoparticles on water quality, growth performance, survival rate and meat quality of the African catfish, *Clarias gariepinus*. *Nanoscience, 1*(1), 12–25.

Wijesena, R. N., Tissera, N., Kannangara, Y. Y., Lin, Y., Amaratunga, G. A. J., & De Silva, K. M. N. (2015). A method for top-down preparation of chitosan nanoparticles and nanofibers. *Carbohydrate Polymers, 117*(2015), 731–738. https://doi.org/10.1016/j.carbpol.2014.10.055

Wisdom, K. S., Bhat, I. A., & Kumar, P. (2018). Fabrication of chitosan nanoparticles loaded with aromatase inhibitors for the advancement of gonadal development in *Clarias magur* (Hamilton, 1822). *Aquaculture, 497*, 125–133.

Wisdom, K. S., Bhat, I. A., & Pathan, M. A. (2022). Teleost nonapeptides, isotocin and vasotocin administration released the milt by abdominal massage in male catfish, *Clarias magur*. *Frontiers in Endocrinology, 13*, 899463. https://doi.org/10.3389/fendo.2022.899463

Wu, Y., Rashidpour, A., Almajano, M. P., & Metón, I. (2020). Chitosan-based drug delivery system: Applications in fish biotechnology. *Polymers, 12*(5), 1177.

Ziaei-nejad, S., Naser-Abad, F. E., Ghaedi, A., Seyyedi, M., & Moradian, S. H. (2020). Effects of enriched food with selenium nanoparticles on breeding efficiency of female rainbow trout (*Oncorhynchus mykiss*). *Iranian Fisheries Scientific Journal, 28*(6), 79–88.

Application of Nanochitosan in Fish Detoxification/Nano-Based Depuration

Patrick Omoregie Isibor

Contents

1 Introduction

Detoxification and depuration are two essential processes in fish farming that focus on ensuring the quality and safety of the fish intended for human consumption. These processes help reduce the potential risks associated with contaminants and environmental pollutants in the fish, ensuring that the fish are safe for consumption.

Detoxification in fish farming refers to the process of reducing or eliminating harmful substances and contaminants that may be present in the fish or their

P. O. Isibor (✉)
Department of Biological Sciences, College of Science and Technology, Covenant University, Ota, Ogun State, Nigeria
e-mail: patrick.isibor@covenantuniversity.edu.ng

P. O. Isibor et al. (eds.), *Nanochitosan-Based Enhancement of Fisheries and Aquaculture*, https://doi.org/10.1007/978-3-031-52261-1_11

environment. These harmful substances can include heavy metals, organic pollutants, and other toxins that may accumulate in fish tissues due to their diet, water quality, or exposure to environmental pollutants (El-Sayed, 2019). Detoxification is crucial for ensuring that the fish are safe for human consumption. Detoxification mechanisms in fish can involve various processes, including metabolism, excretion, accumulation, and depuration. Regarding metabolism, fish may metabolize and transform some harmful substances into less toxic or non-toxic compounds (Goff, 2018). Fish may also excrete certain contaminants through their excretory organs, such as the liver and kidneys. In some cases, fish may accumulate contaminants in specific organs or tissues, which can then be removed through proper processing.

Depuration on the other hand is a specific process within aquaculture that involves the removal of contaminants or undesirable substances from fish by placing them in clean and controlled environments, typically known as depuration facilities. These facilities are designed to provide the fish with clean water and controlled conditions to flush out any contaminants that may have accumulated in their tissues. Depuration helps to ensure that the fish meet the safety standards for human consumption. Key aspects of depuration include clean water, controlled environment, duration, and monitoring. To ascertain clean water, fish are transferred to tanks or ponds with pristine water quality, free from contaminants. For an adequately controlled environment, temperature, salinity, and other environmental conditions are carefully controlled to promote the depuration process (Seyedmohammadi et al., 2016). The duration of depuration can vary depending on the contaminants present in the fish and the specific requirements for safety standards. It can last from a few days to several weeks. The fish are closely monitored during the depuration process to ensure that contaminant levels decrease to safe levels.

Both detoxification and depuration are integral parts of responsible aquaculture and fisheries management. They are critical for meeting food safety standards and ensuring that the fish products are safe for consumers. By implementing these processes, fish farmers can mitigate potential health risks associated with contaminants in the fish and deliver high-quality, safe seafood products to the market (Wassmur, 2012).

Water quality is of paramount importance in aquaculture and fisheries for several reasons, as it directly impacts the health and productivity of aquatic organisms. Aquatic organisms, including fish, shrimp, and shellfish, depend on suitable water quality conditions for their survival and growth. Poor water quality, with factors such as low oxygen levels, high ammonia concentrations, or pH imbalances, can stress or harm the animals, leading to reduced growth rates and even mortality. High-quality water reduces the stress on aquatic animals, making them less susceptible to diseases and infections. Clean water minimizes the prevalence of pathogens and parasites, decreasing the need for antibiotics and other disease treatments, which can be costly and environmentally harmful. Adequate oxygen levels in the water are essential for fish and other aquatic organisms to breathe and carry out their metabolic processes. Insufficient oxygen levels can lead to fish suffocation, reducing production and causing fish kills. Proper water quality management helps control the accumulation of excess nutrients, such as nitrogen and phosphorus, in the water. Excessive nutrients can lead to algal blooms and water quality deterioration, which can negatively affect aquatic ecosystems and fish health. Water quality affects temperature, which is crucial for fish metabolism and behaviour. Maintaining appropriate temperature ranges helps ensure that fish can grow and reproduce optimally (Setiyorini et al., 2022).

The pH level of the water influences various chemical reactions within the aquatic environment. Maintaining stable pH and alkalinity levels is important for the health of aquatic organisms and the buffering capacity of the water. Water clarity is essential for the health of photosynthetic organisms such as algae and phytoplankton, which serve as the base of the aquatic food chain. Poor water quality can lead to reduced primary productivity and, consequently, lower food availability for higher trophic levels. Proper water quality management is essential for the sustainable development of aquaculture operations. Inadequate treatment of effluent and wastewater can lead to environmental pollution and harm surrounding ecosystems. Many countries have established water quality standards and regulations for aquaculture and fisheries to protect the environment and ensure the sustainability of these industries. Non-compliance can result in penalties and legal consequences. Consumers increasingly demand sustainably produced and environmentally friendly seafood products. Good water quality practices in aquaculture and fisheries are essential to meet these market demands and maintain the industry's reputation.

Maintaining high water quality in aquaculture and fisheries is vital for the well-being of aquatic organisms, the sustainability of these industries, and the protection of aquatic ecosystems. Effective water quality management practices are critical for both economic success and environmental responsibility in these sectors (Divya & Jisha, 2018).

Detoxification and depuration are critical processes in fish farming, aiming to enhance the safety and quality of fish products for human consumption. As aquaculture continues to play a vital role in meeting the global demand for seafood, concerns arise regarding the presence of contaminants, pollutants, and potentially harmful substances in farmed fish. While detoxification and depuration are well-established practices, there remains a need to address specific challenges and gaps in knowledge related to these processes. Fish cultivated in various aquaculture systems may accumulate contaminants, including heavy metals, organic pollutants, and environmental toxins, which can pose health risks to consumers. The extent of contamination and the effectiveness of detoxification mechanisms require further investigation. The efficiency and effectiveness of depuration facilities in removing contaminants from fish have not been comprehensively examined. Questions persist regarding optimal depuration conditions, the duration required, and the ability to consistently meet food safety standards. The existing regulatory framework for detoxification and depuration in fish farming varies globally, with limited harmonization of standards and practices. A lack of uniformity may hinder the ability to ensure the safety and quality of farmed fish products across different regions.

Detoxification and depuration processes can consume significant resources, such as water and energy. The environmental sustainability of these practices, particularly in the context of increasing aquaculture production, remains an important concern. Ensuring the safety of fish products is essential to maintaining consumer confidence in the aquaculture industry. Uncertainty surrounding the adequacy of detoxification and depuration processes can erode trust in farmed fish as a safe and healthy food and water source. While detoxification and depuration are recognized components of responsible aquaculture, there is a need for further research to optimize these processes, enhance their efficiency, and develop best practices that can be applied across various fish species and production systems.

This chapter aims to address the issues surrounding detoxification and depuration in fish farming to enhance the safety and quality of farmed fish products, promote sustainable aquaculture practices, and contribute to the overall success and consumer confidence in the industry by exploring the application of nanochitosan in fish detoxification and depuration.

2 Conventional Fish Detoxification

Naturally, the liver of the fish detoxifies ingested and absorbed toxicants from the surrounding water and underlying soil into less toxic metabolites (Fig. 1). But when the toxicant burden goes beyond the natural background level, it may overwhelm the physiological modulation capacity of the fish.

Fish detoxification, or the process of reducing or eliminating harmful substances and contaminants in fish, has been a topic of concern and research for several decades. Various methods and approaches have been employed to address this issue, but they come with their challenges. One of the earlier approaches to fish detoxification involved adjusting the diet of fish to promote the metabolism and excretion of certain contaminants. For example, dietary additives or specialized feeds were developed to enhance the detoxification of heavy metals such as mercury and organic pollutants. Proper management of water quality in aquaculture systems has

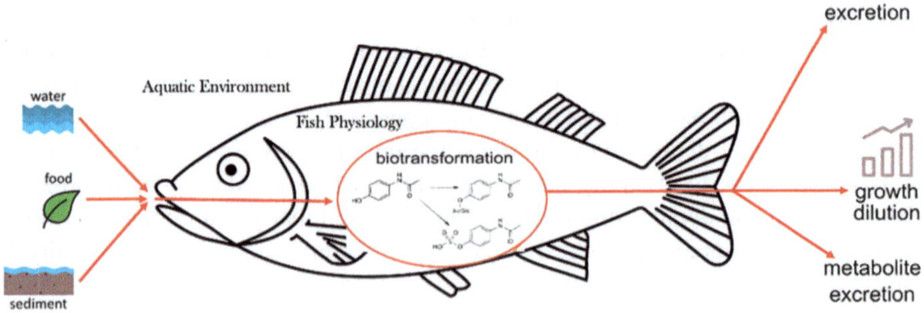

Fig. 1 Adsorption, digestion, metabolism, and excretion effects in fish

long been recognized as a key factor in fish detoxification (Chauhan et al., 2012). Maintaining clean water with adequate oxygen levels and controlling ammonia and nitrate concentrations can reduce stress on fish and help them detoxify more effectively. Some aquaculture systems incorporate bioremediation techniques, where specific microorganisms or plants are introduced to help break down or immobilize contaminants in the water or sediment. This approach can aid in detoxifying the aquatic environment. Depuration facilities are used to remove contaminants from fish by placing them in clean and controlled environments. These facilities offer a controlled setting for fish to purge contaminants from their tissues, enhancing the safety of the final product (Chen et al., 2011).

However, the challenges in the conventional fish depuration and detoxification processes are species variability, bioaccumulation, complexity of contaminants, cost and resource intensiveness, environmental impact, and regulatory compliance. Different fish species have varying abilities to detoxify and eliminate contaminants. Some species are more efficient in detoxification processes than others, making it challenging to develop uniform detoxification strategies across diverse aquaculture practices. Contaminants can accumulate in fish tissues over time, making it difficult to prevent and address long-term exposure. This is especially true for persistent organic pollutants and certain heavy metals. Contaminants found in aquatic environments can be highly diverse and complex. Detoxification methods need to be tailored to the specific contaminant types, and a one-size-fits-all approach is often not feasible. Some detoxification methods, such as depuration facilities, can be costly to implement and may require significant resources, including water, infrastructure, and energy. This can pose financial challenges for aquaculture operations. The disposal of wastewater from depuration facilities and the use of certain detoxification methods may have environmental consequences. Ensuring that these processes are environmentally sustainable is a challenge. Meeting regulatory standards for safe levels of contaminants in fish products is essential, but these standards can vary across regions and may change over time. Compliance with these regulations can be challenging, and non-compliance can lead to product rejection or economic losses (Setiyorini et al., 2022).

Nanochitosan holds the promise of addressing these challenges due to its unique properties. Advancing research in nanochitosan-based fish detoxification is crucial to ensure the safety and quality of farmed fish products while promoting sustainable and responsible aquaculture practices.

3 Novelty of Nanochitosan-Based Detoxification

3.1 High Surface Area

Nanochitosan, a nanoscale version of chitosan derived from chitin, possesses a distinctive property that sets it apart in the field of fish detoxification: its exceptionally high surface area. This unique characteristic arises from its minute particle size, which falls within the nanometer range. In the context of nanomaterials, size matters significantly, and the nanoscale dimensions of nanochitosan grant it a vastly expanded surface area compared to larger chitosan particles (Benettayeb et al., 2023).

This increased surface area is of paramount importance when considering its application in fish detoxification. The primary reason for this is related to the process of adsorption, which refers to the binding or adherence of molecules or contaminants to the surface of a solid material. In the case of nanochitosan, its extensive surface area provides a multitude of sites for interactions with various contaminants present in the aquatic environment and fish tissues (Sun et al., 2006).

This unique property of nanochitosan translates into several key advantages for fish detoxification. Nanochitosan offers an enhanced adsorption capacity due to the heightened surface area that allows it to adsorb a greater quantity of contaminants. Essentially, it provides more 'space' for these substances to attach themselves to the nanochitosan particles (Chiou & Li, 2003; Divya & Jisha, 2018). This means that nanochitosan can capture a larger number of contaminants, including heavy metals, organic pollutants, and toxins, from the surrounding environment. The increased adsorption capacity directly contributes to the efficiency of the detoxification process. Nanochitosan can efficiently and effectively remove a wide range of contaminants from fish tissues, as it can capture and bind with more of these substances, reducing their bioavailability and potential harm to the fish.

The ability of nanochitosan to adsorb diverse types of contaminants underscores its versatility. This versatility makes it a valuable tool for addressing various environmental pollutants found in aquaculture settings, as it can target and capture a wide spectrum of harmful substances (Dutta et al., 2004; El-Naggar et al., 2022a). In addition to the high adsorption capacity, nanochitosan can also be selectively tailored to capture specific contaminants while sparing essential nutrients and compounds in fish tissues. This selectivity ensures that the detoxification process minimally impacts fish health and nutritional quality (El-Naggar et al., 2022b).

The exceptional surface area of nanochitosan, owing to its nanoscale size, significantly enhances its adsorption and binding capacity. This property allows nanochitosan to efficiently capture and remove a broad range of contaminants from fish tissues. This characteristic is instrumental in reducing the accumulation of harmful substances in farmed fish, thereby improving the safety and quality of fish products for human consumption. Nanochitosan's ability to address the diverse challenges associated with fish detoxification underscores its potential as an innovative and valuable tool in the field of aquaculture (El-Naggar et al., 2022a).

3.2 Adsorption Capacity

Nanochitosan, due to its unique properties at the nanoscale, exhibits remarkable adsorption capabilities (Abd-Elhakeem et al., 2016). This means that it has a strong ability to attract and bind various contaminants, which makes it highly effective in the context of fish detoxification. These adsorption properties are particularly beneficial for removing a range of contaminants commonly found in fish, including heavy metals, organic pollutants, and other toxins.

The term 'adsorption' refers to the adhesion or binding of molecules or substances to a solid surface, and nanochitosan's exceptional adsorption properties result from its tiny particle size and a large number of available binding sites. Contaminants, such as heavy metals (e.g. mercury, lead, and cadmium), organic pollutants (e.g. pesticides, polychlorinated biphenyls), and various toxins, are attracted to and held by nanochitosan surfaces (Seyedmohammadi et al., 2016).

When nanochitosan is introduced into the aquatic environment, it acts like a magnet, drawing contaminants towards its surface (Shekhawat et al., 2022). These contaminants can be present in the water, suspended particles, or dissolved compounds (Gamage & Shahidi, 2007; Fu & Wang, 2011). As they come into contact with nanochitosan, the contaminants bind to the numerous active sites on the nanochitosan particles. Furthermore, nanochitosan may prevent contaminant uptake by fish by binding strongly to the contaminants which are thus effectively sequestered and immobilized. This has a crucial benefit for fish, as it prevents the contaminants from being absorbed by fish tissues. In other words, nanochitosan acts as a barrier, shielding the fish from harmful substances that might otherwise accumulate in their organs and muscles. The ability of nanochitosan to reduce the bioavailability of contaminants is a pivotal aspect of its utility in fish detoxification. Bioavailability refers to the portion of a substance that is capable of being absorbed and having an effect within an organism. By binding contaminants, nanochitosan decreases their bioavailability, meaning these substances are less accessible and less likely to harm the fish (Guibal, 2004). The safety of fish and consumers is thus guaranteed by reducing the uptake of contaminants by fish. Hence, nanochitosan contributes to the overall health and well-being of the aquatic organisms. Moreover, it enhances the safety and quality of fish products intended for human consumption, as the concentration of contaminants in the fish is significantly lowered.

The versatility of nanochitosan in adsorbing a wide range of contaminants is a key advantage. It can capture heavy metals, organic pollutants, and toxins simultaneously, making it a versatile solution for addressing the complex and diverse challenges posed by environmental pollutants in aquaculture settings (El-Naggar et al., 2022a). In essence, nanochitosan's strong adsorption properties are a pivotal factor in its effectiveness as a quintessential detoxification agent in fish farming. Its ability to capture and immobilize a variety of contaminants not only ensures the safety of the fish themselves but also safeguards the quality of seafood products for consumers. This property underscores the potential of nanochitosan as a valuable tool in the endeavor to make aquaculture more sustainable and environmentally responsible (El-Naggar et al., 2021).

Nanochitosan's adsorption capacity is characterized by its unique selectivity, a property that allows it to focus on specific contaminants while sparing essential nutrients and compounds in fish. This selectivity is a crucial advantage of nanochitosan in the context of fish detoxification, and it has significant implications for fish health and nutritional quality. Nanochitosan has the potential for tailored contaminant removal. It can be engineered or customized to target particular contaminants of concern (Mohanasrinivasan et al., 2014). For instance, it can be designed to have a higher affinity for heavy metals, organic pollutants, or toxins that are prevalent in a specific aquaculture environment. This tailored approach allows for precise and effective removal of the contaminants that pose the greatest risk. In the process of binding to contaminants, nanochitosan distinguishes between harmful substances and essential nutrients or compounds that are vital for fish health and growth. This is achieved through careful engineering and adjustment of nanochitosan's properties to ensure that it interacts preferentially with the undesired substances. By leaving essential nutrients and compounds in fish unaffected, nanochitosan safeguards the nutritional quality of fish (Marwa et al., 2022). This is of paramount importance, especially in aquaculture where fish are bred for human consumption. Nutrient-rich seafood products are highly valued for their health benefits, and the maintenance of these essential nutrients ensures that the final product meets the standard nutritional requirements. Selective adsorption not only benefits the nutritional quality of the fish but also contributes to the overall health and well-being of the aquatic organisms. The avoidance of essential nutrients and compounds being bound by nanochitosan prevents adverse effects on fish health, growth, and development. The challenge in fish detoxification is to effectively remove contaminants while maintaining the nutritional integrity of the fish. Nanochitosan's selectivity provides a balance between these objectives, allowing for efficient detoxification without compromising the nutritional value of the fish (Mohanasrinivasan et al., 2014).

Researchers and aquaculture practitioners can customize nanochitosan formulations to address specific contaminant profiles and concerns in various aquaculture settings. This adaptability is valuable in tailoring solutions to the unique challenges presented by different fish species and environments (Kaleem & Sabi, 2021). Nanochitosan's selective adsorption capacity is a key feature that distinguishes it as an effective tool for fish detoxification. It enables the precise targeting of contaminants while preserving essential nutrients and compounds in fish, which is essential for maintaining fish health and the nutritional quality of seafood products. This selectivity makes nanochitosan a promising and versatile solution for addressing the complex challenges associated with environmental pollutants in aquaculture.

3.3 Notable Biocompatibility

The biocompatibility of nanochitosan in the context of fisheries and aquaculture is a key attribute that underpins its potential as a quintessential detoxification and depuration tool. Biocompatibility refers to the ability of a material to coexist harmoniously with living organisms without causing harm or adverse reactions (Nalini et al., 2007). In the case of nanochitosan, its biocompatibility with fisheries and aquaculture has several remarkable importance. Nanochitosan is generally well-tolerated by aquatic organisms, including fish and other species commonly raised in aquaculture systems. When introduced into the aquatic environment, nanochitosan is unlikely to induce adverse physiological responses or elicit toxicity in the organisms (Hussein et al., 2012). The use of biocompatible materials like nanochitosan minimizes the risk of stress or negative effects on the health and well-being of aquatic species. This is crucial for the success and sustainability of aquaculture operations. The biodegradability of nanochitosan is a critical factor in its biocompatibility. It can break down naturally in the aquatic environment, reducing the potential for long-term environmental impact.

This property aligns with environmental responsibility in aquaculture, where minimizing the environmental footprint of operations is a priority. The use of materials that do not accumulate in the ecosystem contributes to sustainable practices and satisfactorily aligns with sustainable development goal 14 (life below water). Nanochitosan can be engineered and formulated to enhance its biocompatibility with specific species and aquaculture systems (Setiyorini et al., 2022). Customized formulations can be developed to ensure compatibility with the unique requirements of different fisheries and aquaculture operations. When used appropriately, nanochitosan and its derivatives do not accumulate in fish tissues. They are primarily used as a temporary medium for capturing and immobilizing contaminants. This ensures that fish and other aquatic organisms do not retain nanochitosan in their bodies, preventing any adverse effects on their physiology. Nanochitosan, when employed in fish detoxification or water treatment processes, helps aquaculture operations meet food safety standards. The use of biocompatible materials contributes to the production of safe and high-quality seafood products for human consumption. The safety and biocompatibility of nanochitosan-based methods enhance consumer confidence in the safety and quality of fish and seafood products. This is essential for marketability and reputation within the aquaculture industry. The biocompatibility of the novel biopolymer can be tailored to address specific species and aquaculture practices (Wang & Chen, 2005). Customized approaches ensure that it is well-suited for the unique requirements of different fish species and environments. Furthermore, its biocompatibility extends to various applications beyond detoxification, such as disease control, water quality improvement, and feed additives. This versatility makes it a valuable asset in fisheries and aquaculture. The biocompatibility of nanochitosan with fisheries and aquaculture underscores its potential as a safe and effective solution for addressing various challenges within these industries. Its harmonious coexistence with aquatic organisms, minimal environmental impact, and ability to meet food safety standards contribute to responsible and sustainable aquaculture practices while enhancing the safety and quality of seafood products for consumers (Yu et al., 2013).

3.4 Reduction of Bioavailability

Nanochitosan, through its remarkable property of reducing the bioavailability of contaminants in fish, plays a pivotal role in enhancing the safety and quality of fish products in aquaculture. This process involves nanochitosan binding to contaminants and effectively creating a protective barrier that makes these harmful substances less accessible to the fish's digestive system (Salaah et al., 2021).

As nanochitosan is introduced into the aquatic environment or incorporated into fish feed, it functions as an efficient and versatile adsorbent (Thilagar & Samuthirapandian, 2020). It attracts and immobilizes a diverse array of contaminants, including heavy metals, organic pollutants, and toxins. This adsorption is driven by nanochitosan's high surface area and numerous active binding sites. Once nanochitosan captures contaminants, it acts as a shield that guards against their ingestion by fish. Contaminants that would otherwise be ingested by fish are now firmly bound to the nanochitosan particles, rendering them effectively inaccessible to the fish's digestive system. This inaccessibility is a boon for fish health, as it significantly reduces the bioavailability of harmful substances. Fish are less likely to absorb or accumulate these bound contaminants, which, in turn, lowers the risk of toxic effects and adverse health outcomes. By protecting fish from contamination, nanochitosan contributes to the overall well-being of aquatic organisms (Salaah et al., 2021). Beyond health, the reduction in bioavailability has a direct impact on the quality of fish products intended for human consumption. By mitigating the accumulation of contaminants in fish tissues, nanochitosan ensures that the edible portions of the fish remain safe and meet the stringent safety standards required for seafood products. The positive effects extend beyond the fish. By sequestering contaminants, nanochitosan limits the potential harm these substances may cause to the broader aquatic ecosystem (Wan Ngah & Fatinathan, 2010). This not only safeguards other aquatic organisms but also contributes to the environmental sustainability of aquaculture practices. This reduction in bioavailability aligns with responsible and sustainable practices in aquaculture. It helps maintain the ecological balance and integrity of the aquatic environment, an important consideration as the aquaculture industry strives to minimize its environmental footprint. Nanochitosan's mechanism ensures a balance between efficient detoxification and the preservation of essential nutrients and compounds in fish tissues. It allows fish to maintain their nutritional quality while being effectively protected from contaminants.

Nanochitosan's ability to reduce the bioavailability of contaminants in fish by binding to them and creating a protective barrier is instrumental in fish detoxification (Wang & Li, 2011). This process not only safeguards fish health but also ensures the production of safe and high-quality seafood products for consumers. Nanochitosan's multifaceted benefits contribute to the responsible and sustainable advancement of aquaculture practices.

3.5 Controlled Release

The controlled release feature of nanochitosan in fish detoxification is a valuable and strategic aspect of its application. This property allows nanochitosan to slowly release bound contaminants over time, providing several advantages in the detoxification process while ensuring the well-being of the fish.

One of the key benefits of nanochitosan is that it can be engineered and tailored to meet specific detoxification requirements. This means that the release rate of bound contaminants can be adjusted to align with the unique needs of different aquaculture systems and fish species (Zaki et al., 2015). Customization ensures that detoxification is optimized for specific conditions and contaminants of concern. The controlled release mechanism allows nanochitosan to slowly and steadily eliminate contaminants. Rather than rapidly expelling all bound contaminants at once, nanochitosan releases them over a prolonged period. This gradual elimination minimizes the risk of potential stress or shock to fish that could result from abrupt changes in their environment.

Fish are highly sensitive to changes in their internal environment. Abrupt shifts in water chemistry, or the sudden removal of contaminants can be stressful and detrimental to their health. The controlled release of contaminants by nanochitosan ensures that fish are not subjected to sudden, adverse alterations in their surroundings. By avoiding abrupt changes, the detoxification process becomes less stressful for fish. Stress can weaken the immune system of fish and make them more susceptible to diseases. Controlled release helps maintain a stable and favourable environment for the fish, reducing the risk of stress-related health issues. The gradual elimination of contaminants also contributes to the safety of fish products intended for human consumption. It ensures that fish tissues do not release accumulated contaminants too quickly during processing, which could lead to food safety concerns (Zhao et al., 2018).

The gradual release aligns with environmental sustainability principles by preventing the sudden introduction of concentrated contaminants into the aquatic ecosystem. It reduces the risk of harm to other aquatic organisms and helps maintain ecological balance. The controlled release strategy does not compromise the efficiency of the detoxification process. Contaminants are still effectively removed from fish tissues, but the pace of removal is adjusted to minimize any potential adverse effects on the fish.

Nanochitosan's ability to engineer a controlled release of bound contaminants is an intelligent and versatile approach in fish detoxification. It enables the customization of detoxification processes, promotes fish well-being, and ensures the safety of seafood products for consumers. By gradually eliminating contaminants, nanochitosan contributes to efficient and responsible aquaculture practices while upholding environmental sustainability and fish health.

3.6 Sustainability and Ecofriendliness

The utilization of chitosan, and its nanoscale derivative nanochitosan, as a key component in fish detoxification processes underscores a commitment to sustainability and environmental responsibility in aquaculture. This source is renewable and sustainable, as it can be obtained as a byproduct of the seafood industry, reducing waste and enhancing resource efficiency.

Chitosan is considered an eco-friendly material because it is biodegradable and biocompatible. It can be integrated into aquaculture practices without leaving a lasting environmental footprint. Its use aligns with the principles of responsible resource management and waste reduction. The choice of chitosan in fish detoxification aligns with the growing demand for sustainable and environmentally responsible practices in aquaculture; in the strive to attain Sustainable Development Goal 14 (Life Under Water). Stakeholders in the industry, including consumers, regulators, and environmental advocates, increasingly emphasize the importance of minimizing the environmental impact of aquaculture operations (Abd-Elhakeem et al., 2016; Thilagar & Samuthirapandian, 2020).

One of the notable contributions of nanochitosan in fish detoxification is its ability to reduce the environmental impact of aquaculture. By capturing and immobilizing contaminants, nanochitosan prevents these harmful substances from entering the surrounding aquatic ecosystem. This reduction in contaminant release is particularly significant in areas where aquaculture practices may affect natural water bodies. It ensures that the local environment remains less impacted by aquaculture activities, maintaining the health and integrity of natural water systems. Minimizing the release of contaminants is vital for protecting aquatic ecosystems. These ecosystems often serve as habitats for various wildlife and play a critical role in maintaining biodiversity. Reducing contamination helps safeguard these ecosystems and the species they support (Zareie et al., 2019). Meeting regulatory standards and environmental guidelines is an essential aspect of responsible aquaculture. By using nanochitosan to reduce the release of contaminants, aquaculture operations can enhance their compliance with environmental regulations and ensure they operate within acceptable limits.

Sustainable and environmentally responsible practices are not only ethical but can also lead to economic benefits (Eliaz et al., 2006). They can enhance the reputation of aquaculture operations and attract environmentally conscious consumers who prioritize sustainably sourced seafood products. The use of chitosan and nanochitosan in fish detoxification exemplifies a commitment to sustainability and environmental responsibility in aquaculture. It is driven by the sustainable sourcing of chitosan from chitin, which is a renewable resource found in crustacean shells. By reducing the environmental impact through the controlled capture of contaminants, this approach aligns with the demands of eco-friendly and responsible aquaculture practices, protecting natural water bodies and their ecosystems.

4 Future Perspectives

The adaptability and versatility of nanochitosan are pivotal aspects of its application in fish detoxification and environmental remediation. Researchers can tailor the properties of nanochitosan to suit specific applications, making it a versatile and customizable tool to address varying detoxification needs and target specific contaminants of concern. Additionally, ongoing advancements in the fields of nanotechnology and materials science promise to further enhance the effectiveness and efficiency of nanochitosan-based detoxification methods (Fadlaoui et al., 2019). Nanochitosan can be engineered and customized to meet the precise requirements of different detoxification scenarios. This tailoring process allows researchers to optimize nanochitosan's properties, such as its particle size, surface chemistry, and adsorption capacity, to align with the specific contaminants and environmental conditions they aim to address.

The versatility of nanochitosan means it can be adapted for a wide range of detoxification needs. Whether the objective is to remove heavy metals, organic pollutants, toxins, or a combination of contaminants, nanochitosan can be fine-tuned to effectively capture and immobilize these substances. Different aquaculture systems, geographic locations, and fish species may present varying detoxification challenges, and the adaptability of nanochitosan allows it to offer tailored solutions (El-Naggar et al., 2020). Nanochitosan can be specifically designed to target contaminants of concern in a given aquaculture or environmental context. Researchers can optimize its selectivity to focus on the most problematic substances while avoiding interference with essential nutrients and compounds in the aquatic environment (El-Naggar et al., 2019).

The ability to tailor nanochitosan properties translates into enhanced detoxification efficiency (Benettayeb et al., 2023). By optimizing its properties to match the contaminants and conditions at hand, researchers can achieve more effective and thorough removal of harmful substances from the aquatic environment. This targeted approach ensures that the use of nanochitosan is both resource-efficient and cost-effective. The dynamic field of nanotechnology continues to evolve, offering opportunities for further refinement and innovation in the design and application of nanomaterials like nanochitosan. Researchers can harness these advancements to create more sophisticated and effective nanochitosan-based detoxification systems. Innovations in materials science complement nanotechnology advancements (Cheung et al., 2015). These innovations may lead to the development of novel formulations and modifications of nanochitosan that enhance its performance, stability, and safety for both aquatic organisms and the environment. The pursuit of improved nanochitosan-based detoxification methods aligns with broader goals of environmental responsibility and sustainability. By making the process more efficient and effective, researchers contribute to the reduction of environmental contaminants and the protection of aquatic ecosystems.

5 Conclusion

Nanochitosan's versatility, customization potential, and the promise of ongoing advancements in nanotechnology and materials science position it as a dynamic and powerful tool in the field of fish detoxification and environmental remediation. Researchers have the opportunity to fine-tune nanochitosan properties for specific applications, ensuring that it remains a valuable asset in addressing the ever-evolving challenges posed by environmental pollutants in aquaculture and natural water bodies.

References

Abd-Elhakeem, M. A., Ramadan, M. M., & Basaad, F. S. (2016). Removing of heavy metals from water by chitosan nanoparticles. *Journal of advances in Chemistry, 11*(7), 3765–3771. https://doi.org/10.24297/jac.v11i7.2200

Benettayeb, A., Seihoub, F. Z., Pal, P., Ghosh, S., Usman, M., Chia, C. H., Usman, M., & Sillanpää, M. (2023). Chitosan nanoparticles as potential nano-sorbent for removal of toxic environmental pollutants. *Nanomaterials, 13*(3), 447. https://doi.org/10.3390/nano13030447

Chauhan, D., Jaiswal, M., & Sankararamakrishnan, N. (2012). Removal of cadmium and hexavalent chromium from electroplating waste water using thiocarbamoyl chitosan. *Carbohydrate Polymers, 88*(2), 670–675. https://doi.org/10.1016/j.carbpol.2012.01.014

Chen, L. F., Liang, H. W., Lu, Y., Cui, C. H., & Yu, S. H. (2011). Synthesis of an attapulgite clay@ carbon nanocomposite adsorbent by a hydrothermal carbonization process and their application in the removal of toxic metal ions from water. *Langmuir, 27*(14), 8998–9004. https://doi.org/10.1021/la2017165

Cheung, R. C. F., Ng, T. B., Wong, J. H., & Chan, W. Y. (2015). Chitosan: An update on potential biomedical and pharmaceutical applications. *Marine Drugs, 13*, 5156–5186. https://doi.org/10.3390/md13085156

Chiou, M. S., & Li, H. Y. (2003). Adsorption behavior of reactive dye in aqueous solution on chemical cross-linked chitosan beads. *Chemosphere, 50*(8), 1095–1105. https://doi.org/10.1016/S0045-6535(02)00636-7

Divya, K., & Jisha, M. S. (2018). Chitosan nanoparticles preparation and applications. *Environmental Chemistry Letters, 16*, 101–112. https://doi.org/10.1007/s10311-017-0670-y

Dutta, P. K., Dutta, J., & Tripathi, V. S. (2004). Chitin and chitosan: Chemistry, property and application. *Journal of Scientific and Industrial Research, 63*, 20–31.

Eliaz, I., Weil, E., & Wilk, B. (2006). Integrative medicine and the role of modified citrus pectin/alginates in heavy metal chelation and detoxification-five case reports. *Forschende Komplementärmedizin, 14*(6), 358–364. https://doi.org/10.1159/000109829

El-Naggar, M. M., Abou-Elmagd, W. S. I., Suloma, A. M., El-Shabaka, H. A., Khalil, M. T., & Abd El-Rahman, F. A. A. (2019). Optimization and physicochemical characterization of chitosan and chitosan nanoparticles extracted from the crayfish *Procambarus clarkii* wastes. *Journal of Shellfish Research, 38*(2), 385–395. https://doi.org/10.2983/035.038.0220

El-Naggar, M. M., Haneen, D. S. A., Mehany, A. B. M., & Khalil, M. T. (2020). New synthetic chitosan hybrids bearing some heterocyclic moieties with potential activity as anticancer and apoptosis inducers. *International Journal of Biological Macromolecules, 150*, 1323–1330. https://doi.org/10.1016/j.ijbiomac.2019.10.142

El-Naggar, M. M., Salah, S., El-Shabaka, H. A., Abd El-Rahman, F. A. A., Khalil, M. T., & Suloma, A. (2021). Efficacy of dietary chitosan and chitosan nanoparticles supplementation on health status of Nile tilapia, *Oreochromis niloticus* (L.). *Aquaculture Reports, 19*, 100628. https://doi.org/10.1016/j.aqrep.2021.100628

El-Naggar, M. M., Medhat, F., & Taha, A. (2022a). Applications of chitosan and chitosan nanoparticles in fish aquaculture. *Egyptian Journal Of Aquatic Biology And Fisheries, 26*(1), 23–43. https://doi.org/10.21608/ejabf.2022.213365

El-Naggar, M. M., Salaah, S., Suloma, S., Khalil, M. T., & Emam, W. W. M. (2022b). Impact of chitosan and chitosan nanoparticles on reducing heavy metals in the Nile tilapia, *Oreochromis niloticus. Egyptian Journal of Aquatic Biology & Fisheries Zoology Department, Faculty of Science, Ain Shams University, Cairo, Egypt., 26*(2), 859–875.

El-Sayed, A.-F. M. (2019). *Tilapia culture*. Academic. https://doi.org/10.1016/C2017-0-04085-5

Fadlaoui, S., El Asri, O., Mohammed, L., Sihame, A., Omari, A., & Melhaoui, M. (2019). Isolation and characterization of chitin from shells of the freshwater crab potamon algeriense. *Progress on Chemistry and Application of Chitin and its Derivatives, XXIV*, 23–35. https://doi.org/10.15259/PCACD.24.002

Fu, F., & Wang, Q. (2011). Removal of heavy metal ions from wastewaters: A review. *Journal of Environmental Management, 92*(3), 407–418. https://doi.org/10.1016/j.jenvman.2010.11.011

Gamage, A., & Shahidi, F. (2007). Use of chitosan for the removal of metal ion contaminants and proteins from water. *Food Chemistry, 104*(3), 989–996. https://doi.org/10.1016/j.foodchem.2007.01.004

Goff, J. P. (2018). Invited review: Mineral absorption mechanisms, mineral interactions that affect acid-base and antioxidant status, and diet considerations to improve mineral status. *Journal of Dairy Science, 101*(4), 2763–2813. https://doi.org/10.3168/jds.2017-13112

Guibal, E. (2004). Interactions of metal ions with chitosan-based sorbents: A review. *Separation and Purification Technology, 38*(1), 43–74. https://doi.org/10.1016/j.seppur.2003.10.004

Hussein, M. H. M., El-Hady, M. F., Sayed, W. M., & Hefni, H. (2012). Preparation of some chitosan heavy metal complexes and study its properties. *Polymer Science, Series A, 54*(2), 113–124. https://doi.org/10.1134/S0965545X12020046

Kaleem, O., & Sabi, A.-F. B. S. (2021). Overview of aquaculture systems in Egypt and Nigeria, prospects, potentials, and constraints. *Aquaculture and Fisheries, 6*(6), 535–547. https://doi.org/10.1016/j.aaf.2020.07.017

Marwa, M. E., Sally, S., Ashraf, S., Magdy, T. K., & Wiame, W. M. E. (2022). Impact of chitosan and chitosan nanoparticles on reducing heavy metals in the Nile tilapia, *Oreochromis niloticus. Egyptian Journal of Aquatic Biology & Fisheries Zoology. Department, Faculty of Science, Ain Shams University, Cairo, Egypt, 26*(2), 859–875. ISSN 1110–6131.

Mohanasrinivasan, V., Mishra, M., Paliwal, J. S., Singh, S. K., Selvarajan, E., Suganthi, V., & Devi, C. S. (2014). Studies on heavy metal removal efficiency and antibacterial activity of chitosan prepared from shrimp shell waste. *3 Biotech, 4*, 167–175.

Nalini, S., Ajit, K. S., & Rashmi, S. (2007). Novel chitosan derivative for the removal of cadmium in the presence of cyanide from electroplating wastewater. *Journal of Hazardous Materials, 148*(1–2), 353–359.

Salaah, S. M., El-Gaar, D. M., & Gaber, H. S. (2021). Potential effects of dietary chitosan against lead-induced innate immunotoxicity and oxidative stress in Nile tilapia (*Oreochromis niloticus*). *The Egyptian Journal of Aquatic Research, 48*(2), 123–129. https://doi.org/10.1016/j.ejar.2021.10.004

Setiyorini, Y., Anggraeni, A., & Pintowantoro, S. (2022). In-vivo study of nano chitosan as therapeutic agent for toxic metal implant. *Results in Engineering, 13*, 100352. https://doi.org/10.1016/j.rineng.2022.100352

Seyedmohammadi, J., Motavassel, M., Maddahi, M. H., & Nikmanesh, S. (2016). Application of nanochitosan and chitosan particles for adsorption of Zn(II) ions pollutant from aqueous solution to protect environment. *Modelling Earth Systems and Environment, 2*, 165. https://doi.org/10.1007/s40808-016-0219-2

Shekhawat, A., Kahu, S., Saravanan, D., & Jugade, R. (2022). Bi-functionalized Ionic Liquid-Thiourea Chitosan for effective decontamination of Cd(II) and Hg(II) from water bodies. *Current Research in Green and Sustainable Chemistry, 5*, 100246.

Sun, S., Wang, L., & Wang, A. (2006). Adsorption properties of crosslinked carboxymethyl-chitosan resin with Pb(II) as template ions. *Journal of Hazardous Materials, 136*(2006), 930–937.

Thilagar, G., & Samuthirapandian, R. (2020). Chitosan from crustacean shell waste and its protective role against lead toxicity in *Oreochromis mossambicus. Toxicology Reports, 7*, 296–303. https://doi.org/10.1016/j.toxrep.2020.02.006

Wan Ngah, W. S. W., & Fatinathan, S. (2010). Pb (II) biosorption using chitosan and chitosan derivatives beads: Equilibrium, ion exchange and mechanism studies. *Journal of Environmental Sciences, 22*(3), 338–346. https://doi.org/10.1016/S1001-0742(09)60113-3

Wang, S. H., & Chen, J. C. (2005). The protective effect of chitin and chitosan against *Vibrio alginolyticus* in white shrimp *Litopenaeus vannamei. Fish & Shellfish Immunology, 19*(2005), 191–204. https://doi.org/10.1016/j.fsi.2004.11.003

Wang, Y., & Li, J. (2011). Effects of chitosan nanoparticles on survival, growth, and meat quality of tilapia, *Oreochromis nilotica. Nanotoxicology, 5*(3), 425–431. https://doi.org/10.3109/17435390.2010.530354

Wassmur, B. (2012). *Detoxification mechanisms in fish – Regulation and function of biotransformation and efflux in fish exposed to pharmaceuticals and other pollutants* (pp. 66). ISBN 978-91-628-8540-3. http://hdl.handle.net/2077/30452

Yu, K., Ho, J., Mccandlish, E., Buckley, B., Patel, R., Li, Z., & Shapley, N. C. (2013). Copper ion adsorption by chitosan nanoparticles and alginate microparticles for water purification applications. *Colloids and Surfaces A: Physicochemical and Engineering Aspects, 425*, 31–41. https://doi.org/10.1016/j.colsurfa.2012.12.043

Zaki, M. A., Salem, M. E. S., Gaber, M. M., & Nour, A. M. (2015). Effect of chitosan supplemented diet on survival, growth, feed utilization, body composition & histology of sea bass (*Dicentrarchus labrax*) world. *Journal of Engineering Technology, 3*(2015), 38–47.

Zareie, C., Eshkalak, S. K., Darzi, G. N., Baei, M. S., Younesi, H., & Ramakrishna, S. (2019). Uptake of Pb (II) ions from simulated aqueous solution via nanochitosan. *Coatings, 9*(12), 862. https://doi.org/10.3390/coatings9120862

Zhao, D., Yu, S., Sun, B., Gao, S., Guo, S., & Zhao, K. (2018). Biomedical applications of chitosan and its derivative nanoparticles. *Polymers, 10*(4), 462–479. https://doi.org/10.3390/polym10040462

Economic and Social Implications of Nanochitosan

Solomon Uche Oranusi, Emmanuel Ojochegbe Mameh, Samuel Adeniyi Oyegbade, Daniel Oluwatobiloba Balogun, Austine Atokolo, Victoria-grace Onyekachi Aririguzoh, and Oluwapelumi Shola Oyesile

Contents

S. U. Oranusi (✉) · O. S. Oyesile
Department of Biological Sciences, College of Science and Technology, Covenant University, Ota, Ogun State, Nigeria
e-mail: solomon.oranusi@covenantuniversity.edu.ng

E. O. Mameh · S. A. Oyegbade · D. O. Balogun · A. Atokolo · V.-g. O. Aririguzoh
Department of Biological Sciences, College of Science and Technology, Covenant University, Ota, Ogun State, Nigeria

Covenant Applied Informatics and Communication Africa Centre of Excellence (CApIC-ACE), Covenant University, Ota, Ogun State, Nigeria
e-mail: emmanuel.mamehpgs@stu.cu.edu.ng

P. O. Isibor et al. (eds.), *Nanochitosan-Based Enhancement of Fisheries and Aquaculture*, https://doi.org/10.1007/978-3-031-52261-1_12

1 Introduction

Nanotechnology is the study of functional materials, devices, and systems as well as their design, synthesis, fabrication, manipulation, and application for the matter harnessing of unique phenomena and characteristics. Nanotechnology is the controlled modification of atomic and molecular components at scales between 1 and 100 nm and the understanding and controlling of the behaviour of matter at dimensions typically ranging from 1 to 100 nm, where different phenomena offer novel applications. Globally, there has been a notable surge in the quantity of nanoscience applications and inventive patent authorizations (Fitridge et al., 2012). The fields of electronics, materials science, human health, animal husbandry, aquaculture, and biological and biomedical sciences all stand to benefit greatly from nanotechnology. Applications in these fields include biomolecule analysis, and the development of vectors that are not virus-like for clinical diagnostics, medicines, gene therapy, and cancer treatment. Research in nanomedicine, nanoelectronics, and nanopharmaceuticals receives the lion's share of funding for nanomaterials studies. There aren't many nanotechnologies and nanoproducts accessible in the fields of agriculture, aquaculture, and animal husbandry because the sector has invested less in nanotechnology than the other industries mentioned. Nonetheless, there is a great deal of promise for nanotechnology in this crucial field, which is needed for a significant population-wide socioeconomic status revolution. With an estimated yearly production of over 50 million tons and a value of US$ 80 billion, aquaculture is a significant global sector (Tacon & Metian, 2015). Since 1995, the average yearly growth rate of aquaculture production worldwide has been 6.6%. The business handled 76.6 million aquatic animals and 29.4 million aquatic plants in 2015. As stated by FAO (2016), the industry of aquaculture is projected to rival fish farming shortly and play a crucial part in fostering food safety, reducing poverty, and fostering economic growth. Aquaculture has the potential to produce high-quality food for a growing population and to create jobs. However, there are numerous challenges related to environmental concerns (Martinez-Porchas & Martinez-Cordova, 2012). The aquaculture industry faces various challenges, such as the devastation of naturally occurring ecological systems (Rajitha et al., 2007); pollution of waterways (Avnimelech & Kochba, 2009); salinization and acidification of earth crust (Martinez-Porchas & Martinez-Cordova, 2012); nutrient enrichment and nitrification of ecosystems receiving effluents (Fenaroli et al., 2014; Martinez-Porchas & Martinez-Cordova, 2012); contaminants from chemicals (Justino et al., 2016); biological contamination resulting from the entry of unnatural species (Bashir et al., 2020); and alteration of landscapes and detrimental effects on fisheries (Osman et al., 2022). Numerous transdisciplinary applications of nanotechnology can be found in the aquaculture and agricultural sectors. Nanotechnology has the potential to transform the aquaculture sector and the fishing industry by providing new methods for quick disease detection and improving fish absorption of medications, vaccinations, and nutrition, among other things. Numerous uses of nanotechnology in aquaculture are now under development. One of the finest industries to use and market nanotechnological goods may be the highly integrated fish farming sector (Okeke et al., 2022). Additionally, by strengthening processing protection, the use of nanotechnology in fish processing can also be used to detect fish infectious diseases in packaging and enhance product safety. Even though further investigation and development are required in this field, there are various signs that aquaculture health management, water treatment, fish animal breeding, harvest, and postharvest technologies might all benefit from nanotechnology.

Fish aquaculture is seen as a significant agricultural endeavour that can eliminate nutritional deficits and help reduce poverty (Kaleem & Sabi, 2021). Aquaculture has advanced significantly during the last few decades, surpassing the growth of catch fisheries. However, it is anticipated to become the main source of aquatic animal requirements in the future years (Witeska et al., 2022). As a result, there is a growing global demand for fish, which is driving the development of intensive aquaculture systems, particularly in underdeveloped nations where fish constitute a major source of animal protein and are used in artificial feeding. According to recent estimates, aquaculture contributes approximately 75% of Egypt's annual fish production. To ensure sustained expansion in aquaculture, novel and non-conventional fish meals that support it are necessary for improving the physiology and biological functions of the cultured fish. Thus, to achieve aquaculture development, researchers must look for sustainable, biodegradable, economically viable, and environmentally favourable nanomaterials (Abdel-Tawwab et al., 2019). According to Krishnani et al. (2022), chitosan (CS) and chitosan nanoparticles (CSNP) are both naturally occurring cationic biopolymers that are safe and boost fish development and immunological response of chitin, an essential part of the land and freshwater crustaceans' exoskeletons, including those of shrimp, crabs, and crayfish, and the cell walls of certain microbial agents (Kumar et al., 2020). However, as one of the fastest-growing industries, nanotechnology offers fresh nano-enabled goods with creative and distinctive applications. Nanoparticles, which have a size range of 1–100 nm in at least a single dimension, are crucial to the success of the nanoscale industry. In contrast to their bulk materials, the nanoparticles have unique physicochemical properties. Their bigger surface area to volume ratio sets them apart and contributes to their notable features and increased reactivity (Khosravi-Katuli et al., 2017). Applications for CSNP and CS include cancer treatment, agriculture, and water purification. Additionally, they serve as safe, natural feed additives in aquaculture, improving fish development, boosting immunity, and thwarting intestinal microbial infections (Krishnani et al., 2022).

A biocompatible, non-toxic substance, CS (Poliglusam) [β-(1-4)-N-acetyl-D-glucosamine] is a renewable polymeric substance that is soluble in an acidic aqueous solution. Characterized by unique qualities, for example, non-toxicity, decomposition, biocompatibility, and increased solubility (Shard et al., 2014), CS may be readily and widely modified, making it appropriate for a range of applications in aquaculture, cosmetics, cancer therapy, administering drugs, and water purification (Ahmed et al., 2019; El-Naggar et al., 2021; Abd-Elghany & Salem, 2020).

N-acetyl-D-glucosamine and -(14)-linked D-glucosamine are the two chemical building blocks that contribute to making up the molecular structure of CS. It is made from chitin, which is the second-most common polysaccharide after cellulose and is found in the exoskeletons of aquatic animals including fish, shrimp, crabs, and some insects as well as in the cell walls of some fungi.

2 Cost-Effectiveness of Nanochitosan in Aquaculture

Waste products like the remains after consuming crustaceans, shrimp, and crabs are used for the production of bio-polymer chitosan. The processing of crustaceans produces a large number of underused by-products. About 35–40% of the overall weight of this biowaste is made up primarily of heads and shells (de la Caba et al., 2019). Because all arthropods, including littoral crustaceans, have chitin-containing exoskeletons, the second most prevalent organic substance on earth, chitin, is being consumed through these waste materials. Chitin is a valuable substance. Arthropods comprise the bulk of all species on earth, thus supply shouldn't be an issue. Except for the acetamide groups that replace the hydroxyl groups at position C_2, chitin and cellulose are chemically similar. Functional groups in chitin render it an insoluble polymer and restrict its application. However, this by-product's deacetylation produces chitosan, which has processability and is soluble in acidic solutions. The chemical approach dominates large-scale production owing to the increasing demand. Even though chitosan is a naturally occurring substance, using organic solvents raises environmental issues because they may have detrimental effects on the ecosystem. In light of this, alternative strategies are investigated. When compared to conventional chemical processes, deep electrostatic solvents, ionic liquids, and ultrasonic extraction all provide better process control, energy efficiency, cost-effectiveness (Rodrigues et al., 2021), and microwave-assisted extraction (Mohan et al., 2022). Particularly, pretreatment and polymer manufacturing, including chitin extraction, have been used in developing deep eutectic solvents (DESs) and ionic liquids that are examples of green solvents in a variety of applications, including biomass for purifying and separating it (Pellis et al., 2022). Nanochitosan is a valuable substance with applications in the culinary, pharmaceutical, cosmetic, agricultural, and chemical industries, among other fields. Nanochitosan is highly valued for its several biological properties, including its anti-inflammatory, antibacterial, anti-coagulant, and anti-tumour properties (Flórez et al., 2022). These properties eventually account for chitosan's widespread recognition. In terms of the legal framework, the European Union has started implementing law (EC 450/2009) on active and intelligent materials intended to come into contact with food substrates (European Union, 2009; Flórez et al., 2022). Nanochitosan has been approved as GRAS (generally recognized as safe) by the China National Standards (GB 29941-2013), the European Union (No. 749/2012) regulation, and the United States Food and Drug Administration (USFDA). Depending on the origin of derivatives, chitosan can be applied in various fields, which further solidifies its position as a suitable and sustainable candidate for food packaging that is biodegradable is being developed (Wang et al., 2021).

Technological developments and the circular economy approach have proven to be extremely effective ways to reduce the amount of clean natural resources used as well as the final waste burden. To put it simply, it's the state in which everything is wholesome fodder for something else. According to Saleeger et al. (2020), this strategy tends to enhance the efficiency of products, which enables a quality of life about environmental soundness. Additionally, the circular economy concept's 3 R approach, which is reduce, reuse, and recycle, can foster social and economic advantages and may be lessen the load on the environment (de la Caba et al., 2019). From now on, it will be crucial to manage garbage in a way that is both environmentally responsible and methodical in the future. Numerous objectives and targets in the United Nations 2030 Agenda for Sustainable Development Goals (SDGs) seem to have included the importance of the circular economy. The production of chitosan from the waste of crustaceans aligns with the circular economy concept, which cycles resources at a high value while adhering to strict environmental and human health protection standards. It seems like a win–win situation to use this marine waste biopolymer to preserve food quality and, in the end, reduce food loss. According to Abdollahzadeh et al. (2023), chitosan nanoemulsions provide all the characteristics and potentials inherent in chitosan, but in a more enhanced and advanced way. Because fish-based products are always susceptible to microbiological deterioration, the fish processing sector must cope with related issues, which can lead to financial losses and health risks. The antibacterial property of chitosan-based nanoemulsions, which results from the cationic nature of biopolymeric substances, is one of the most important characteristics in the industrial sector. Regarding the underlying mechanism, not many have been documented. One of these has to do with the positively charged amino groups found in biopolymers. These groups react with negatively charged microbial membranes to test the strength of the membrane and release important components. According to some reports, chitosan may also act as a chelating agent. To increase cell membrane permeability and permit the outflow of physiologically vital components, the purpose of chitosan nanoemulsions is to precisely chelate a variety of ions from the lipo-polysaccharide layer that constitutes bacterial outer membrane (Khalid & Arif, 2022).

Chitosan emulsions have been shown to have antioxidant activity in addition to their antibacterial potency, which supports their application in preventing fish deterioration from oxidation. Fish lipid and protein oxidation can lead to the production of free radicals, which can eventually attract colour loss, off-flavours, and the creation of hazardous chemical compounds that can create major concerns for humans. Reactive oxygen species (ROS) can be efficiently countered and inactivated by chitosan nanoemulsions, which also provide protection against oxidative deterioration and almost entirely preserve food quality. Abd El-Hack et al. (2020) claim that chitosan nanoemulsions currently can replace synthetic antioxidants (BHA, BHT) and address the problem that results from their use. Haemoglobin and metal ions in high quantities have been known to oxidize, particularly in fish and other shellfish. According to several studies, a biopolymer's molecular weight (MW), concentration, and viscosity are all closely connected with its effectiveness as an antioxidant (Abd El-Hack et al., 2020). By blocking the conversion of ferrous ions into ferric ions, chitosan is thought to inhibit oxidation in lipids by chelating ferrous ions present in fishes and thereby removing their prooxidant activity. Better antioxidant activity is found in chitosan emulsions with lower molecular weights because short chains are more or less likely to establish intramolecular hydroxyl bonds, which leads to more amino groups and activated hydroxyl thereby increasing the activity of scavenging free radicals (Ozogul et al., 2021). By utilizing cutting-edge technologies, the properties of the biopolymer can be altered. The matrix's barricade features aim for optimal optimization that best suits the goal of preserving the product. By purposefully interacting with the chitosan, functional components can be included to improve water vapour permeability (WVP) by decreasing the areas where water molecules can interact. In the end, this results in a rapid increase in the hydrophobicity of the nanochitosan particles, which gives the manufactured matrices superior WVP properties. Above all, when the system is reduced to the nanoscale, functional chemicals spread and occlude open spaces in the microstructure more effectively, leaving just a tiny, restricted zone for movement, which stops moisture migration (Wu et al., 2021). The WVP is further influenced by the chitosan's degree of deacetylation, the pH of the solvent, and the type of acid used to produce chitosan. In a similar vein, the polymer's capacity to permeabilize oxygen is another important aspect in determining its effectiveness in food storage and preservation, particularly in fish and meat products, which have rich matrices that make them particularly susceptible to excessive oxidative putrefaction. The structural framing of nanoscale chitosan emulsions is hydrogen woven, preventing the food system from leaking oxygen. Because particles are arranged in a nanometric pattern, the amount of available space is further reduced, which stops the gas from moving through the manufactured film or coating. Upgraded crystallinity of the matrices is the outcome of these confirmed strong interactions among components (Yan et al., 2016).

3 Socioeconomic Impacts of Nanochitosan on Fisheries and Aquaculture

Because of their minimal negative effects and capacity to improve immunological response, antioxidant activity, and fish growth performance (GP), CS and CSNP have been extensively employed in a variety of applications (Abd El-Naby et al., 2019; Krishnani et al., 2022). Furthermore, drugs can be transported via CS and CSNP (Tardy et al., 2021). Additionally, because CS and its derivatives may chelate heavy metals, they are employed in the treatment of water (Janani et al., 2022). According to Abd El-Naby et al. (2019), CSNP and CS have been shown to improve water quality in this regard.

3.1 CSNP and CS as Feed Additives

According to Nathanailides et al. (2021), fish nutrition plays a major role in determining a fish's capacity for growth and reproduction. One major obstacle facing aquaculture is the exorbitant cost of fish meal, which is a major component of fish feed. Nathanailides et al. (2021) recommended utilizing the ideal dosages of feed additives or supplements to boost fish growth while consuming the least amount of fish meal possible, which would lower fish mortality. Feed additives are consumable materials that are added in small amounts to animal feed to improve feed quality. This improves fish development performance and lowers their mortality rate (Abou-Hadeed et al., 2021). According to Tardy et al. (2021), it is advised to look for alternate protein sources, especially from plant sources because fish meal is becoming more and more expensive. The authors further identified that certain vital nutritional elements are absent from plant-based diets, which has an impact on fish survival rates as well as growth performance. In this particular context, the author suggested supplementing plant-based diets with CS and CSNP to improve Oreochromis niloticus's growth performance and immunological response. According to Abd El-Naby et al. (2019), several research studies suggested using CSNP and CS as growth-promoting and immune-stimulating agents.

3.2 Effect of CSNP and CS on the Growth Performance of Fish

The appropriate amount of CSNP and CS to have the most growth-enhancing effect relies primarily on the species of animal (Abdel-Ghany & Salem, 2020). The addition of chitosan in diets did not affect sea bream growth, and no change in fish development performance was observed in Paramigurunus dabryanus when CS was added along with selenium (Se) at a level of (0.6, 1, 2, and 1.8) mg CS per kg food for 60 days (Victor et al., 2019). One noteworthy finding is that adding CS as a supplement to the feed of *O. niloticus* reduced growth performance. That being said, *Cyprinus carpio, Dicentrachus labrax*, and *Paralichthys olivaceus* all showed improved growth following CS treatment (Zaki et al., 2015). Zaki et al. (2015) discussed the CS effect in terms of its capacity to stimulate the proliferation of microvilli, increasing their absorption surface, as well as their capacity to generate healthy intestinal epithelium. Alternatively, the intestinal obstruction caused by an excess of microvilli caused by an increase in CS concentrations in the diets resulted in a decrease in growth performance. The results showed that the amount of CS and CSNP supplied to fish predominantly impacted how well they developed. In the case of *Oreochromis niloticus*, Wang and Li (2011) reported that CS supplementation at a dosage of 0.5 g/kg diet increased the growth performance. Moreover, enriched diets with varying amounts of chitosan (1800, 4000, 7500, 10,000, and 20,000 milligram per kilogram of food for 75 days) were used for *Carasius auratus* juveniles (Butt et al., 2021). They discovered that the ideal amount of CS supplementation was 4000 mg per kilogram of food. Through its effects on the formation of goblet cells and microvilli in the intestinal epithelia, it caused fish to grow to their maximum potential. Furthermore, by raising feed intake and feed utilization, the application of CSNP supplementation at various doses in *O. niloticus* increased growth efficiency (Abdel-Tawwab et al., 2019; Abd El-Naby et al., 2019). Besides, fish disease resistance to infections was enhanced, and overall health was improved when 0.5 g/kg of CSNP was added to the diet of *O. niloticus* (El-Naggar et al., 2021).

3.3 Nanochitosan-Based Food Enhancement

Nanoparticles are presently one of the key components for many different fields. A lot more processes and applications can be made to work better, more efficiently, and more affordably consequent nanoparticles. Chitosan nanoparticles are used in agriculture to load herbicides, insecticides, and pesticides to enhance crop cultivation and are also utilized in food packaging (Prasad et al., 2022). According to recent studies, nanotechnology has the potential to revolutionize several areas of the food industry, including post-harvest management, food storage and packaging, biosensors, medicine, the delivery of nutrients and nutraceuticals, ingredients and additives, and food bioprocessing (Dholariya et al., 2021).

Nanochitosan-based improvement is rooted in the rapidly evolving field of nanotechnology and the adaptable qualities of chitosan, a naturally occurring biopolymer produced from chitin, typically found in the shells of crustaceans (Adetunji et al., 2023). The distinctive properties of chitosan at the nanoscale serve as the basis for this invention's science. Chitosan acquires extraordinary qualities, including a large surface area and improved reactivity, when it is broken down into nanoparticles. It can interact molecularly with other compounds attributable to these qualities (Adetunji et al., 2023; Dholariya et al., 2021).

Nanochitosan is used in the food manufacturing industry to enhance both the safety and quality of edible items. It functions as a naturally occurring preservative, prolonging shelf life, and minimizing the need for synthetic modifications. It can also be utilized to encapsulate bioactive substances, improving their transport and bioavailability. Nanochitosan-based formulations for agriculture encourage plant growth, offer disease protection, and lessen environmental pollution. Utilizing the possibilities of nanostructures for the improvement of food supply chains and ecological sustainability, the science of nanochitosan-based enhancement provides promising solutions for sustainable agriculture and food safety (Prasad et al., 2022).

3.4 Chitosan Utilization in Food Processing and Preservation

One of the most crucial strategies for extending the shelf life of fruits and vegetables is covering them. Anindita et al. (2022) reported that strawberries' shelf life would be greatly extended when chitosan nanoparticles (CHNPs) are used in food processing. Strawberries treated with CHNPs kept their quality at 6–2 °C for 16 days and 25–3 °C at 6-day intervals, respectively. A different approach involving study findings points to a potential remedy for *Botrytis*, a greyish fungus that affects the growth of strawberries. Irradiation chitosan was more bioactive and had a lower molecular weight when compared to natural fungal chitosan. Chitosan-Zano or chitosan/copper complexes have piqued the curiosity of researchers due to their possible application in the preservation of vegetables and fruits (Anindita et al., 2022).

As a useful addition to food processing and preservation, nanochitosan is a chitosan derivative that is nanoscale in size and provides several important benefits to the food industry. Nanochitosan has the following four major benefits:

(i) Natural preservative: Nanochitosan is used in the food sector as a natural preservative. Its antibacterial qualities contribute to increasing the shelf life of numerous food goods by preventing the growth of spoilage organisms. In line with customer expectations for clearer labelling and healthier food alternatives, this enables food makers to decrease or perhaps completely remove the need for synthetic chemical preservatives (Saputra et al., 2022).

(ii) Enhanced food safety: By lowering the risk of foodborne diseases, nanochitosan can increase food safety. It is an effective method for maintaining the safety of food items since it can stop the growth of germs like *Salmonella* and *E. coli*. Nanochitosan leads to decreased contamination rates and, eventually, a safer food supply by limiting the growth of harmful pathogens (Anindita et al., 2022; Saputra et al., 2022).

(iii) Food quality improvement: Nanochitosan has the power to raise the general standard of food items. It can assist in preserving the flavour, colour, and texture of food while halting oxidation and rotting. The retention of sensory qualities guarantees that consumers obtain goods that are not only more attractive and fresher but also safer, enhancing their pleasure with food selections (Prasad et al., 2022; Anindita et al., 2022; Saputra et al., 2022).

(iv) Reduced chemical preservatives: Using nanochitosan in the manufacture of food can help cut down on the use of synthetic chemical additives. The move towards greener, more natural solutions is in line with customer expectations for healthier, less processed food. Additionally, the decrease in chemical additions can benefit food labelling by allowing for shorter ingredient lists and clearer product information (Prasad et al., 2022; Saputra et al., 2022).

According to reports in 2016, there were 10.3 million fatalities and 229.1 million DALYs (disability-adjusted life-years) around the globe, and poor eating habits are the second-leading risk factor (Mertens et al., 2019). The globalization of the food industry and an overall improvement in living standards have caused dietary habits, which are influenced by cultural, environmental, technical, and economic variables, to grow more similar over time (Traill et al., 2014). The preservation of vital nutrients and bioactive molecules can be facilitated by nanochitosan's capacity to encapsulate and shield bioactive components in food.

4 Economic Impacts of Nanochitosan-Based Food Enhancement

The monetary implications of food improvement with nanochitosan are becoming obvious in the food sector. A nanoscale derivative of chitosan has demonstrated tremendous promise for enhancing food quality, increasing shelf life, and lowering food waste (Traill et al., 2014). Due to its capacity to function as a natural preservative by preventing the growth of bacteria and fungi that can degrade food, this cutting-edge food additive has attracted interest. Nanochitosan helps minimize the need for early disposal and disposal expenses by extending the shelf life of food goods, which is advantageous to both consumers and producers (Fig. 1). Furthermore, because of its biocompatibility and sustainability, there may be less of a need for synthetic chemical preservatives, which would save money and be better for the environment. With the use of food enhancements based on nanochitosan, producers may notice a reduction in spoilage losses, which will ultimately boost their profitability (Dholariya et al., 2021; Saputra et al., 2022; Traill et al., 2014). Some of the economic impacts on the use of nanochitisan for food enhancement include:

(i) *Increased agricultural yields:* The majority of the world's population relies on the agricultural industry as their primary source of income. Chitosan is a

Fig. 1 Applications of nanochitosan to agriculture. (Ingle et al., 2022)

renewable substance that may be utilized during an array of operations in industries to replace harmful and non-biodegradable substances in their nanoform (Ingle et al., 2022). ChNPs have potential applications in agriculture as antimicrobial agents against agricultural produce, disease-causing microorganisms such as fungi, bacteria, and other insect pests such as *Aphis gossypii, Callosobruchus chinensis,* and *Callosobruchus maculatus,* as well as stimulants for plant growth have a ripple effect in determining the health of plant species (Sahab et al., 2015).

(ii) *Enhanced product quality:* Modern agriculture's principal focus is the production of high-quality food in sufficient quantity to fulfil the world's population growth while limiting environmental concerns. As a result, scientists began to consider nanotechnology in agriculture. Although several research studies have described the application of chitosan in farming, there is much research to be done on ChNP use in agriculture. Different complexes that might be beneficial for agricultural use are produced by the negatively charged polymers and the amine group in chitosan. To encourage the growth of plants, several forms of chitosan are applied to the soil (Ingle et al., 2022; Sahab et al., 2015; Traill et al., 2014).

(iii) *Reduced post-harvest losses:* Post-harvest losses, or the degradation and wasting of crops and food items after they have been harvested, have long been a major problem in the agriculture and food sectors. In addition to having an intense economic impact, these losses also add to environmental problems and food hunger across the world. To reduce post-harvest losses by prolonging the shelf life of various agricultural goods, nanochitosan has emerged as a viable alternative (Ingle et al., 2022). Throughout storage, a nanochitosan coating with a 0.5% chitosan concentration greatly improved fruit quality and reduced weight loss. These findings supported the potential advantages of covering extremely sensitive apples with nanochitosan to increase shelf life and preserve quality (Sahraei Khosh Gardesh et al., 2016).

5 Social Impacts of Aquacultural Nanochitosan

The social effects of food improvement with nanochitosan are substantial. It helps to improve food security by prolonging the shelf life of food products while minimizing waste, guaranteeing a steadier supply of food. By reducing the need for synthetic chemical preservatives, this technique can help meet the rising demand for natural and healthier food alternatives. Additionally, as less food waste results from nanochitosan applications, this can aid in the fight against hunger and support sustainability. These social advantages collectively result in a more reliable food supply, healthier options, and a proportionate allocation of resources, all of which have a favourable effect on society (Prasad et al., 2022; Dholariya et al., 2021; Anindita et al., 2022; Ingle et al., 2022). Some of the social impacts of the use of nanochitosan include:

1. *Accessibility and affordability of food:* Nanochitosan can be used to greatly improve food cost and accessibility. This ground-breaking technique can help provide access to food for a wider variety of customers by increasing the shelf life of food goods and lowering spoilage. It lowers the need for pricey chemical preservatives and decreases post-harvest losses, which eventually lowers production and distribution expenses. As a result, customers could get access to more reasonably priced and durable food alternatives (Prasad et al., 2022).
2. *Health and nutrition outcomes:* The possible applications of chitosan in medicine and biomedicine include drug delivery and pharmaceutical formulation (for antibiotics, anti-inflammatories, vaccines, peptides, proteins, and growth factors), Antimicrobial applications, gene therapy and gene delivery and wound healing and burn recovery, used in dermatology, ophthalmology, dentistry, and cancer (treatment, therapy, and diagnostic approach), as well as in a variety of additional uses, such as magnetic bioimaging support for enzymes that are immobilized, and magnetic resonance imaging medical care for animals (Morin-Crini et al., 2019). Friedman and Juneja (2010) highlighted the exceptional antibacterial properties of chitosan in powders, films, coatings, and solutions. Chitosan with a low molecular weight produced better outcomes. The main emphasis of study findings, as shown in Fig. 1, is on new chitosan derivatives and oligomers that can be employed as antibacterial agents against food microorganisms. Derivatives like this look encouraging, particularly for applications in nutraceuticals. Chitosan and its derivatives, as shown by Kardas et al. (2012), have a wide range of unique applications in the food industry, including packaging food, preserving food against microbial deterioration, prolonging shelf life, and producing renewable films.

6 Importance of Sustainable Practices

Nanotechnology is widely acknowledged as an emerging method that is effectively being used to enhance plant resistance to abiotic stress as well as improve crop productivity, quality, and nutrient usage efficiency. By minimizing resource, input, expense, and energy waste, the use of nanoscale fertilizers is one of the conceivable practices of precision farming that makes crop production systems more effective, sustainable, and ecologically safe (Ashraf et al., 2022). Based on these, many sustainable methods for using nanochitosan are adopted.

6.1 Environmental Impact Reduction

Sustainable environmental outcomes result from the usage of nanochitosan in food production. This method decreases food waste and the need for continuous manufacturing and distribution, preserving energy and natural resources in the process. Perishable products' periods of preservation are extended. Reducing the use of artificial chemical preservatives, which can be harmful to the environment, is in line with sustainable agricultural methods. Nanochitosan helps make food production more environmentally friendly by lowering the impact of the food sector on the environment. Reducing the amount of food that ends up in landfills reduces methane emissions while also conserving resources such as water, electricity, and agricultural inputs that are utilized in the production of tossed food. By lessening the burden on ecosystems and the pollution caused by food waste, this in turn has a good effect on the environment. Nanochitosan helps create a more ecologically conscious food system by reducing waste and encouraging resource conservation (Traill et al., 2014; Prasad et al., 2022; Kardas et al., 2012; Friedman & Juneja, 2010).

6.2 Nanochitosan Use for Resource Efficiency

In farming, it improves the absorption of nutrients and water by plants, optimizing the use of resources. By extending the shelf life of food, it can cut down on food waste and conserve production-related resources (Abou El-Enin et al., 2023). According to Abou El-Enin et al. (2023), a two-year field study was undertaken to evaluate the efficiency of nanochitosan-loaded N (CS-NNPs) for lowering the amount of mineral N used in maize-based maize-soybean intercropping and increasing land productivity. Therefore, adding N to maize that is interplanted with soybeans remains acceptable. Without any doubt, this will help maize producers economically since the agricultural environment would be protected.

The goal of sustainable agriculture is to increase resource utilization efficiency (Zulfiqar et al., 2019; Alghamdi et al., 2023) and meet current and future generations' needs. Through effective management, agricultural ecosystems may be made more diverse and a healthy environment can be attained (Mousavi & Eskandari, 2011; Mekdad et al., 2022). According to Brooker et al. (2015), intercropping is an example of a sustainable agricultural method that aids in achieving agroecological balance, effectively utilizing the nutrients, water, and sunlight available for the growth of plants, enhancing output per unit of land, and reducing yield losses from disease, insect pests, and weeds.

7 Cost-Effectiveness of Nanochitosan in Various Industries

The application of nanochitosan offers diverse cost-efficiency in various sectors, saving costs while maximizing profits. Products containing chitosan have been employed since the 1990s to protect crops against bacteria causing decay and other harmful effects on crop production both during the growing season as well as following the harvest season (Yin & Du, 2011). Nanochitosan has bacteriostatic (preventing bacterial growth) or bactericidal (killing bacteria) activities. However, the exact procedure is still not well-known. Additionally, chitosan hinders viroids and plant viruses from proliferating. It has the possibility of being used as a biological pesticide, also used as a spray, for root application, and seed-soaking agent. These actions have a significant impact on plant disease management and stress tolerance (Morin-Crini et al., 2019).

In the development of cost-effective methods for production, effective technologies for recovering chitin and its byproducts, such as proteins and pigments, are necessary for chitin's application in the industry. The recovery of chitinous substances from waste is a well-known extra revenue source. Significant levels of unsynthesized carotenoids found in crustacean shells are sold as a fish meal ingredient in aquaculture, mostly for salmon. Alishahi and Ader (2012) described the usage of chitosan and its related products in aquaculture. Functional foods, dietary supplements (synbiotics), or a few uses for it include the release of drugs, encapsulating of pathogens or nucleic acids, and pollution removal from water and wastewater.

Several enzymes were immobilized by chitosan, including lysozyme, urease, amylases, and cells of the bacterium *Escherichia coli*. They become enveloped and absorbed by the macromolecule in chains. Chitosan is mostly used in cross-linking processes in biochemistry to sustain enzymes. Using chitosan and its substitutes, technological advancement has also employed biological sensors and other biodevices. The depolymerization and de-N-acetylation of chitin by chitinases and deacetylases results in an array of alternatives, including chitooligosaccharides, which have numerous applications in biotechnology (Grifoll-Romero et al., 2018).

As it relates to the usage of nanochitosan, ethical practices discuss the sociological and ethical implications (SEI) of the development of nanotechnology/science and how it affects people, society, and the environment. It is critical to consider the immediate and prospective benefits of nanotechnology as well as its drawbacks, possible dangers, and hazards as science and nanotechnology evolve (*Nanotechnology and Ethics*, 2020).

8 Ethical Implications of Unregulated Use

There are serious ethical considerations raised by the uncontrolled usage of nanochitosan. Without adequate regulation, there is a probability that it will be misused such as by being used excessively in agriculture, which might be harmful to ecosystems and human health. Uncontrolled use may also result in uneven access, with populations that are economically underprivileged perhaps being more negatively affected than other areas. It is essential to provide competent regulation and control of nanochitosan applications to resolve these moral dilemmas and strike a balance between innovation, equality, and safety.

There are several roles ethical considerations play in the use of nanochitosan. The following roles are employed to avoid the challenges of innovative technology:

(i) Environmental impact and sustainability: The possible environmental effects of nanochitosan require careful ethical consideration. Its introduction into many industries, especially agriculture, as a revolutionary nanomaterial, must emphasize sustainability. Its possible effects on ecosystems and water quality give rise to ethical considerations. Comprehensive environmental risk assessments, ongoing monitoring, and the adoption of sustainable agriculture methods that limit harm to ecosystems, product biodiversity, and assure resource conservation are required for the appropriate use of nanochitosan, impact on the environment, and sustainability.

(ii) Consumer safety and health: It is important to maintain the health and safety of the public whilst ensuring that the nanochitosan does not provide unanticipated health hazards, careful assessment is needed for its usage in food preservation, and packaging is employed. To ascertain that customers are informed of its existence in food items, transparency in labelling and information sharing is essential. To utilize nanochitosan responsibly, strict safety guidelines and extensive toxicity analyses must be followed, taking into account any potential health risks.

(iii) Equity and accessibility: It is ethically right to guarantee that all individuals have access to nanochitosan benefits. Existing inequalities in resource usage or food security shouldn't be made worse by this technology. It raises ethical issues if it becomes a technology that is predominantly used by larger agricultural enterprises, thus putting smaller, resource-constrained farmers at a disadvantage. To promote a fair and inclusive nanochitosan deployment that benefits a wider variety of stakeholders and promotes social and economic fairness, policymakers and industry stakeholders must address these equity challenges.

9 Conclusion

The cost-effectiveness of nanochitosan applications in agriculture is evident through improved disease control and water quality management, ultimately leading to enhanced productivity. These economic benefits ripple through to fisheries and aquaculture communities, contributing to livelihoods and food supply stability. However, the socioeconomic impacts of nanochitosan should be approached with ethical and sustainable practices including equitable access to these innovations while minimizing environmental impacts, which is crucial for a responsible, ethical, and sustainable application of nanochitosan across the agriculture and food supply chain.

References

Abd El-Hack, M. E., El-Saadony, M. T., Shafi, M. E., Zabermawi, N. M., Arif, M., Batiha, G. E., Khafaga, A. F., Abd El-Hakim, Y. M., & Al-Sagheer, A. A. (2020). Antimicrobial and antioxidant properties of chitosan and its derivatives and their applications: A review. *International Journal of Biological Macromolecules, 164*, 2726–2744. https://doi.org/10.1016/j.ijbiomac.2020.08.153

Abd-Elghany, H.M. & Salem, M.E.S. (2020). Effects of dietary chitosan supplementation on farmed fish; a review. Rev. Aquacult., *12*(1): 438–452

Abd El-Naby, F. S., Naiel, M. A. E., Al-Sagheer, A. A., & Negm, S. S. (2019). Dietary chitosan nanoparticles enhance the growth, production performance, and immunity of *Oreochromis niloticus. Aquaculture (Amsterdam, Netherlands), 501*, 82–89. https://doi.org/10.1016/j.aquaculture.2018.11.014

Abdel-Tawwab, M., Razek, N. A., & Abdel-Rahman, A. M. (2019). Immunostimulatory effect of dietary chitosan nanoparticles on the performance of Nile tilapia, *Oreochromis niloticus* (L.). *Fish & Shellfish Immunology, 88*, 254–258. https://doi.org/10.1016/j.fsi.2019.02.063

Abdollahzadeh, M., Elhamirad, A. H., Shariatifar, N., Saeidiasl, M., & Armin, M. (2023). Effects of nano-chitosan coatings incorporating with free/nano-encapsulated essential oil of Golpar (*Heracleum persicum* L.) on quality characteristics and safety of rainbow trout (*Oncorhynchus mykiss*). *International Journal of Food Microbiology, 385*(109996), 109996. https://doi.org/10.1016/j.ijfoodmicro.2022.109996

Abou El-Enin, M. M., Sheha, A. M., El-Serafy, R. S., Ali, O. A. M., Saudy, H. S., & Shaaban, A. (2023). Foliage-sprayed nano-chitosan-loaded nitrogen boosts yield potentials, competitive ability, and profitability of intercropped maize-soybean. *International Journal of Plant Production, 17*(3), 517–542. https://doi.org/10.1007/s42106-023-00253-4

Abou-Hadeed, A. H., Mohamed, A. T., Hegab, D. Y., & Ghoneim, M. H. (2021). Ethoxyquin and butylated hydroxy toluene distrub the hematological parameters and induce structural and functional alterations in liver of rats. *Archives of Razi Institute, 76*(6), 1765–1776. https://doi.org/10.22092/ari.2021.356439.1844

Adetunji, C. O., Olaniyan, O. T., Dare, A., Adeniyi, M. J., & Ajayi, O. O. (2023). Significance of nanochitosan for improving bioavailability, stability and shelf life of sensitive ingredients in aquatic and animal-based foods. In *Next generation nanochitosan* (pp. 421–430). https://doi.org/10.1016/b978-0-323-85593-8.00014-x

Ahmed, F., Soliman, F. M., Adly, M. A., Soliman, H. A. M., El-Matbouli, M., & Saleh, M. (2019). Recent progress in biomedical applications of chitosan and its nanocomposites in aquaculture: A review. *Research in Veterinary Science, 126*, 68–82. https://doi.org/10.1016/j.rvsc.2019.08.005

Alghamdi, S. A., Alharby, H. F., Abdelfattah, M. A., Mohamed, I. A., Hakeem, K. R., Rady, M. M., & Shaaban, A. (2023). *Spirulina platensis*-inoculated humified compost boosts rhizosphere soil hydro-physico-chemical properties and *Atriplex nummularia* forage yield and quality in an arid saline calcareous soil. *Journal of Soil Science and Plant Nutrition, 23*, 2215. https://doi.org/10.1007/s42729-023-01174-x

Alishahi, A., & Aïder, M. (2012). Applications of chitosan in the seafood industry and aquaculture: A review. *Food and Bioprocess Technology, 5*, 817–830. https://doi.org/10.1007/s11947-011-0664-x

Anindita, R. C., Swachchhatoya, G., Anapurba, K., Anuradha, M., & Subhasis, D. (2022). Antimicrobial properties of chitosan nanoparticles and their role in post-harvest shelf life extension. *Journal of Environmental Science, Toxicology and Food Technology, 16*, 51–60. https://doi.org/10.9790/2402-1606015160

Ashraf, U., Zafar, S., Ghaffar, R., Sher, A., Mahmood, S., Noreen, Z., Maqbool, M. M., Saddique, M., & Ashraf, A. (2022). Impact of nano chitosan-NPK fertilizer on field crops. In *Role of chitosan and chitosan-based nanomaterials in plant sciences* (pp. 165–183). Elsevier.

Avnimelech, Y., & Kochba, M. (2009). Evaluation of nitrogen uptake and excretion by tilapia in bio floc tanks, using 15N tracing. *Aquaculture (Amsterdam, Netherlands), 287*(1–2), 163–168. https://doi.org/10.1016/j.aquaculture.2008.10.009

Bashir, I., Lone, F. A., Bhat, R. A., Mir, S. A., Dar, Z. A., & Dar, S. A. (2020). Concerns and threats of contamination on aquatic ecosystems. In *Bioremediation and biotechnology* (pp. 1–26). Springer.

Brooker, R. W., Bennett, A. E., Cong, W., Daniell, T. J., George, T. S., Hallett, P. D., Hawes, C., Iannetta, P. P. M., Jones, H. G., Karley, A. J., Li, L., McKenzie, B. M., Pakeman, R. J., Paterson, E., Schob, C., Shen, J., Squire, G., Watson, C. A., Zhang, C., & White, P. J. (2015). Improving intercropping: A synthesis of research in agronomy, plant physiology and ecology. *New Phytologist, 206*, 107–117. https://doi.org/10.1111/nph.13132

Butt, U. D., Lin, N., Akhter, N., Siddiqui, T., Li, S., & Wu, B. (2021). Overview of the latest developments in the role of probiotics, prebiotics and synbiotics in shrimp aquaculture. *Fish & Shellfish Immunology, 114*, 263–281. https://doi.org/10.1016/j.fsi.2021.05.003

de la Caba, K., Guerrero, P., Trung, T. S., Cruz-Romero, M., Kerry, J. P., Fluhr, J., Maurer, M., Kruijssen, F., Albalat, A., Bunting, S., Burt, S., Little, D., & Newton, R. (2019). From seafood waste to active seafood packaging: An emerging opportunity of the circular economy. *Journal of Cleaner Production, 208*, 86–98. https://doi.org/10.1016/j.jclepro.2018.09.164

de Sadeleer, I., Brattebø, H., & Callewaert, P. (2020). Waste prevention, energy recovery or recycling – Directions for household food waste management in light of circular economy policy. *Resources, Conservation and Recycling, 160*(104908), 104908. https://doi.org/10.1016/j.resconrec.2020.104908

Dholariya, P. K., Borkar, S., & Borah, A. (2021). Prospect of nanotechnology in food and edible packaging: A review. *The Pharma Innovation, 10*(5), 197–203. https://doi.org/10.22271/tpi.2021.v10.i5c.6202

European Union. (2009). Lisbon Treaty: The Treaty of the Functioning of the European Union (TFEU). European Union. https://eur-lex.europa.eu/legal-content/EN/TXT/?uri=celex%3A12012E%2FTXT

El-Naggar, M., Salaah, S., El-Shabaka, H., El-Rahman, F. A., Khalil, M., & Suloma, A. (2021). Efficacy of dietary chitosan and chitosan nanoparticles supplementation on health status of Nile tilapia, *Oreochromis niloticus* (L.). *Aquaculture Reports, 19*(100628), 100628. https://doi.org/10.1016/j.aqrep.2021.100628

FAO (2016). Food and Agriculture Organization of the United Nations. Fao.org. Available at: https://www.fao.org/news/archive/news-by-date/2016/en/?page=8&ipp=10&tx_dynalist_pi1%5Bpar%5D=YToxOntzOjE6IkwiO3M6MToiNiI7fQ%3D%3D (Accessed: January 23, 2024).

Fenaroli, F., Westmoreland, D., Benjaminsen, J., Kolstad, T., Skjeldal, F. M., Meijer, A. H., van der Vaart, M., Ulanova, L., Roos, N., Nyström, B., Hildahl, J., & Griffiths, G. (2014). Nanoparticles as drug delivery system against tuberculosis in zebrafish embryos: Direct visualization and treatment. *ACS Nano, 8*(7), 7014–7026. https://doi.org/10.1021/nn5019126

Fitridge, I., Dempster, T., Guenther, J., & de Nys, R. (2012). The impact and control of biofouling in marine aquaculture: A review. *Biofouling, 28*(7), 649–669. https://doi.org/10.1080/08927014.2012.700478

Flórez, M., Guerra-Rodríguez, E., Cazón, P., & Vázquez, M. (2022). Chitosan for food packaging: Recent advances in active and intelligent films. *Food Hydrocolloids, 124*(107328), 107328. https://doi.org/10.1016/j.foodhyd.2021.107328

Friedman, M., & Juneja, V. K. (2010). Review of antimicrobial and antioxidative activities of chitosans in food. *Journal of Food Protection, 73*, 1737–1761.

Grifoll-Romero, L., Pascual, S., Aragunde, H., Biarnes, X., & Planas, A. (2018). Chitin deacetylases: Structures, specifcities, and biotech applications. *Polymers, 10*, 352. https://doi.org/10.3390/polym10040352

Ingle, P. U., Shende, S. S., Shingote, P. R., Mishra, S. S., Sarda, V., Wasule, D. L., Rajput, V. D., Minkina, T., Rai, M., Sushkova, S., Mandzhieva, S., & Gade, A. (2022). Chitosan nanoparticles (ChNPs): A versatile growth promoter in modern agricultural production. *Heliyon, 8*(11), e11893. https://doi.org/10.1016/j.heliyon.2022.e11893

Janani, R., Gurunathan, B., Sivakumar, K., Varjani, S., Ngo, H. H., & Gnansounou, E. (2022). Advancements in heavy metals removal from effluents employing nano-adsorbents: Way towards cleaner production. *Environmental Research, 203*(111815), 111815. https://doi.org/10.1016/j.envres.2021.111815

Justino, C. I. L., Duarte, K. R., Freitas, A. C., Panteleitchouk, T. S. L., Duarte, A. C., & Rocha-Santos, T. A. P. (2016). Contaminants in aquaculture: Overview of analytical techniques for their determination. *Trends in Analytical Chemistry: TRAC, 80*, 293–310. https://doi.org/10.1016/j.trac.2015.07.014

Kaleem, O. & Bio Singou Sabi, A.-F. (2021) "Overview of aquaculture systems in Egypt and Nigeria, prospects, potentials, and constraints," Aquaculture and fisheries, 6(6), pp. 535–547. Available at: https://doi.org/10.1016/j.aaf.2020.07.017

Kardas, I., Struszczyk, M. H., Kucharska, M., van den Broek, L. A. M., van Dam, J. E. G., & Ciechańska, D. (2012). Chitin and chitosan as functional biopolymers for industrial applications. In P. Narvard (Ed.), *The European polysaccharide network of excellence (EPNOE)* (pp. 329–373). Springer. https://doi.org/10.1007/978-3-7091-0421-7_11

Khalid, M. Y., & Arif, Z. U. (2022). Novel biopolymer-based sustainable composites for food packaging applications: A narrative review. *Food Packaging and Shelf Life, 33*(100892), 100892. https://doi.org/10.1016/j.fpsl.2022.100892

Khosravi-Katuli, K., Prato, E., Lofrano, G., Guida, M., Vale, G., & Libralato, G. (2017). Effects of nanoparticles in species of aquaculture interest. *Environmental Science and Pollution Research International, 24*(21), 17326–17346. https://doi.org/10.1007/s11356-017-9360-3

Krishnani, K. K., Boddu, V. M., Chadha, N. K., Chakraborty, P., Kumar, J., Krishna, G., & Pathak, H. (2022). Metallic and non-metallic nanoparticles from plant, animal, and fisheries wastes: Potential and valorization for application in agriculture. *Environmental Science and Pollution Research International, 29*(54), 81130–81165. https://doi.org/10.1007/s11356-022-23301-4

Kumar, D., Gihar, S., Shrivash, M. K., Kumar, P., & Kundu, P. P. (2020). A review on the synthesis of graft copolymers of chitosan and their potential applications. *International Journal of Biological Macromolecules, 163*, 2097–2112. https://doi.org/10.1016/j.ijbiomac.2020.09.060

Martinez-Porchas, M., & Martinez-Cordova, L. R. (2012). World aquaculture: Environmental impacts and troubleshooting alternatives. *The Scientific World Journal, 2012*, 1–9. https://doi.org/10.1100/2012/389623

Mekdad, A. A., El-Sherif, A., Rady, M. M., & Shaaban, A. (2022). Culture management and application of humic acid in favor of *Helianthus annuus* L. oil yield and nutritional homeostasis in a dry environment. *Journal of Soil Science and Plant Nutrition, 22*, 71–86. https://doi.org/10.1007/s42729-021-00636-4

Mertens, E., Kuijsten, A., Dofková, M., Mistura, L., D'Addezio, L., Turrini, A., Dubuisson, C., Favret, S., Havard, S., Trolle, E., van't Veer, P, & Geleijnse, J. M. (2019). Geographic and socioeconomic diversity of food and nutrient intakes: A comparison of four European countries. *European Journal of Nutrition, 58*(4), 1475–1493. https://doi.org/10.1007/s00394-018-1673-6

Mohan, K., Ganesan, A. R., Ezhilarasi, P. N., Kondamareddy, K. K., Rajan, D. K., Sathishkumar, P., Rajarajeswaran, J., & Conterno, L. (2022). Green and eco-friendly approaches for the extraction of chitin and chitosan: A review. *Carbohydrate Polymers, 287*(119349), 119349. https://doi.org/10.1016/j.carbpol.2022.119349

Morin-Crini, N., Lichtfouse, E., Torri, G., & Crini, G. (2019). Applications of chitosan in food, pharmaceuticals, medicine, cosmetics, agriculture, textiles, pulp and paper, biotechnology, and environmental chemistry. *Environmental Chemistry Letters, 17*(4), 1667–1692. https://doi.org/10.1007/s10311-019-00904-x

Mousavi, S. R., & Eskandari, H. (2011). A general overview on intercropping and its advantages in sustainable agriculture. *Journal of Applied Environmental and Biological Sciences, 1*, 4482–4482. https://doi.org/10.3389/fsufs.2021.634361

Nanotechnology and Ethics. (2020). *Nanotechnology in STEM*. https://serc.carleton.edu/msu_nanotech/nano_ethics.html

Nathanailides, C., Kolygas, M., Choremi, K., Mavraganis, T., Gouva, E., Vidalis, K., & Athanassopoulou, F. (2021). Probiotics have the potential to significantly mitigate the environmental impact of freshwater fish farms. *Fishes, 6*(4), 76. https://doi.org/10.3390/fishes6040076

Okeke, E. S., Chukwudozie, K. I., Nyaruaba, R., Ita, R. E., Oladipo, A., Ejeromedoghene, O., Atakpa, E. O., Agu, C. V., & Okoye, C. O. (2022). Antibiotic resistance in aquaculture and aquatic organisms: A review of current nanotechnology applications for sustainable management. *Environmental Science and Pollution Research International, 29*(46), 69241–69274. https://doi.org/10.1007/s11356-022-22319-y

Osman, A. I., Fawzy, S., Farghali, M., El-Azazy, M., Elgarahy, A. M., Fahim, R. A., Maksoud, M. I. A. A., Ajlan, A. A., Yousry, M., Saleem, Y., & Rooney, D. W. (2022). Biochar for agronomy, animal farming, anaerobic digestion, composting, water treatment, soil remediation, construction, energy storage, and carbon sequestration: A review. *Environmental Chemistry Letters, 20*(4), 2385–2485. https://doi.org/10.1007/s10311-022-01424-x

Ozogul, F., Cagalj, M., Šimat, V., Ozogul, Y., Tkaczewska, J., Hassoun, A., Kaddour, A. A., Kuley, E., Rathod, N. B., & Phadke, G. G. (2021). Recent developments in valorisation of bioactive ingredients in discard/seafood processing by-products. *Trends in Food Science & Technology, 116*, 559–582. https://doi.org/10.1016/j.tifs.2021.08.007

Pellis, S.M., Pellis, V.C., Ham, J.R. & Achterberg, E.J.M. (2022) "The rough-and-tumble play of rats as a natural behavior suitable for studying the social brain," Frontiers in behavioral neuroscience, 16. Available at: https://doi.org/10.3389/fnbeh.2022.1033999

Prasad, R. D., Abhilash, J., Chaitanya, V., Khdabadi, S. S., & Mukund, T. (2022). Current applications of chitosan nanoparticles. A multifaceted review journal in the field of pharmacy systematic review. *The Systematic Reviews in Pharmacy, 13*(10), 685–693. https://doi.org/10.31858/0975-8453.13.10.685-693

Rajitha, K., Mukherjee, C. K., & Vinu Chandran, R. (2007). Applications of remote sensing and GIS for sustainable management of shrimp culture in India. *Aquaculture Engineering, 36*(1), 1–17. https://doi.org/10.1016/j.aquaeng.2006.05.003

Rodrigues, L. A., Radojčić Redovniković, I., Duarte, A. R. C., Matias, A. A., & Paiva, A. (2021). Low-phytotoxic deep eutectic systems as alternative extraction media for the recovery of chitin from brown crab shells. *ACS Omega, 6*(43), 28729–28741. https://doi.org/10.1021/acsomega.1c03402

Sahab, A. F., Waly, A. I., Sabbour, M. M., & Nawar, L. S. (2015). Synthesis, antifungal and insecticidal potential of chitosan (CS)-g-poly (acrylic acid) (PAA) nanoparticles against some seed borne fungi and insects of soybean. *International Journal of ChemTech Research, 8*(2), 589–598.

Sahraei Khosh Gardesh, A., Badii, F., Hashemi, M., Ardakani, A. Y., Maftoonazad, N., & Gorji, A. M. (2016). Effect of nanochitosan based coating on climacteric behavior and postharvest shelf-life extension of apple cv. *Golab Kohanz. Lebensmittel-Wissenschaft Und Technologie [Food Science and Technology], 70*, 33–40. https://doi.org/10.1016/j.lwt.2016.02.002

Saputra, D., Ula, F. R., Fadhila, A. B. N., Sijabat, Y. Y., Romadoni, A. A., & Windarto, S. (2022). Nano-chitosan spray as a preservative and food security of fishery products in the middle of the covid-19 pandemic. *Jurnal Ilmiah Perikanan Dan Kelautan, 14*(1), 71–82. https://doi.org/10.20473/jipk.v14i1.30121

Shard, C., McMichael, G., Girirajan, S., Moreno-De-Luca, A., Gecz, J., Nguyen, L.S., Nicholl, J., Gibson, C., Haan, E., Eichler, E., Martin, C.L. & MacLennan, A. (2014) "Rare copy number variation in cerebral palsy," European journal of human genetics: EJHG, 22(1), pp. 40–45. Available at: https://doi.org/10.1038/ejhg.2013.93

Tacon, A.G.J. & Metian, M. (2015) "Feed matters: Satisfying the feed demand of aquaculture," Reviews in fisheries science & aquaculture, 23(1), pp. 1–10. Available at: https://doi.org/10.1080/23308249.2014.987209

Tardy, B. L., Mattos, B. D., Otoni, C. G., Beaumont, M., Majoinen, J., Kämäräinen, T., & Rojas, O. J. (2021). Deconstruction and reassembly of renewable polymers and biocolloids into next generation structured materials. *Chemical Reviews, 121*(22), 14088–14188. https://doi.org/10.1021/acs.chemrev.0c01333

Traill, W. B., Mazzocchi, M., Shankar, B., & Hallam, D. (2014). Importance of government policies and other influences in transforming global diets. *Nutrition Reviews, 72*(9), 591–604. https://doi.org/10.1111/nure.12134

Victor, H., Zhao, B., Mu, Y., Dai, X., Wen, Z., Gao, Y., & Chu, Z. (2019). Effects of Se-chitosan on the growth performance and intestinal health of the loach *Paramisgurnus dabryanus* (Sauvage). *Aquaculture (Amsterdam, Netherlands), 498*, 263–270. https://doi.org/10.1016/j.aquaculture.2018.08.067

Wang, H., Ding, F., Ma, L., & Zhang, Y. (2021). Edible films from chitosan-gelatin: Physical properties and food packaging application. *Food Bioscience, 40*(100871), 100871. https://doi.org/10.1016/j.fbio.2020.100871

Wang, Y., & Li, J. (2011). Effects of chitosan nanoparticles on survival, growth and meat quality of tilapia, *Oreochromis nilotica*. *Nanotoxicology, 5*(3), 425–431. https://doi.org/10.3109/17435390.2010.530354

Witeska, M., Kondera, E., Ługowska, K., & Bojarski, B. (2022). Hematological methods in fish – Not only for beginners. *Aquaculture (Amsterdam, Netherlands), 547*(737498), 737498. https://doi.org/10.1016/j.aquaculture.2021.737498

Wu, F., Misra, M., & Mohanty, A. K. (2021). Challenges and new opportunities on barrier performance of biodegradable polymers for sustainable packaging. *Progress in Polymer Science, 117*(101395), 101395. https://doi.org/10.1016/j.progpolymsci.2021.101395

Yan, N., Capezzuto, F., Lavorgna, M., Buonocore, G. G., Tescione, F., Xia, H., & Ambrosio, L. (2016). Borate cross-linked graphene oxide–chitosan as robust and high gas barrier films. *Nanoscale, 8*(20), 10783–10791. https://doi.org/10.1039/c6nr00377j

Yin, H., & Du, Y. (2011). Chapter 41: Mechanism and application of chitin/chitosan and their derivatives in plant protection. In S. K. Kim (Ed.), *Chitin, chitosan, oligosaccharides and their derivatives: Biological activities and applications* (pp. 605–618). CRC Press/Taylor & Francis Group LLC. ISBN 9781439816035.

Zaki, M. A., Salem, M. E.-S., Gaber, M. M., & Nour, A. M. (2015). Effect of chitosan supplemented diet on survival, growth, feed utilization, body composition & histology of sea bass (*Dicentrarchus labrax*). *World Journal of Engineering and Technology, 03*(04), 38–47. https://doi.org/10.4236/wjet.2015.34c005

Zulfiqar, F., Navarro, M., Ashraf, M., Akram, N. A., & Munné-Bosch, S. (2019). Nano fertilizers use for sustainable agriculture: Advantages and limitations. *Plant Science, 289*, 1–11. https://doi.org/10.1016/j.plantsci.2019.110270

Prospects and Challenges of Nanochitosan Application in Aquaculture

Patrick Omoregie Isibor, Ifeoluwa Ihotu Kayode-Edwards, and Ogochukwu Oritseweyinmi Onwaeze

Contents

1 Introduction

Advancements in aquaculture, the farming of aquatic organisms such as fish, shellfish, and aquatic plants, have been significant in recent years. These advancements are driven by the growing demand for seafood, environmental concerns, and the

P. O. Isibor (✉) · I. I. Kayode-Edwards · O. O. Onwaeze
Department of Biological Sciences, College of Science and Technology, Covenant University, Ota, Ogun State, Nigeria
e-mail: patrick.isibor@covenantuniversity.edu.ng

P. O. Isibor et al. (eds.), *Nanochitosan-Based Enhancement of Fisheries and Aquaculture*, https://doi.org/10.1007/978-3-031-52261-1_13

need for sustainable food production (Mustapha et al., 2021). Nanochitosans, also known as nanoscale chitosan particles, are molecules that have been broken down into nanoparticles, which are particles with sizes typically in the range of 1–100 nm (Rampino et al., 2013). They are derived from chitosan, a biopolymer that is obtained from chitin, the second most abundant natural polysaccharide after cellulose (Ahmad et al., 2020). This reduction in size imbues chitosan with remarkable properties that significantly enhance its versatility. As a result, nanochitosans have gained attention for their potential applications in various fields, including fishery and aquaculture, environmental remediation, pharmaceutics, and agriculture.

One of the most significant prospects of nanochitosan lies in their ability to promote sustainable aquaculture. Traditional aquaculture methods face difficulties due to resource depletion and environmental impact as the world's demand for seafood rises (Mustapha et al., 2021). By enhancing resource efficiency, reducing waste, and minimizing environmental harm via efficient pollutant adsorption and nutrient cycling, nanochitosan can support more environmentally conscious and conscientious aquaculture (Reid et al., 2019). Mustapha et al. (2021) also report that the future of aquaculture operations with effective resource utilization and maintenance largely depends on innovative technologies, one of which is the utilization of nanochitosan for their unique properties, including high surface area and reactivity, potent sorption, rapid dissolution, possession of a large number of functional groups, electric and optical properties, and improved active sites (Maleki et al., 2015; Haripriyan et al., 2022; Isibor et al., 2023).

2 Potential Advancements and Innovative Applications of Nanochitosan in Fishery and Aquaculture Systems

Nanochitosan is a transformative biocatalytic tool with immense potential in fisheries and aquaculture. Their versatility, sustainability, and biocompatibility make them invaluable in addressing the challenges faced by these industries. From improving feed quality and disease management to wastewater treatment and environmental impact reduction, nanochitosan plays a multifaceted role in enhancing the sustainability and efficiency of fisheries and aquaculture operations.

Some significant areas for potential advancement of nanochitosan include cost reduction strategies, optimization of processing method, environmental impacts, regulatory considerations, scale-up studies, mechanistic understanding, nanochitosan modification, combination with other technologies and regeneration and reusability of nanochitosan (Isibor et al., 2023).

2.1 Biomedical Applications of Nanochitosan

Nanochitosans serve as effective drug delivery systems in fishery and aquaculture, offering targeted and controlled drug administration to aquatic organisms (Abimbola et al., 2023). These nanoparticles, derived from chitosan, can encapsulate drugs, vaccines, or therapeutic agents for precise delivery to aquatic organisms (Patra et al., 2018). This targeted approach enhances disease management and overall aquatic organism health by minimizing dosage requirements and reducing the risk of off-target effects (Sulthana et al., 2023). Also, according to Sulthana et al. (2023), nanochitosan improves drug solubility, stability, and controlled release to maximize drug delivery and produce more effective treatments with fewer side effects. Additionally, they support ethical and sustainable aquaculture methods that protect fish populations and lessen the negative effects of conventional treatments on the environment. By reducing the negative effects of treatments on the environment, nanochitosan supports sustainable aquaculture methods. Additionally, because it is biocompatible and biodegradable, it supports responsible fish health management techniques (Dar et al., 2020).

Nanochitosan can be applied to cell and gene therapy in aquaculture and fisheries, in addition to improving drug delivery methods (Dar et al., 2020). Aquatic species can effectively receive genetically modified cells or therapeutic genes via these chitosan-derived nanoparticles. Because of this, it is possible to precisely manipulate genetic features, increase disease resistance, and improve the general well-being and productivity of fish populations (Han et al., 2022). According to Tripathi et al. (2023), nanochitosan offers a targeted and safe delivery strategy that minimizes off-target effects and increases the efficacy of gene and cell therapies.

Aquaculture and fisheries both use nanochitosan for a variety of biomedical purposes. They help treat diseases and lessen stress in aquatic organisms by acting as drug carriers for targeted medication delivery (Dar et al., 2020). Furthermore, gene therapy makes use of nanochitosan to modify specific genetic features in fish populations to enhance growth and resistance to disease (Ansari, 2023).

With the creation of stimuli-responsive carriers, advances in the application of nanochitosan as drug carriers in fisheries and aquaculture systems are being made (Selvasudha et al., 2022). These nanocarriers have the ability to release medications intelligently in response to particular physiological cues from aquatic species. For example, when pH levels fluctuate or when certain enzymes are present in the fish's body, nanochitosans may release therapeutic compounds (Adewuyi et al., 2019; Vyshnava et al., 2023). The precise and customized drug delivery provided by this technology guarantees that prescriptions are released when and where they are most needed. These developments lessen their negative effects on the environment, increase the efficacy of therapies, and support the sustainable management of aquaculture and fisheries (Bilal et al., 2019).

2.2 Application of Nanochitosan in Environmental Remediation

A large amount of organic matter and nutrient-rich effluent is produced by aquaculture systems. Therefore, efficient wastewater treatment is necessary to reduce contamination of the environment and preserve the system's water quality.

In fisheries and aquaculture systems, nanochitosan offers substantial promise for environmental rehabilitation. According to Isa et al. (2022), these nanoparticles are efficient at adsorbing organic contaminants, pollutants, and heavy metals from aquaculture effluent. Nanochitosan effectively removes dangerous compounds, which helps to preserve aquatic ecosystems, lessen environmental pollution, and maintain water quality (Ahuekwe et al., 2023c). Their biodegradable nature ensures minimal long-term impact (Isibor et al., 2023). Additionally, nanochitosan can support remediation efforts in cases of accidental spills or pollution events, aiding in the restoration of affected aquatic environments (Ahankari et al., 2023; Rather et al., 2023). These applications underscore nanochitosans' vital role in minimizing the environmental footprint of fishery and aquaculture practices while safeguarding aquatic habitats.

Nanochitosans possess exceptional adsorption capabilities, efficiently binding heavy metals, organic pollutants, and toxins present in aquaculture wastewater (Chaudhary et al., 2023). This helps in removing harmful substances, reducing the risk of water contamination. The goal of ongoing research is to advance nanochitosan's efficiency and selectivity for heavy metal adsorption. This includes improvements in the surface properties of nanochitosan and the creation of hybrid materials with enhanced adsorption capacities (Ferreira et al., 2022; Isibor et al., 2023).

In addition to their adsorption capabilities for heavy metals, nanochitosan can also inhibit algal blooms by adsorbing excess nutrients, like phosphates and nitrates, which are often the culprits behind water eutrophication and oxygen depletion (Ahuekwe et al., 2023c).

Functionalized nanochitosan can also serve as carriers for beneficial microorganisms that aid in the biological removal of nutrients like ammonia and nitrate, thereby maintaining balanced water chemistry (Ahuekwe et al., 2023a). On the other hand, nanochitosan can mitigate the spread of waterborne pathogens in aquaculture systems and aid in the control of biofouling (Olisaka et al., 2023). They do this by binding and immobilizing bacteria and viruses, reducing the risk of disease outbreaks (Zou et al., 2016). In addition, nanochitosan-based materials are being explored for removing emerging contaminants such as pharmaceuticals, personal care products, and microplastics from water sources (Isibor et al., 2023).

By eliminating suspended particles and impurities, nanochitosan contributes to clearer water. This enhanced clarity not only benefits fish health but also supports efficient fish monitoring and harvesting (Tumwesigye et al., 2022). Implementing nanochitosan in water treatment aligns with sustainable aquaculture practices by reducing the reliance on chemical treatments that can harm the environment and aquatic life (Rebello et al., 2023). Also, effective pollutant removal and improved water quality support water recycling initiatives, reducing the demand for freshwater and the discharge of contaminated wastewater (Tayel et al., 2019).

Nanochitosan also holds promise in air purification for fishery and aquaculture operations. These nanoparticles, derived from chitosan, can be used to remove airborne pollutants, odors, and harmful gases generated within aquaculture facilities (Zhang et al., 2017). Nanochitosan-coated indoor air filters for heating, ventilation, and air conditioning (HVAC) systems can be utilized to capture particulate matter, allergens, and volatile organic compounds, improving indoor air quality in the aquaculture facility (Lou et al., 2023).

Nanochitosan-based sediment capping techniques are becoming involved in fishery and aquaculture management. These techniques involve applying nanochitosan to the sediment bed, creating a barrier that immobilizes pollutants, heavy metals, and contaminants, preventing their release into the water column (Inobeme et al., 2023; Rather et al., 2023). This method promotes fish and other aquatic organisms' health, safeguards aquatic ecosystems, and helps maintain water quality. Fishery and aquaculture operations can lessen their environmental impact while maintaining the safety and welfare of aquatic life by using nanochitosan-based sediment capping (Inobeme et al., 2023).

2.3 *Application of Nanochitosan in Food Processing*

Numerous aspects of fish processing also make use of nanochitosan. To maximize the use of fish resources and reduce waste, efficient fish processing is necessary. In order to prolong the shelf life of seafood products without compromising their quality or safety, nanochitosan can also be added to fish preservation methods (Abdollahzadeh et al., 2023).

Nanochitosan can be utilized for enhanced food packaging in fishery and aquaculture. These nanoparticles create biodegradable and edible packaging materials, reducing plastic waste and ensuring seafood safety (Nilsen-Nygaard et al., 2021). Nanochitosan-based edible films can extend the shelf life and quality of perishable seafood products and reduce food waste by providing a protective barrier against moisture loss and contamination (Gardesh et al., 2016). Nanochitosan-based coatings can also be tailored to release bioactive compounds, such as antioxidants or antimicrobials, further improving seafood preservation (Ray et al., 2022). Additionally, these coatings reduce plastic waste and are biodegradable, aligning with sustainability goals in the industry. Nanochitosan in edible films and coatings signifies a leap towards responsible seafood packaging and quality assurance in fishery and aquaculture (Bilal et al., 2019).

Smart packaging refers to the integration of advanced technologies and features into packaging materials and designs to enhance the functionality, safety, and user experience of products (Schaefer & Cheung, 2018). Smart packaging incorporating nanochitosan is transforming in the fishery and aquaculture industries. Nanochitosan-based sensors embedded in food packaging can provide real-time information on food freshness and safety (Dong et al., 2023). These sensors detect changes in temperature, pH, and microbial activity, providing valuable data on seafood conditions during storage and transportation. Smart packaging equipped with nanochitosan helps reduce food waste by alerting consumers and suppliers to potential spoilage (Yu et al., 2023). Moreover, it contributes to sustainable seafood practices by ensuring the safety and integrity of seafood products. In fishery and aquaculture, nanochitosan-based smart packaging enhances monitoring, transparency, and efficiency while promoting responsible seafood consumption (Jamróz, 2021).

2.4 Application of Nanochitosan to Boost and Monitor Aquatic Health

Another important contribution of nanochitosan is in aquatic health monitoring. Monitoring the health of fish populations in an aquaculture system is crucial as the productivity of the system depends on it (Føre et al., 2018). Nanochitosan plays a role in diagnostic assays and monitoring techniques. Enzyme-linked immunosorbent assays (ELISA) and polymerase chain reaction (PCR) techniques can be enhanced using nanochitosan as carriers for biomolecules (Elbhnsawi et al., 2023). These assays help detect pathogens and monitor the health of fish populations.

Nanochitosans are also utilized as feed additives to optimize the nutritional value of fish diets. They can encapsulate and deliver essential nutrients, probiotics, and enzymes, improving nutrient absorption and overall fish health (Adetunji et al., 2023). This leads to enhanced growth rates and immune health, reduced feed waste, and better utilization of resources, making aquaculture more sustainable and economically viable (El-Naggar et al., 2022a; Uyanga et al., 2023). Hidangmayum and Dwivedi (2022) report that there have been citings of chitosan-based nanoformulation exhibiting a defense response to stress, however, the exact mechanism of action is completely known. Hence, nanochitosan-based feed additives can contain stress-relieving compounds, aiding fish in adapting to changing conditions and minimizing stress-related health issues during transport and handling of the fish.

Nanochitosan can also serve as effective growth stimulants in fishery and aquaculture. These nanoparticles can be used to develop bio-stimulant formulations that enhance nutrient absorption and metabolism in aquatic organisms, promoting faster and healthier growth (Ingle et al., 2022; Selvaraj et al., 2022).

2.5 Application of Nanochitosan for Pesticides

Nanochitosans are at the forefront of eco-friendly pesticide applications in fishery and aquaculture. These nanoparticles, derived from chitosan, serve as potent carriers for natural and biopesticides (Campos et al., 2018). They lessen the need for chemical pesticides that can damage aquatic ecosystems by improving the targeted delivery of pest-controlling agents (Bandara et al., 2020). Pesticide stability is increased by nanochitosan-based formulations, which ensure effective pest control while reducing environmental impact (Fatima et al., 2021). By promoting ethical and sustainable fishery and aquaculture methods, this strategy protects aquatic life and maintains the equilibrium of the ecosystem. In the aquaculture and fishery sectors, nanochitosan holds great promise for improving pest control while reducing environmental impact.

2.6 Application of Nanochitosan in Material Science

Within the field of material science, nanochitosan exhibits promise for diverse uses in aquaculture and fishery. These nanoparticles provide innovative ways to create sophisticated structures and materials. By adding nanochitosan to edible and biodegradable packaging materials, seafood packaging can use less plastic (Gardesh et al., 2016; Korsah et al., 2023). Additionally, according to Dutta et al. (2012), nanochitosan aids in the development of specialty coatings, films, and membranes for aquaculture facilities, enabling the establishment of controlled environments that maximize fish growth and health. These uses in material science demonstrate how important nanochitosan is to improving efficiency, sustainability, and responsible fishery and aquaculture methods.

To create multifunctional materials with improved mechanical, thermal, and barrier properties, nanochitosan can be incorporated into a variety of materials, including polymer, ceramic, or metal matrices (Shapi'i et al., 2022). Nanochitosan-based nanocomposites are used in aquaculture to create sturdy, ecologically friendly structures including cages, nets, and tanks (Olisaka et al., 2023). They offer lower maintenance costs, increased durability, and resistance to environmental stresses. These nanocomposites can also be engineered to release nutrients or antimicrobial agents gradually, which supports environmentally friendly methods of raising fish (Shwetha et al., 2020). Nanochitosan-based nanocomposites support ethical seafood production while reducing environmental impact, enhancing the sustainability and efficiency of fisheries and aquaculture operations.

With regard to 3D printing for uses in aquaculture and fishery, nanochitosan is opening up new possibilities. By incorporating these nanoparticles into 3D printing materials—such as filaments and inks—new structures customized to meet the demands of the market can be produced (Diwan & Sah, 2023; Lam et al., 2023). They make it possible to develop intricate, personalized designs for fish feeders, aquaculture tanks, and even elaborate fish habitats. Materials for 3D printing based on nanochitosan provide improved biodegradability, biocompatibility, and strength (Rihayat et al., 2022; Siripongpreda et al., 2022). This lessens the environmental impact while promoting effective and sustainable fish farming methods.

Applications for nanochitosan are also being found in energy storage systems for aquaculture and fishery. Supercapacitors and energy-dense materials for underwater sensors, monitoring systems, and remote aquaculture facilities can be created with these nanoparticles (Zhang et al., 2019; Goda, 2022). Nanochitosan-based materials have large surface areas, which enable effective energy release and storage (Bandara et al., 2020). In keeping with sustainable practices in these industries, they also aid in the development of environmentally friendly and biodegradable energy storage devices (Zhang et al., 2019; Goda, 2022). Fishery and aquaculture operations can improve their monitoring capabilities, increase energy efficiency, and lessen their environmental impact by harnessing nanochitosan for energy storage. This will ultimately support the production of seafood in a responsible and environmentally friendly manner.

2.7 Application of Nanochitosan in Biocatalysis

The process of using biological catalysts, like enzymes, to speed up chemical reactions is known as biocatalysis (Bilal et al., 2020). These biological catalysts lower energy consumption and environmental impact by allowing certain reactions to happen in milder conditions (Bilal et al., 2020). With a variety of applications that support the efficiency and sustainability of these sectors, biocatalysis is important to the fishery and aquaculture industries. Biocatalysis can be used in controlled breeding programs to create hormone treatments that improve fish reproduction and facilitate artificial fertilization (Singh et al., 2018). Biocatalysis also helps the fishery industry create edible and biodegradable packaging materials, which cuts down on plastic waste and increases the sustainability of seafood packaging (Gardesh et al., 2016). In line with its preservation activities, biocatalytic processes can be employed to develop natural preservatives that extend the shelf life of seafood products while maintaining their quality and safety (Chellaram et al., 2014).

By optimizing feed conversion and waste management through nanochitosan-based biocatalysis, aquaculture systems can minimize their environmental impact, including issues like eutrophication and habitat degradation, while enhancing overall health and productivity of the aquaculture systems (Ahuekwe et al., 2023c).

Nanochitosans play important roles in the various applications of biocatalysis in fisheries and aquaculture. Nanochitosan can break down complex carbohydrates and proteins in feed ingredients, making them more digestible for fish (Bashar et al., 2021). This decreases the quantity of undigested feed and waste in the aquaculture system while simultaneously increasing nutrient absorption. Fish can be guaranteed to receive the right nutrition at the right time by controlling and optimizing the release of vital nutrients and vitamins by encasing them in nanochitosan particles (Luo et al., 2012; Azevedo et al., 2014). Moreover, feed additive enzymes can be stabilized by nanochitosan (Zhang et al., 2020). Enzymes such as phytases and proteases can be added as a result, improving nutrient utilization and lowering feed expenses (Pragya et al., 2021; Filippovich et al., 2023). According to Dar et al. (2020), nanochitosan improves the nutritional content and digestibility of feed ingredients, which eventually increases feed conversion efficiency and lessens the environmental impact of aquaculture operations.

Enzyme immobilization could improve catalytic efficiency and stability by using nanochitosan matrices. The process of immobilizing enzymes, which increases their stability and reusability, depends extensively on nanochitosan (Ajayi et al., 2023). These nanoparticles' large surface area and biocompatibility make them the perfect medium for enzyme attachment. Because of their increased catalytic activity, enzymes anchored onto nanochitosan are highly valuable in a variety of industries, such as environmental remediation, biomedicine, and food processing (Ajayi et al., 2023). In addition to extending the lifespan of the enzymes, this immobilization process enables their recovery and reuse, which lowers expenses and has a positive environmental impact (Zhong et al., 2020). Ansari and Husain (2012) highlight the importance of nanochitosan in biocatalysis and biotechnological developments by highlighting its capacity to maximize enzyme performance.

Because of its special qualities, nanochitosan also functions as a very powerful biosensor. These nanoparticles improve the sensitivity and specificity of biosensors due to their biocompatibility, adaptability, and capacity to immobilize biomolecules such as enzymes, antibodies, or DNA probes (Ansari & Husain, 2012; Baranwal et al., 2018). Their remarkable precision allows them to detect a wide range of analytes, from environmental pollutants to pathogens (Jampílek & Kráľová, 2018; Ahuekwe et al., 2023b). With their quick and accurate detection techniques, nanochitosan-based biosensors are used in environmental monitoring and food safety (Dholariya et al., 2021; Rather et al., 2023). They are essential tools in the development of next-generation biosensing technologies because of their versatility, affordability, and environmental friendliness. Furthermore, nanochitosan can be used in bioprocessing to enhance the yield and stability of biopharmaceuticals and bio-based products (Arya et al., 2022).

Utilizing nanochitosan in biocatalysis offers high selectivity, efficiency, and sustainability, making it a green alternative as opposed to using traditional chemical methods (Anwar et al., 2023).

3 Challenges in Scalability, Cost-Effectiveness, and Regulatory Considerations

The cost-effectiveness and feasibility of using nanochitosan in an aquaculture system depend on factors including production cost, adsorption capacity, and regeneration, integration into existing systems, durability, scalability, and environmental impact (Isibor et al., 2023). Efficient production methods can optimize production costs, while higher adsorption capacity and regeneration reduce the need for frequent replacement. Integration into existing systems minimizes infrastructure costs, and environmental benefits align with sustainability objectives.

As nanochitosan applications expand, regulatory bodies will need to establish guidelines and safety assessments to ensure the responsible development and commercialization of these innovations (Subhan & Subhan, 2022). Safety, biocompatibility, and environmental impact assessments will be crucial in the regulations of nanochitosan.

3.1 Challenges of Nanochitosan Scalability

The scalability of nanochitosan production is a fundamental challenge as its production methods are limited. Traditional methods for chitosan extraction involve labour-intensive processes, including chemical deacetylation and purification, which are not well-suited for large-scale production (El Knidri et al., 2018; Pakizeh et al., 2021). Converting chitosan into nanochitosan typically involves techniques such as ionic gelation, coacervation, or precipitation, which may not be easily scalable due to their dependence on precise conditions and extensive processing steps (Kou et al., 2021).

Additionally, scalability often comes with increased environmental concerns. Traditional chitosan extraction methods generate substantial waste (El Knidri et al., 2016). The biodegradability property of nanochitosan, however, contributes to its cost-effectiveness by reducing its potential environmental impact while still meeting the growing demand without compromising the environment (Isibor et al., 2023).

3.2 Cost-Effectiveness of Nanochitosan

The cost of chitosan, the primary raw material for nanochitosan production, can significantly affect the cost (Isibor et al., 2023). Cost is also greatly impacted by the industrial processes that chitosan undergoes. To increase the competitiveness of nanochitosan-based products in the market, alternative chitosan sources must be obtained.

Furthermore, the processes involved in producing nanochitosan can be energy-intensive, especially if high-energy methods such as sonication or high-pressure homogenization are used (Yanat & Schroën, 2021). Reducing production costs and increasing the cost-effectiveness of products based on nanochitosan requires finding energy-efficient ways or improving current processes (Isibor et al., 2023).

Moreover, major infrastructure and equipment investments are frequently needed to scale up the production of nanochitosan, which raises the initial costs. To guarantee that nanochitosan stays commercially feasible and available for a wider range of industries and applications, cost-effective scaling techniques and strategies must be developed (Shegokar & Nakach, 2020).

3.3 Regulatory Considerations for Nanochitosan

The regulatory environment surrounding nanomaterials is always changing, with an emphasis on risk management and safety evaluation. Similar to other nanomaterials, nanochitosan needs to be thoroughly tested to determine any possible effects on human health and the environment. To comply with regulatory requirements, it is essential to establish standardized protocols and methodologies for the safety evaluation of nanochitosan (Isigonis et al., 2020; Subhan & Subhan, 2022).

Additionally, registration and approval from regulatory agencies may be necessary for nanochitosan products, depending on the intended use and jurisdiction (Rauscher et al., 2017). Products containing nanomaterials are often subject to regulations requiring comprehensive documentation. Manufacturers are required to give detailed information about the composition, characteristics, and possible risks of products based on nanochitosan (Quinn, 2021). Ensuring compliance with documentation requirements is essential to prevent regulatory issues (Subhan & Subhan, 2022).

Furthermore, navigating the global regulatory landscape can be challenging due to differences in regulations and standards across countries and regions. Efforts to harmonize international regulations for nanomaterials like nanochitosan are essential to streamline market access and reduce compliance complexities (Dave et al., 2021; Subhan & Subhan, 2022).

4 Addressing Challenges in Scalability, Cost-Effectiveness, and Regulatory Considerations

One way to address scalability and environmental concerns is by developing sustainable production methods. Research efforts are focused on finding alternative sources of chitosan, such as fungal-derived chitosan, which may offer a more consistent and cost-effective supply (Islam et al., 2023). Furthermore, optimizing extraction processes to minimize waste and energy consumption can contribute to sustainability.

Optimizing nanochitosan production processes, including improving extraction and purification techniques, particle size control, and energy efficiency, can also enhance its cost-effectiveness (El-Naggar et al., 2022b; Isibor et al., 2023). Innovations like microfluidic technology and continuous-flow reactors offer promising avenues for process improvement (Hara & Singh, 2021). Process optimization not only reduces production costs but also enhances the overall quality and consistency of nanochitosan.

Collaboration between industry, academia, and regulatory bodies is also crucial to addressing regulatory challenges. Establishing standardized protocols for nanochitosan safety assessment and harmonizing international regulations can facilitate market access and regulatory compliance (Dave et al., 2021).

Although nanochitosan has great potential for use in many different industries, its full potential can only be achieved when the issues of scalability, affordability, and regulatory compliance are resolved. Researchers, manufacturers, regulators, and other stakeholders must work together to overcome these challenges. Nanochitosan-based technologies can be widely adopted and their benefits for society unlocked by creating sustainable production methods, streamlining procedures, encouraging cooperation, and standardizing practices—all while guaranteeing safety and regulatory compliance. The unique properties of nanochitosan, such as their large surface area, biocompatibility, and ability to be tailored for specific applications, have garnered significant attention (Isibor et al., 2023). Interdisciplinary collaborations between researchers, industries, and regulatory bodies are essential for achieving their full potential. These cooperative endeavours foster the sharing of information, assets, and skills, propelling progress in the research, development, and application of nanochitosan.

Through collaborations, experts from different fields can pool their resources, such as funding, knowledge, equipment, and research materials, to accelerate progress and save costs. Furthermore, collaborative efforts improve the quality of research because they enable thorough peer review and validation of research findings, which produce outcomes that are more robust and dependable (Hoffman, 2022). Additionally, access to specialized research facilities and equipment that may not be available at their home institutions is frequently granted to collaborators (van Rijnsoever & Hessels, 2021).

Collaborations between academia and industry also bridge the gap between fundamental research and practical applications. This ensures that nanochitosan-based innovations meet real-world needs. Regulatory bodies come into play by ensuring the safety and efficacy of nanochitosan applications. Nanochitosan has the potential to address pressing global challenges, such as water pollution, healthcare, and sustainable agriculture. Collaborations allow for a collective response to these challenges (Dusdal & Powell, 2021). Collaborations between water treatment companies, research institutions, and environmental agencies have led to the development of advanced nanochitosan-based membranes that exhibit superior filtration and fouling resistance properties (Isibor et al., 2023). These membranes could be used in large-scale water treatment plants worldwide, exemplifying the impact of interdisciplinary partnerships on addressing water pollution. Collaborative efforts between agricultural research organizations, agrochemical companies, and government bodies have resulted in nanochitosan-based formulations that reduce the environmental impact of agricultural practices (de Oliveira et al., 2021). These products enhance crop yields, reduce chemical runoff, and minimize soil contamination, showcasing the potential of industry-academia collaborations for sustainable agriculture. Numerous other collaborations exist between various institutions, bodies, companies, groups, agencies, and researchers to uncover and explore the potential advancements and innovative applications in nanochitosan (Anwar et al., 2023).

Successful collaborative efforts in nanochitosan research and implementation often employ specific models and strategies:

1. Public-Private Partnerships (PPPs): Public research institutions collaborate with private industries and government bodies to jointly fund and conduct research projects. PPPs can accelerate the translation of research into practical applications (Koschatzky & Stahlecker, 2010).
2. Interdisciplinary Research Centers: Establishing interdisciplinary research centres that bring together experts from various fields can foster close collaboration. These centres often facilitate the cross-pollination of ideas and resources, accelerating nanochitosan research and development (Oluwasanu et al., 2019).
3. Consortia and Networks: Forming consortia or networks involving multiple research institutions, industries, and regulatory agencies allows for shared goals, resources, and expertise. Such collaborative structures are well-suited for addressing complex challenges like environmental remediation or healthcare innovations (Garousi et al., 2016).
4. Open Innovation Platforms: Open innovation platforms invite external stakeholders, including startups, research institutions, and individuals, to contribute their expertise and ideas (Locatelli et al., 2021). These platforms can drive innovation by tapping into a diverse talent pool.

5 Challenges and Future Directions

While collaborative efforts in nanochitosan research and implementation offer substantial benefits, several challenges must be addressed to ensure their success and sustainability (Singh et al., 2021).

First, in funding and resource allocation, it is important to ensure equitable resource distribution among collaborators, especially when imbalances in funding or infrastructure exist, remains a challenge and must be sorted out during collaborative research studies to ensure seamless progression (Bromham et al., 2016). Second, developing clear agreements regarding intellectual property rights and commercialization strategies is also crucial to prevent disputes (Somaya et al., 2011). Third, collaborative projects involving nanochitosan often span multiple jurisdictions. Harmonizing regulatory standards and approvals can be complex but is essential for global implementation. Fourth, ensuring open access to research data while respecting privacy and confidentiality can be a delicate balance in collaborative efforts (Hoffman, 2022). Finally, maintaining long-term collaborations requires ongoing commitment from all stakeholders. Addressing changing priorities and personnel turnover is essential (Margerum & Robinson, 2015).

As nanochitosan continues to find innovative applications, collaborative efforts will remain essential to maximize its potential and address pressing global challenges in fishery and aquaculture, environmental sustainability, agriculture, and materials science (ABIO, 2021). By fostering a culture of cooperation, embracing interdisciplinary perspectives, and addressing challenges proactively, collaborative initiatives will play a pivotal role in shaping the future of nanochitosan research and application.

References

Abdollahzadeh, M., Elhamirad, A. H., Shariatifar, N., Saeidiasl, M., & Armin, M. (2023). Effects of nano-chitosan coatings incorporating with free/nano-encapsulated essential oil of Golpar (*Heracleum persicum* L.) on quality characteristics and safety of rainbow trout (*Oncorhynchus mykiss*). *International Journal of Food Microbiology, 385*, 109996.

Abimbola, O. F., Olotu, T. M., Adetunji, C. O., & Alabetutu, A. (2023). Applications of nano-chitosan in the detection and control of aquatic diseases. In *Next generation nanochitosan* (pp. 221–234). Academic Press.

ABIO. (2021). The challenges facing research collaborations. *Nature, 594*, 316–319.

Adetunji, C. O., Olaniyan, O. T., Dare, A., Adeniyi, M. J., & Ajayi, O. O. (2023). Significance of nanochitosan for improving bioavailability, stability and shelf life of sensitive ingredients in aquatic and animal-based foods. In *Next generation nanochitosan* (pp. 421–430). Academic Press.

Adewuyi, S., Cruz-Maya, I., Ejeromedoghene, O., & Guarino, V. (2019). Design of nano-chitosans for tissue engineering and molecular release. In *Sustainable agriculture reviews 36: Chitin and chitosan: Applications in food, agriculture, pharmacy, medicine and wastewater treatment* (pp. 315–334). Springer.

Ahankari, S. S., Mohanty, A. K., & Misra, M. (Eds.). (2023). *Nanomaterials from renewable resources for emerging applications*. CRC Press.

Ahmad, S. I., Ahmad, R., Khan, M. S., Kant, R., Shahid, S., Gautam, L., Hasan, G. M., & Hassan, M. I. (2020). Chitin and its derivatives: Structural properties and biomedical applications. *International Journal of Biological Macromolecules, 164*, 526–539.

Ahuekwe, E. F., Isibor, P. O., Oziegbe, O., Salami, A. O., Akinyosoye, A. D., Akinhanmi, F., Oyewale, J. O., Taiwo, O. S., Akinwunmi, R. A., Ajiboye, I., Adekeye, B. T., Akinpelu, S. O., Kuye, A. D., Bello, A. O., George, D. Z., Ojo-Omoniyi, O. A., Popoola, T. S., Akinyemi, O. D., Adebayo, G. P., Oniha, M. I., Iheagwam, F. N., Adelodun, C. A., Orukotan, K. E., Bilewu, O. F., Onibokun, E. A., Fasuyi, N. O., Akinduti, P. A., Onuselogu, C. C., Oshamika, O. O., Oyesola, O. L., Ichor, T., Ezekiel, O. M., Nwinyi, O. C., Ayanda, I. O., Akinnola, O. O., Oranusi, S. U., Eni, A. O., Popoola, J. O., Omonhinmin, C. A., Olasehinde, G. I., & Obembe, O. O. (2023a). Nanochitosan derived from marine bacteria. In *Next generation nanochitosan* (pp. 147–168). Academic Press.

Ahuekwe, E. F., Isibor, P. O., Akinduti, P. A., Salami, A. O., Onuselogu, C. C., Oshamika, O. O., Oyesola, O. L., Ichor, T., Ezekiel, O. M., Oziegbe, O., Akinyosoye, A. D., Akinhanmi, F., Oyewale, J. O., Taiwo, O. S., Akinwunmi, R. A., Ajiboye, I., Adekeye, B. T., Akinpelu, S. O., Kuye, A. D., Bello, A. O., George, D. Z., Ojo-Omoniyi, O. A., Popoola, T. S., Akinyemi, O. D., Adebayo, G. P., Oniha, M. I., Iheagwam, F. N., Adelodun, C. A., Orukotan, K. E., Bilewu, O. F., Onibokun, E. A., Fasuyi, N. O., Nwinyi, O. C., Ayanda, I. O., Omonhinmin, C. A., Akinnola, O. O., Oranusi, S. U., Eni, A. O., Popoola, J. O., Olasehinde, G. I., & Obembe, O. O. (2023b). Application of nanochitosan in tagging and nano-barcoding of aquatic and animal meats. In *Next generation nanochitosan* (pp. 483–498). Academic Press.

Ahuekwe, E. F., Isibor, P. O., Akinhanmi, F., Ajiboye, I., Salami, A. O., Adekeye, B. T., Akinpelu, S. O., Kuye, A. D., Bello, A. O., Oyewale, J. O., George, D. Z., Ojo-Omoniyi, O. A., Popoola, T. S., Akinyemi, O. D., Adebayo, G. P., Oziegbe, O., Akinduti, P. A., Akinyosoye, A. D., Oshamika, O. O., Oniha, M. I., Taiwo, O. S., Akinwunmi, R. A., Iheagwam, F. N., Adelodun, C. A., Orukotan, K. E., Bilewu, O. F., Onibokun, E. A., Fasuyi, N. O., Onuselogu, C. C., Oyesola, O. L., Ichor, T., Ezekiel, O. M., Akinnola, O. O., Oranusi, S. U., Eni, A. O., Popoola, J. O., Omonhinmin, C. A., Nwinyi, O. C., Ayanda, I. O., Olasehinde, G. I., & Obembe, O. O. (2023c). Utilization of nanochitosan in the sterilization of ponds and water treatment for aquaculture. In *Next generation nanochitosan* (pp. 301–338). Academic Press.

Ajayi, E. I., Oladele, J. O., & Babalola, A. D. (2023). Application of nanochitosan in enzyme immobilization. In *Next generation nanochitosan* (pp. 235–272). Academic Press.

Ansari, M. A. (2023). Nanotechnology in food and plant science: Challenges and future prospects. *Plants, 12*(13), 2565.

Ansari, S. A., & Husain, Q. (2012). Potential applications of enzymes immobilized on/in nano materials: A review. *Biotechnology Advances, 30*(3), 512–523.

Anwar, A., Imran, M., & Iqbal, H. M. (2023). Smart chemistry and applied perceptions of enzyme-coupled nano-engineered assemblies to meet future biocatalytic challenges. *Coordination Chemistry Reviews, 493*, 215329.

Arya, S. S., Rookes, J. E., Cahill, D. M., & Lenka, S. K. (2022). Chitosan nanoparticles and their combination with methyl jasmonate for the elicitation of phenolics and flavonoids in plant cell suspension cultures. *International Journal of Biological Macromolecules, 214*, 632–641.

Azevedo, M. A., Bourbon, A. I., Vicente, A. A., & Cerqueira, M. A. (2014). Alginate/chitosan nanoparticles for encapsulation and controlled release of vitamin B2. *International Journal of Biological Macromolecules, 71*, 141–146.

Bandara, S., Du, H., Carson, L., Bradford, D., & Kommalapati, R. (2020). Agricultural and biomedical applications of chitosan-based nanomaterials. *Nanomaterials, 10*(10), 1903.

Baranwal, A., Kumar, A., Priyadharshini, A., Oggu, G. S., Bhatnagar, I., Srivastava, A., & Chandra, P. (2018). Chitosan: An undisputed bio-fabrication material for tissue engineering and biosensing applications. *International Journal of Biological Macromolecules, 110*, 110–123.

Bashar, A., Hasan, N. A., Haque, M. M., Rohani, M. F., & Hossain, M. S. (2021). Effects of dietary silica nanoparticle on growth performance, protein digestibility, hematology, digestive morphology, and muscle composition of Nile tilapia, Oreochromis niloticus. *Frontiers in Marine Science, 8*, 706179.

Bilal, M., Cui, J., & Iqbal, H. M. (2019). Tailoring enzyme microenvironment: State-of-the-art strategy to fulfill the quest for efficient bio-catalysis. *International Journal of Biological Macromolecules, 130*, 186–196.

Bilal, M., Barceló, D., & Iqbal, H. M. (2020). Nanostructured materials for harnessing the power of horseradish peroxidase for tailored environmental applications. *Science of the Total Environment, 749*, 142360.

Bromham, L., Dinnage, R., & Hua, X. (2016). Interdisciplinary research has consistently lower funding success. *Nature, 534*(7609), 684–687.

Campos, E. V., Proença, P. L., Oliveira, J. L., Pereira, A. E., de Morais Ribeiro, L. N., Fernandes, F. O., Gonçalves, K. C., Polanczyk, R. A., Pasquoto-Stigliani, T., Lima, R., Melville, C. C., Vechia, J. F. D., Andrade, D. J., & Fraceto, L. F. (2018). Carvacrol and linalool co-loaded in β-cyclodextrin-grafted chitosan nanoparticles as sustainable biopesticide aiming pest control. *Scientific Reports, 8*(1), 7623.

Chaudhary, P., Ahamad, L., Chaudhary, A., Kumar, G., Chen, W. J., & Chen, S. (2023). Nanoparticle-mediated bioremediation as a powerful weapon in the removal of environmental pollutants. *Journal of Environmental Chemical Engineering, 11*, 109591.

Chellaram, C., Murugaboopathi, G., John, A. A., Sivakumar, R., Ganesan, S., Krithika, S., & Priya, G. (2014). Significance of nanotechnology in food industry. *APCBEE Procedia, 8*, 109–113.

Dar, A. H., Rashid, N., Majid, I., Hussain, S., & Dar, M. A. (2020). Nanotechnology interventions in aquaculture and seafood preservation. *Critical Reviews in Food Science and Nutrition, 60*(11), 1912–1921.

Dave, V., Sur, S., & Gupta, N. (2021). Current framework, ethical consideration and future challenges of regulatory approach for nano-based products. In *Nanopharmaceutical advanced delivery systems* (pp. 447–472). Wiley/Scrivener Publishing LLC.

de Oliveira, A. L. B., Cavalcante, F. T. T., da Silva Moreira, K., Lima, P. J. M., de Castro Monteiro, R. R., Pinheiro, B. B., dos Santos, K. P., & dos Santos, J. C. S. (2021). Chitosan nanoparticle: Alternative for sustainable agriculture. In *Nanomaterials and nanotechnology: Biomedical, environmental, and industrial applications* (pp. 95–132). Springer.

Dholariya, P. K., Borkar, S., & Borah, A. (2021). Prospect of nanotechnology in food and edible packaging: A review. *The Pharma Innovation Journal, 10*(5), 197–203.

Diwan, H., & Sah, M. K. (2023). Exploring the potential of keratin-based biomaterials in orthopedic tissue engineering: A comprehensive review. *Emergent Materials, 6*, 1–20.

Dong, S., Zhang, Y., Lu, D., Gao, W., Zhao, Q., & Shi, X. (2023). Multifunctional intelligent film integrated with purple sweet potato anthocyanin and quercetin-loaded chitosan nanoparticles for monitoring and maintaining freshness of shrimp. *Food Packaging and Shelf Life, 35*, 101022.

Dusdal, J., & Powell, J. J. (2021). Benefits, motivations, and challenges of international collaborative research: A sociology of science case study. *Science and Public Policy, 48*(2), 235–245.

Dutta, J., Tripathi, S., & Dutta, P. K. (2012). Progress in antimicrobial activities of chitin, chitosan and its oligosaccharides: A systematic study needs for food applications. *Food Science and Technology International, 18*(1), 3–34.

El Knidri, H., El Khalfaouy, R., Laajeb, A., Addaou, A., & Lahsini, A. (2016). Eco-friendly extraction and characterization of chitin and chitosan from the shrimp shell waste via microwave irradiation. *Process Safety and Environmental Protection, 104*, 395–405.

El Knidri, H., Belaabed, R., Addaou, A., Laajeb, A., & Lahsini, A. (2018). Extraction, chemical modification and characterization of chitin and chitosan. *International Journal of Biological Macromolecules, 120*, 1181–1189.

Elbhnsawi, N. A., Elwakil, B. H., Hassanin, A. H., Shehata, N., Elshewemi, S. S., Hagar, M., & Olama, Z. A. (2023). Nano-chitosan/eucalyptus oil/cellulose acetate nanofibers: Manufacturing, antibacterial and wound healing activities. *Membranes, 13*(6), 604.

El-Naggar, M., Medhat, F., & Taha, A. (2022a). Applications of chitosan and chitosan nanoparticles in fish aquaculture. *Egyptian Journal of Aquatic Biology and Fisheries, 26*(1), 23–43.

El-Naggar, N. E. A., Shiha, A. M., Mahrous, H., & Mohammed, A. A. (2022b). Green synthesis of chitosan nanoparticles, optimization, characterization and antibacterial efficacy against multi drug resistant biofilm-forming *Acinetobacter baumannii*. *Scientific Reports, 12*(1), 19869.

Fatima, F., Hashim, A., & Anees, S. (2021). Efficacy of nanoparticles as nanofertilizer production: A review. *Environmental Science and Pollution Research, 28*, 1292–1303.

Ferreira, P. G., Ferreira, V. F., da Silva, F. D. C., Freitas, C. S., Pereira, P. R., & Paschoalin, V. M. F. (2022). Chitosans and nanochitosans: Recent advances in skin protection, regeneration, and repair. *Pharmaceutics, 14*(6), 1307.

Filippovich, S. Y., Isakova, E. P., Gessler, N. N., & Deryabina, Y. I. (2023). Advances in immobilization of phytases and their application. *Bioresource Technology, 379*, 129030.

Føre, M., Frank, K., Norton, T., Svendsen, E., Alfredsen, J. A., Dempster, T., Eguiraun, H., Watson, W., Stahl, A., Sunde, L. M., Schellewald, C., Skøien, K. R., Alver, M. O., & Berckmans, D. (2018). Precision fish farming: A new framework to improve production in aquaculture. *Biosystems Engineering, 173*, 176–193.

Gardesh, A. S. K., Badii, F., Hashemi, M., Ardakani, A. Y., Maftoonazad, N., & Gorji, A. M. (2016). Effect of nanochitosan based coating on climacteric behavior and postharvest shelf-life extension of apple cv. Golab Kohanz. *LWT – Food Science and Technology, 70*, 33–40.

Garousi, V., Petersen, K., & Ozkan, B. (2016). Challenges and best practices in industry-academia collaborations in software engineering: A systematic literature review. *Information and Software Technology, 79*, 106–127.

Goda, E. S. (2022). Bio-nanomaterial for renewable energy storage applications. In *Biorenewable nanocomposite materials, Vol. 1: Electrocatalysts and energy storage* (pp. 91–127). American Chemical Society.

Han, X., Alu, A., Liu, H., Shi, Y., Wei, X., Cai, L., & Wei, Y. (2022). Biomaterial-assisted biotherapy: A brief review of biomaterials used in drug delivery, vaccine development, gene therapy, and stem cell therapy. *Bioactive Materials, 17*, 29–48.

Hara, T. O., & Singh, B. (2021). Electrochemical biosensors for detection of pesticides and heavy metal toxicants in water: Recent trends and progress. *ACS ES&T Water, 1*(3), 462–478.

Haripriyan, U., Gopinath, K. P., & Arun, J. (2022). Chitosan based nano adsorbents and its types for heavy metal removal: A mini review. *Materials Letters, 312*, 131670.

Hidangmayum, A., & Dwivedi, P. (2022). Chitosan based nanoformulation for sustainable agriculture with special reference to abiotic stress: A review. *Journal of Polymers and the Environment, 30*, 1–20.

Hoffman, A. J. (2022). A modest proposal to the peer review process: A collaborative and interdisciplinary approach in the assessment of scholarly communication. *Research Ethics, 18*(1), 84–91.

Ingle, P. U., Shende, S. S., Shingote, P. R., Mishra, S. S., Sarda, V., Wasule, D. L., Rajput, V. D., Minkina, T., Rai, M., Sushkova, S., Mandzhieva, S., & Gade, A. (2022). Chitosan nanoparticles (ChNPs): A versatile growth promoter in modern agricultural production. *Heliyon, 8*(11), e11893.

Inobeme, A., Adetunji, C. O., & Singh, K. R. (2023). Removal of chlorinated compounds using bionanomaterials. In *Nanobiotechnology for bioremediation* (pp. 227–247). Elsevier.

Isa, E. D. M., Shameli, K., Ali, R. R., & Balasundram, V. (2022). Advances in nanocellulose based materials as adsorbent for wastewater treatment. *Journal of Research in Nanoscience and Nanotechnology, 5*(1), 43–64.

Isibor, O. P., Osagie, A. D., & Oritseweyinmi, O. O. (2023). Crustacean nanochitosan-based bioremediation of nanoplastic-polluted aquatic habitat: A review pursuant to SDG 6. *Scientific African, 21*, e01881

Isigonis, P., Afantitis, A., Antunes, D., Bartonova, A., Beitollahi, A., Bohmer, N., Bouman, E., Chaudhry, Q., Cimpan, M. R., Cimpan, E., Doak, S., Dupin, D., Fedrigo, D., Fessard, V., Gromelski, M., Gutleb, A. C., Halappanavar, S., Hoet, P., Jeliazkova, N., Jomini, S., Lindner, S., Linkov, I., Longhin, E. M., Lynch, I., Malsch, I., Marcomini, A., Mauriussen, E., de la Fuente, J. M., Melagraki, G., Murphy, F., Neaves, M., Packroff, R., Pfuhler, S., Puzyn, T., Rahman, Q., Pran, E. R., Semenzin, E., Serchi, T., Steinbach, C., Trump, B., Vrček, I. V., Warheit, D., Wiesner, M. R., Willighagen, E., & Dusinska, M. (2020). Risk governance of emerging technologies demonstrated in terms of its applicability to nanomaterials. *Small, 16*(36), 2003303.

Islam, N., Hoque, M., & Taharat, S. F. (2023). Recent advances in extraction of chitin and chitosan. *World Journal of Microbiology and Biotechnology, 39*(1), 28.

Jampílek, J., & Kráľová, K. (2018). Nanomaterials applicable in food protection. In *Nanotechnology applications in the food industry* (p. 75). CRC Press.

Jamróz, E. (2021). Nanomaterials for packaging application. In *Biopolymeric nanomaterials* (pp. 423–447). Elsevier.

Korsah, M. A., Bulawa, A., Jeevanandam, J., & Danquah, M. K. (2023). Utilization of nanochitosan for enzyme immobilization-based food packages. In *Next generation nanochitosan* (pp. 509–528). Academic Press.

Koschatzky, K., & Stahlecker, T. (2010). New forms of strategic research collaboration between firms and universities in the German research system. *International Journal of Technology Transfer and Commercialisation, 9*(1–2), 94–110.

Kou, S. G., Peters, L. M., & Mucalo, M. R. (2021). Chitosan: A review of sources and preparation methods. *International Journal of Biological Macromolecules, 169*, 85–94.

Lam, W. S., Lam, W. H., & Lee, P. F. (2023). The studies on chitosan for sustainable development: A bibliometric analysis. *Materials (Basel), 16*(7), 2857.

Locatelli, G., Greco, M., Invernizzi, D. C., Grimaldi, M., & Malizia, S. (2021). What about the people? Micro-foundations of open innovation in megaprojects. *International Journal of Project Management, 39*(2), 115–127.

Lou, C. W., Li, S., Fan, Y., Li, T. T., Liu, X., Liu, L., Shiu, B. C., & Lin, J. H. (2023). Preparation of antibacterial composite fiber membrane for air filtration with micro/nano structure. *Journal of Materials Research and Technology, 24*, 9949–9960.

Luo, Y., Teng, Z., & Wang, Q. (2012). Development of zein nanoparticles coated with carboxymethyl chitosan for encapsulation and controlled release of vitamin D3. *Journal of Agricultural and Food Chemistry, 60*(3), 836–843.

Maleki, A., Pajootan, E., & Hayati, B. (2015). Ethyl acrylate grafted chitosan for heavy metal removal from wastewater: Equilibrium, kinetic and thermodynamic studies. *Journal of the Taiwan Institute of Chemical Engineers, 51*, 127–134.

Margerum, R. D., & Robinson, C. J. (2015). Collaborative partnerships and the challenges for sustainable water management. *Current Opinion in Environmental Sustainability, 12*, 53–58.

Mustapha, U. F., Alhassan, A. W., Jiang, D. N., & Li, G. L. (2021). Sustainable aquaculture development: A review on the roles of cloud computing, internet of things and artificial intelligence (CIA). *Reviews in Aquaculture, 13*(4), 2076–2091.

Nilsen-Nygaard, J., Fernández, E. N., Radusin, T., Rotabakk, B. T., Sarfraz, J., Sharmin, N., Sivertsvik, I. S., & Pettersen, M. K. (2021). Current status of biobased and biodegradable food packaging materials: Impact on food quality and effect of innovative processing technologies. *Comprehensive Reviews in Food Science and Food Safety, 20*(2), 1333–1380.

Olisaka, F. N., Obayagbona, N. O., Innocent, I. H., Adetunji, C. O., & Inobeme, A. (2023). Application of nanochitosan in the control of biofouling in aquatic ponds. In *Next generation nanochitosan* (pp. 349–362). Academic Press.

Oluwasanu, A. A., Oluwaseun, F., Teslim, J. A., Isaiah, T. T., Olalekan, I. A., & Chris, O. A. (2019). Scientific applications and prospects of nanomaterials: A multidisciplinary review. *African Journal of Biotechnology, 18*(30), 946–961.

Pakizeh, M., Moradi, A., & Ghassemi, T. (2021). Chemical extraction and modification of chitin and chitosan from shrimp shells. *European Polymer Journal, 159*, 110709.

Patra, J. K., Das, G., Fraceto, L. F., Campos, E. V. R., Rodriguez-Torres, M. D. P., Acosta-Torres, L. S., Diaz-Torres, L. A., Grillo, R., Swamy, M. K., Sharma, S., Habtemariam, S., & Shin, H. S. (2018). Nano based drug delivery systems: Recent developments and future prospects. *Journal of Nanobiotechnology, 16*(1), 1–33.

Pragya, Sharma, K. K., Kumar, A., Singh, D., Kumar, V., & Singh, B. (2021). Immobilized phytases: An overview of different strategies, support material, and their applications in improving food and feed nutrition. *Critical Reviews in Food Science and Nutrition, 63*(22), 5465–5487.

Quinn, B. M. (2021). Substance in nanomaterials regulation. *Nature Nanotechnology, 16*(11), 1172–1175.

Rampino, A., Borgogna, M., Blasi, P., Bellich, B., & Cesàro, A. (2013). Chitosan nanoparticles: Preparation, size evolution and stability. *International Journal of Pharmaceutics, 455*(1–2), 219–228.

Rather, M. A., Bhuyan, S., Chowdhury, R., Sarma, R., Roy, S., & Neog, P. R. (2023). Nanoremediation strategies to address environmental problems. *Science of the Total Environment, 886*, 163998.

Rauscher, H., Rasmussen, K., & Sokull-Klüttgen, B. (2017). Regulatory aspects of nanomaterials in the EU. *Chemie Ingenieur Technik, 89*(3), 224–231.

Ray, A., Ghosh, S., Kanjilal, A., Mukherjee, A., & Datta, S. (2022). Antimicrobial properties of chitosan nanoparticles and their role in post-harvest shelf life extension. *IOSR Journal of Environmental Science, Toxicology and Food Technology, 16*(6), 51–60.

Rebello, S., Sali, S., Jisha, M. S., Reshmy, R., Pugazhendhi, A., Madhavan, A., Binod, P., Awasthi, M. K., Pandey, A., & Sindhu, R. (2023). Chitosan a versatile adsorbent in environmental remediation in the era of circular economy – A mini review. *Sustainable Chemistry and Pharmacy, 32*, 101004.

Reid, G. K., Gurney-Smith, H. J., Flaherty, M., Garber, A. F., Forster, I., Brewer-Dalton, K., Knowler, D., Marcogliese, D. J., Chopin, T., Moccia, R. D., Smith, C. T., & De Silva, S. (2019). Climate change and aquaculture: Considering adaptation potential. *Aquaculture Environment Interactions, 11*, 603–624.

Rihayat, T., Aidy, N., Safitri, A., & Aida, A. (2022). Synthesis of poly lactic acid (PLA)/nanochitosan–based for bioscaffold materials with the addition of Zn-curcumin. *Materials Today: Proceedings, 63*, S526–S531.

Schaefer, D., & Cheung, W. M. (2018). Smart packaging: Opportunities and challenges. *Procedia CIRP, 72*, 1022–1027.

Selvaraj, M., Ramasami, N., & Jagan, E. G. (2022). The effect of chitosan nanoparticles on immune responses in plants. In *Role of chitosan and chitosan-based nanomaterials in plant sciences* (pp. 185–196). Academic Press.

Selvasudha, N., Sweety, J. P., Dhanalekshmi, U. M., & Ruckmani, K. (2022). Chitosan-based nanocomposite as targeted drug delivery carrier. In *Marine biomaterials: Drug delivery and therapeutic applications* (pp. 97–137). Singapore.

Shapi'i, R. A., Othman, S. H., Basha, R. K., & Naim, M. N. (2022). Mechanical, thermal, and barrier properties of starch films incorporated with chitosan nanoparticles. *Nanotechnology Reviews, 11*(1), 1464–1477.

Shegokar, R., & Nakach, M. (2020). Large-scale manufacturing of nanoparticles—An industrial outlook. In *Drug delivery aspects* (pp. 57–77). Elsevier.

Shwetha, H. J., Shilpa, S., Mukherjee, M. B., Ambedkar, R., Raichur, A. M., & Lakshminarayana, R. (2020). Fabrication of chitosan nanoparticles with phosphatidylcholine for improved sustain release, basolateral secretion, and transport of lutein in Caco-2 cells. *International Journal of Biological Macromolecules, 163*, 2224–2235.

Singh, R. P., Singh, P. K., Gupta, R., & Singh, R. L. (2018). Biotechnological tools to enhance sustainable production. In *Biotechnology for sustainable agriculture* (pp. 19–66). Woodhead Publishing.

Singh, H., Sharma, A., Bhardwaj, S. K., Arya, S. K., Bhardwaj, N., & Khatri, M. (2021). Recent advances in the applications of nano-agrochemicals for sustainable agricultural development. *Environmental Science: Processes and Impacts, 23*(2), 213–239.

Siripongpreda, T., Hoven, V. P., Narupai, B., & Rodthongku, N. (2022). Emerging 3D printing based on polymers and nanomaterial additives: Enhancement of properties and potential applications. *European Polymer Journal, 184*, 111806.

Somaya, D., Teece, D., & Wakeman, S. (2011). Innovation in multi-invention contexts: Mapping solutions to technological and intellectual property complexity. *California Management Review, 53*(4), 47–79.

Subhan, M. A., & Subhan, T. (2022). Safety and global regulations for application of nanomaterials. In *Nanomaterials recycling* (pp. 83–107). Elsevier.

Sulthana, A. S., Raguvaran, K., Prabu, P., Mani, R., Sundaramoorthi, C., Padarthi, P. K., Ragavendran, C., Manimegalai, T., & Elangovan, N. (2023). Design, preparation, and in vitro characterizations of chitosan-loaded nanostructured lipid carriers: A promising drug delivery system. *Biomass Conversion and Biorefinery*, 1–18. https://doi.org/10.1007/s13399-023-04608-9

Tayel, A. A., Elsaied, B. E., & Diab, A. M. (2019). Nanotechnology for aquaculture. In *Nanoscience for sustainable agriculture* (pp. 479–544). Springer.

Tripathi, S., Siddiqui, M. H., Kumar, A., & Vimal, A. (2023). Nanoparticles: A promising vehicle for the delivery of therapeutic enzymes. *International Nano Letters, 13*, 209–221.

Tumwesigye, Z., Tumwesigye, W., Opio, F., Kemigabo, C., & Mujuni, B. (2022). The effect of water quality on aquaculture productivity in Ibanda District, Uganda. *Aquaculture Journal, 2*(1), 23–36.

Uyanga, V. A., Ejeromedoghene, O., Lambo, M. T., Alowakennu, M., Alli, Y. A., Ere-Richard, A. A., Min, L., Zhao, J., Wang, X., Jiao, H., Onagbesan, O. M., & Lin, H. (2023). Chitosan and chitosan-based composites as beneficial compounds for animal health: Impact on gastrointestinal functions and biocarrier application. *Journal of Functional Foods, 104*, 105520.

van Rijnsoever, F. J., & Hessels, L. K. (2021). How academic researchers select collaborative research projects: A choice experiment. *The Journal of Technology Transfer, 46*(6), 1917–1948.

Vyshnava, S. S., Swetha, K., & Dowlathabad, M. R. (2023). Smart drug nanoparticles from microorganisms and drug delivery. In *Microbial processes for synthesizing nanomaterials* (pp. 227–259). Singapore.

Yanat, M., & Schroën, K. (2021). Preparation methods and applications of chitosan nanoparticles; with an outlook toward reinforcement of biodegradable packaging. *Reactive and Functional Polymers, 161*, 104849.

Yu, Z., Boyarkina, V., Liao, Z., Lin, M., Zeng, W., & Lu, X. (2023). Boosting food system sustainability through intelligent packaging: Application of biodegradable freshness indicators. *ACS Food Science and Technology, 3*(1), 199–212.

Zhang, B., Zhang, Z. G., Yan, X., Wang, X. X., Zhao, H., Guo, J., Feng, J. Y., & Long, Y. Z. (2017). Chitosan nanostructures by in situ electrospinning for high-efficiency PM2. 5 capture. *Nanoscale, 9*(12), 4154–4161.

Zhang, C., Liu, Y., Yu, W., Zhan, Y., Wang, J., Xiong, C., Shi, Z., & Yang, Q. (2019). Nanopolysaccharides in energy storage applications. In *Advanced functional materials from nanopolysaccharides* (pp. 137–169). Springer.

Zhang, Z., Yang, J., Xie, P., Gao, Y., Bai, J., Zhang, C., Liu, L., Wang, Q., & Gao, X. (2020). Characterization of a thermostable phytase from Bacillus licheniformis WHU and further stabilization of the enzyme through disulfide bond engineering. *Enzyme and Microbial Technology, 142*, 109679.

Zhong, L., Feng, Y., Wang, G., Wang, Z., Bilal, M., Lv, H., Jia, S., & Cui, J. (2020). Production and use of immobilized lipases in/on nanomaterials: A review from the waste to biodiesel production. *International Journal of Biological Macromolecules, 152*, 207–222.

Zou, P., Yang, X., Wang, J., Li, Y., Yu, H., Zhang, Y., & Liu, G. (2016). Advances in characterisation and biological activities of chitosan and chitosan oligosaccharides. *Food Chemistry, 190*, 1174–1181.

Real-World Application of Nanochitosan in Refinery-Produced Water Treatment: A Case Study

Geetha Devi and Khadija Salim Abdullah Al Balushi

Contents

G. Devi (✉) · K. S. A. Al Balushi
Mechanical & Industrial Engineering Department, National University of Science & Technology, Muscat, Oman
e-mail: geethadevi@nu.edu.om

P. O. Isibor et al. (eds.), *Nanochitosan-Based Enhancement of Fisheries and Aquaculture*, https://doi.org/10.1007/978-3-031-52261-1_14

1 Introduction

Environmental pollution is a serious concern all over the world, and green chemistry is receiving considerable attention to solve these problems. The consequences of the generation of wastewater, its treatment, and disposal gradually increase and thereby develop major environmental distress. Nanotechnology is one of the promising areas of research, and its development has contributed a substantial role in meeting the requirement of freshwater and the economy of the country (Magwaza et al., 2020). Recently, particular research interest has been devoted to the development of nanoparticles and polymer-based nanocomposites from synthetic or natural sources (Pedro et al., 2018). Crab shell chitosan is a widely employed biomaterial for the treatment of industrial wastewater in a safe and environment-friendly approach. The main constituents of crustacean shells are chitin, minerals, pigment lipids, and proteins. Crab, shrimp, and prawn are considered staple food rich in chitin and chitosan with excellent nutritional value. Production of natural polymers from crustacean shells could lessen the current dependency on this value-added product. Chitosan is the second most abundant polysaccharide with straight chain natural polysaccharide obtained by the deacetylation of chitin. Chitosan is β-(1,4)-related glucosamine units (2-amino-2-deoxy-β-D-glucopyranose) with N-acetylglucosamine units (2-acetamino-2-deoxy-β-D-glucopyranose) (Pedro et al., 2018). Recent studies show the importance of nanochitosan, and nanoparticles derived from chitosan are recommended as a valuable biomaterial for industrial wastewater treatment applications. The excellent biological and chemical properties make it suitable for water purification (Dash et al., 2011; Luo et al., 2013). The chemical structures of chitin and chitosan are displayed in Fig. 1a, b.

The cell walls of fungi, green algae, cuticles of insects, and exoskeleton of crustaceans are the main sources of chitin. The composition analysis of chitin shows 20–40% calcium, 40% protein, and the rest magnesium carbonate, along with other minor constituents, such as lipids and minerals (Khoushab & Yamabhai, 2010; Ahyat et al., 2017). Both chitin and chitosan derivatives are excellent biosorbents due to their enriched surface properties, nontoxic, cost-effective, and freely available surface functional groups of amino and hydroxyl, which have attractive adsorption capacity to eliminate a variety of pollutants from wastewater (Bhatnagar & Sillanpää, 2009). The outstanding properties of flocculation, coagulation, and

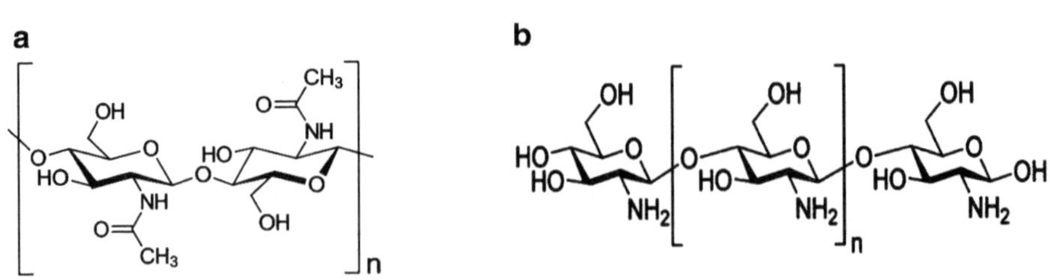

Fig. 1 (**a**) Chemical structure of Chitin. (**b**) Chemical structure of Chitosan

environment-friendly nature of chitosan are the most interesting, making it appropriate for environmental pollution control applications (Pradip et al., 2004; Krajewska, 2005; Kashyap et al., 2005; Prashanth & Tharanathan, 2007; Renault et al., 2009; Al Sagheer et al., 2009; Islam et al., 2011; Dong et al., 2014). The worldwide consumption of crab is on the higher side and huge amounts of crab shells are discarded as solid waste, which creates severe environmental concerns. Hence, it's important to convert the waste crab shell into value-added products. The majority of the processes in the extraction of chitin and chitosan require enormous amounts of poisonous chemicals that result in pollution problems by the generation of an excess quantity of toxic waste and are also harmful to humans and the environment. Over the years, several techniques have been extensively applied in the extraction of biopolymers (chitin and chitosan). In this chapter, the extraction of chitosan from waste crab shells and their performance in the removal of pollutants from refinery wastewater treatment applications has been highlighted. The extraction of chitin and chitosan is carried out at different processing conditions, and the method of extraction is described in this chapter with particular emphasis on applications of chitosan in the batch treatment of refinery wastewater. Herein, we report an eco-friendly, cost-effective, and energy-efficient isolation technique for the preparation of biopolymer from crab shells.

The major steps involved in the preparation of chitosan from crab shells are the isolation of chitin and the conversion of chitin into chitosan by the N-deacetylation process (Pradip et al., 2004; Sowmya et al., 2011). The reported studies show the extraction process of chitosan from crustacean shells, which includes demineralization and deproteinization process, while some research reports included a decolourization step (Sowmya et al., 2011).

2 Materials and Methods

The fresh crabs used in the extraction of chitosan were received from a local fish market in Muscat, sultanate of Oman (Fig. 2). The shells were removed from the body and washed several times with fresh water to eliminate debris and then allowed to dry in a furnace operated at 80 °C for 30 min. The dried samples were crushed and then ground to fine powder followed by sieve analysis to get desired average particle size of 75 μm. This powdered sample was subjected to demineralization and deproteinization processes. Hydrochloric acid (HCl), nitric acid (HNO_3), sodium hydroxide pellets, and acetone are purchased from Chemistry for Life Company, Oman. The characterization techniques employed are SEM, XRD, EDX, FTIR, TGA, XRF, etc. The refinery effluent samples were collected from the outlet of produced water from the Occidental Company, Oman.

The schematic representation of the synthesis of chitosan is shown in Fig. 2. A crab shell can be dried (Fig. 3a) and pulverized crab shell (Fig. 3b).

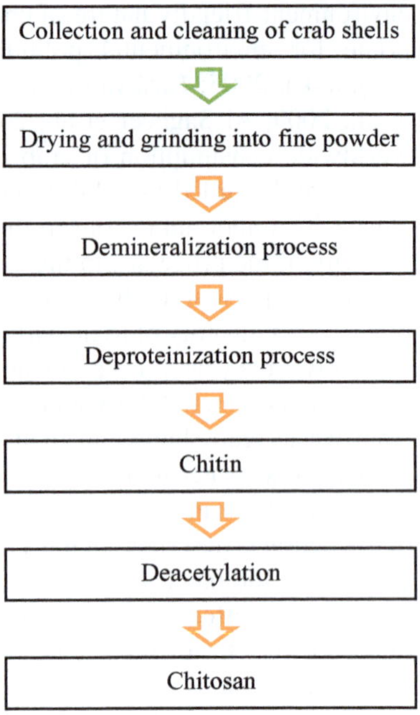

Fig. 2 Schematic representation of the production of chitosan from crab shell

(a) Dried crab shell (b) Crab shell powder

Fig. 3 Production of chitosan. (**a**) Dried crab shell. (**b**) Crab shell powder

3 Synthesis of Chitosan

This section contains a detailed procedure and methodology employed in the extraction of chitin and chitosan from waste crab shells. The comprehensive step-by-step procedures are described with the inclusion of experimental conditions. The chitosan synthesis step consists of three stages, viz, demineralization, deproteinization, and deacetylation process.

3.1 Demineralization Process

The purpose of the demineralization process is to remove the inorganic materials such as calcium phosphate, calcium carbonate, salts, and other minerals present in the crab shells using dilute acids. The demineralization process was carried out at room temperature by treating 100.0 g of crab shell powder with 150 ml of HNO_3 under agitated conditions for 1 h. The reaction mixture was thoroughly washed with millipore water for several cycles until it reached a pH of 7.0 followed by filtration to remove the minerals and then dried in a furnace at 80 °C to form dry powder.

3.2 Deproteinization

The deproteinization process was performed by reacting 1.0 M NaOH with the demineralized product at a ratio of 10:1 (solid/liquid ratio). The reaction temperature was kept at 80 °C for a mixing duration of 3 h. The reaction product was filtrated and washed with distilled water to reach a neutral pH. The mixture was bleached using 1% ethanol for 10 min followed by drying in a furnace operated at 70 °C to form chitin powder.

3.3 Decolorization

The deproteinized powder sample was dispersed in an acetone ratio of 1:10 to remove the colour. The decolourization was performed by dissolving 1.0 g of deproteinized powder in 100 ml of acetone under continuous stirring for 24 h with a stirring speed of 100 rpm. The decolourized product was filtered and washed five times with distilled water to reach a neutral pH and then dried in an oven at 60 °C to form powder.

3.4 Deacetylation Process

The decolourized chitin powder was subjected to a deacetylation process using thermal heating. The deacetylation process removes the acetyl group from the chitin powder thereby enriching the properties of the biopolymer. In the deacetylation process, 1.0 g of chitin powder was dissolved in 20.0 ml of 50% NaOH solution under continuous agitation for 6 h at 110 °C. The excess NaOH was removed after the deacetylation process. The product chitosan was finally washed with distilled water and dried in the oven at 60 °C to get pure chitosan.

4 Characterization Techniques

The main characterization tools employed in the analysis of chitosan are Scanning Electron Microscopy (SEM), X-ray diffraction (XRD), Fourier Transforms Infrared Spectroscopy (FTIR), Thermo Gravimetric Analysis (TGA), Nuclear Magnetic Resonance (NMR), Energy Dispersive Spectrometer (EDS), and X-Ray Fluorescence (XRF). These techniques are used to study the surface morphology, identification of surface functional groups, microstructural features, phase identification, and elemental analysis of the extracted chitosan. The samples are prepared 1 day before SEM analysis. Thermo gravimetric analysis was performed to determine the mass of the sample over a range of 30–900 °C. The following section focuses on the working of various characterization equipment used in the testing and analysis of chitosan.

4.1 Scanning Electron Microscope (SEM)

SEM was employed to study the surface morphological and microstructural characterization of the synthesized chitosan. In SEM analysis, the image of an object is captured using electron beams, and the resulting image is magnified using electromagnetic fields. The resolution of an electron microscope is 200 times that of a light microscope.

Working Principle

In SEM, a narrow beam of electrons is directed towards the test sample to be characterized. The electron gun emits a narrow beam of electrons that falls on the sample surface and scans the surface by releasing secondary electrons along with other radiations from the surface of the specimen (Fig. 4). The intensity of the secondary electrons is different depending on the shape and chemical composition of the powder sample. The detector collects the secondary electrons and subsequently produces electronic signals. A cathode ray tube projects the image of the sample surface. The test samples are prepared by dropping a sample on a stub and dried overnight to remove moisture content from the sample. The sample surface is coated with platinum and placed in a closed chamber with argon as inlet gas to make the sample conducting. It is then placed inside the SEM for further characterization. Figure 4 explains the working of the scanning electron microscope. An electron gun, located at the top of the device, shoots out a beam of highly concentrated electrons. The types of electron guns used are thermionic guns, which heat a filament until electrons stream away. The second type is field emission guns, field emission guns. The microscope is aligned with a couple of lenses within a vacuum chamber. These electrons are directed towards the specimen through lenses to maximize efficiency. The higher the number of electrons allowed to pass through, the better the view of the sample. The SEM characterization is carried out using a vacuum chamber to function the device, and the electron beam must not be obstructed as it passes.

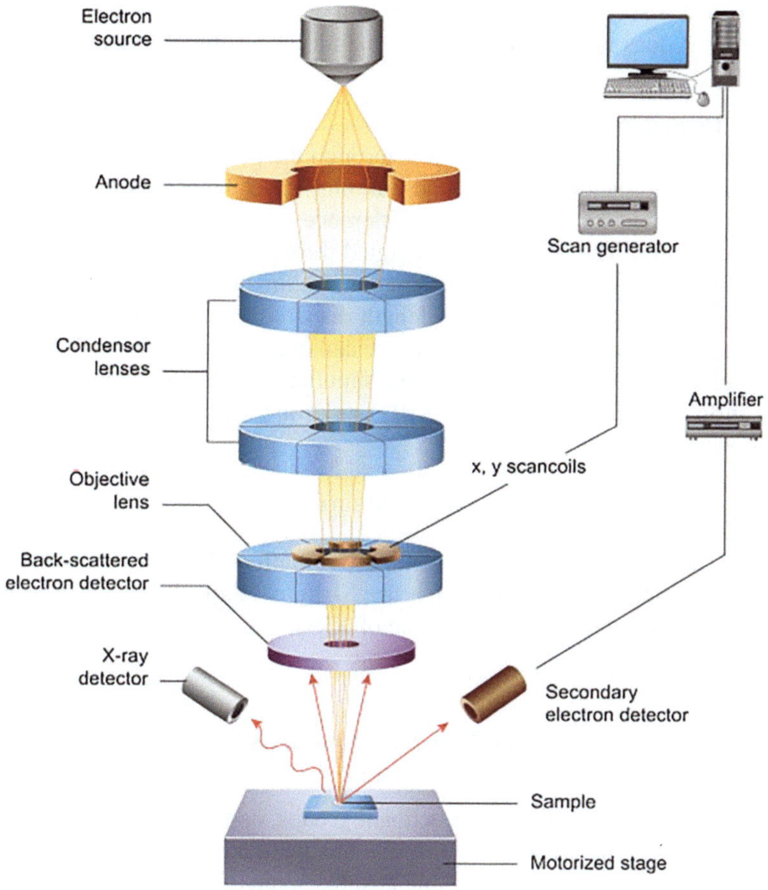

Fig. 4 Working principle of scanning electron microscope

When electron beams hit the sample, X-rays are radiated with primary back-scattered electrons, secondary electrons, and Auger electrons. The SEM employs primary back-scatter electrons and secondary electrons. An electron recorder picks up the rebounding electrons and records their imprint. This information is translated onto a screen, which allows three-dimensional images to be represented clearly.

4.2 X-Ray Diffractometer (XRD)

X-ray diffractometers are used to determine the crystallographic structure, atomic arrangement, unit cell dimensions, grain size, and purity of a material as a function of time. The X-ray diffraction technique is a non-destructive and rapid analytical technique that is suitable for the corroboration of the structure and crystallinity of a sample, but it does not provide any information about the chemical nature of the substance.

Working Principle

XRD works by passing an X-ray source through the powder sample to be analyzed and irradiated by incident X-rays and emitting their characteristic X-rays. X-ray diffraction is based on constructive interference of monochromatic X-rays and a crystalline sample. The X-rays are generated through a cathode ray tube, which is filtered to produce monochromatic radiation, collimated to concentrate, and then directed towards the sample. A wavelength dispersive detector was used to assess the emitted X-ray peaks for the quantitative and qualitative analysis. The X-ray diffraction techniques are also applied in the simultaneous determination of elemental composition and film thickness. The XRD works on the principle of Bragg's Law of diffraction,

$$n\lambda = 2d \sin \theta \tag{1}$$

where θ is the angle of incidence, λ is the wavelength of the incident X-ray beam, d is the interspacing distance (on which the X-ray is incident), and n is an integer. Figure 5 illustrates the working principle of XRD.

4.3 *Fourier Transforms Infrared Spectroscopy (FTIR)*

The FTIR is used to identify the functional groups, composition, and purity of the sample. In FTIR spectroscopic analysis, the incident infrared radiation passes through the sample surface, the sample absorbs some of the infrared radiation, and the remaining passes through it. The molecular fingerprint of the sample was obtained through the spectrum arising from absorption and transmission modes. In FTIR spectroscopy, no two molecular fingerprints can ever be the same. FTIR analysis is used to identify solids, liquids, or gases. Two types of sampling techniques

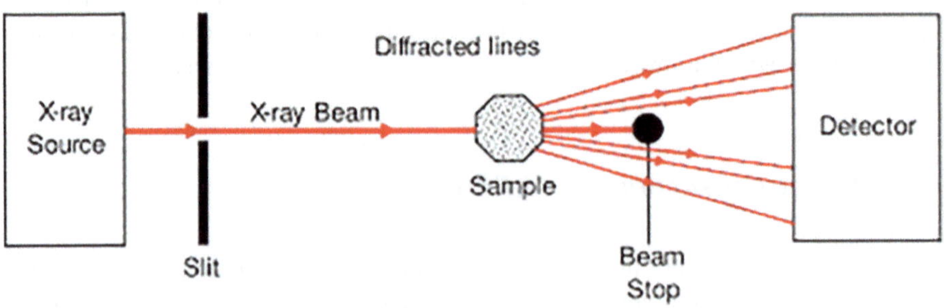

Fig. 5 Working principle of XRD

used in FTIR analysis are Attenuated Total Reflectance (ATR) and Absorbance mode. Among these, ATR mode is the most common FTIR sampling technique.

Working Principle
The main parts of FTIR are source, interferometer, sample compartment, detector, and computer. When the infrared radiation is passed through the sample, it is absorbed and some are transmitted. The resulting spectrum makes a molecular fingerprint of the sample. The beam enters the interferometer where the interference of two beams of light is employed to make precise measurements. The electron beam enters the detector, and the final measured signal will be converted into a digital signal and transferred to the computer, where the Fourier transform takes place. The working principle of FTIR is shown in Fig. 6.

4.4 *Thermo Gravimetric Analysis (TGA)*

Thermo Gravimetric Analysis (TGA) detects the weight changes in a material as a function of temperature (or time) in a controlled atmosphere. TGA is used to measure the thermal stability of a material, filler content in polymers, the quantification of components in a compound, and moisture and solvent content.

Working Principle
The working of TGA is based on thermal analysis in which any changes in chemical and physical properties of a material are quantified as a function of temperature rise, or as a function of time. TGA is performed by gradually increasing the temperature of samples in a furnace as their weight is measured on an analytical balance outside the furnace. When a thermal event involves the loss of a volatile component, weight loss is observed in TGA. Chemical reactions, such as combustion, involve weight loss, and no physical changes, such as melting. The sample weight is plotted with temperature or time to illustrate the thermal transitions of the material such as loss of solvent and plasticizers in polymers, hydration water in inorganic materials, and

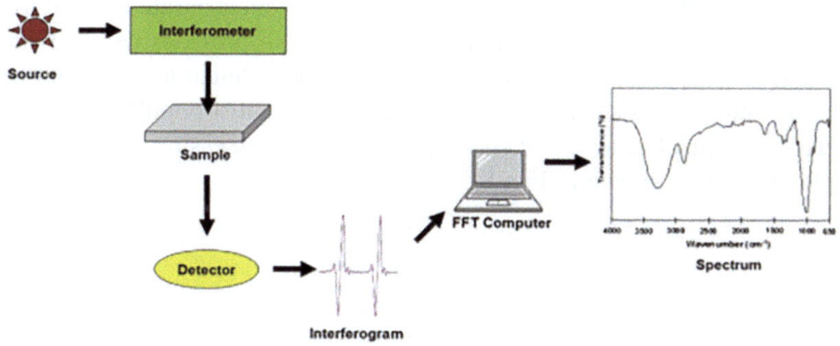

Fig. 6 Working principle of FTIR

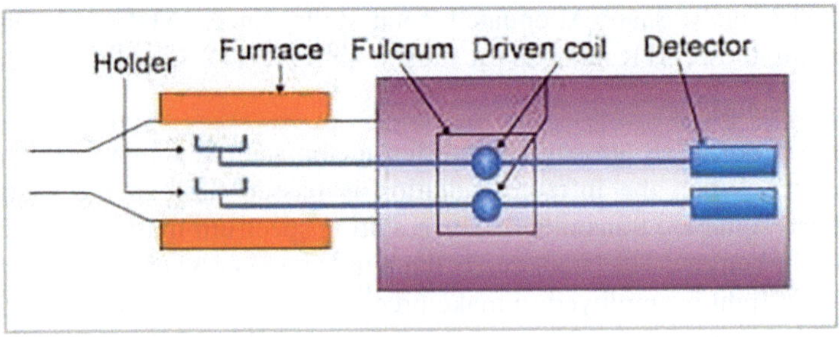

Fig. 7 Working principle of TGA

finally material decomposition. TGA consists of a high-precision balance and sample pan. Figure 7 represents the working principle of TGA.

4.5 Nuclear Magnetic Resonance (NMR) Spectroscopy

NMR spectroscopy is an analytical technique employed in the determination of the purity of a sample by witnessing the local magnetic fields around the nuclei. NMR is based on the absorption by nuclei of atoms of electromagnetic radiation in the radio frequency range of 4–900 MHz. NMR spectroscopic techniques are used to study the physical, chemical, and biological properties of materials. NMR can quantitatively analyze mixtures containing known compounds and also study chemical structure using simple one-dimensional techniques. The structure of more complicated molecules is determined by two-dimensional techniques.

Working Principle

In NMR spectroscopic analysis, the sampling is placed in a magnetic field, and the excitation of the nuclei sample with radio waves produces nuclear magnetic resonance and is detected using radio receivers. The change in resonance frequency will provide the complete electronic structure of the molecule along with the functional groups present in the molecule. NMR spectroscopy is an accurate method to determine the monomolecular organic compounds and also provides detailed statistics about the dynamics, structure, state of reaction, and chemical nature of molecules. The proton and carbon-13 NMR spectroscopy are the most popular types of NMR applied to any type of sample that encompasses nuclei possessing spin. The working principle of NMR is shown in Fig. 8.

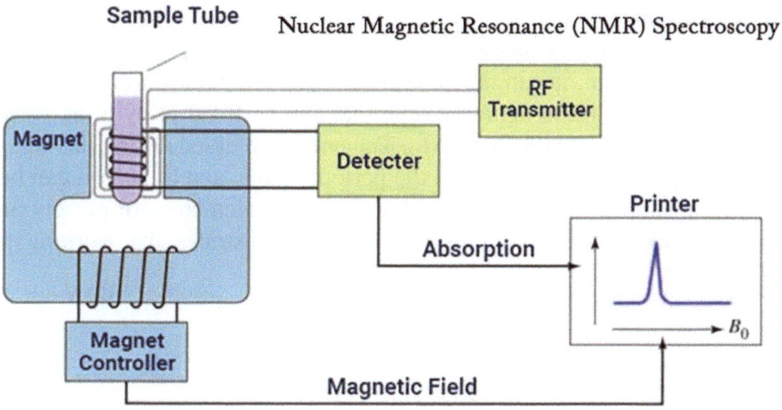

Fig. 8 Working principle of NMR

Fig. 9 working principle of EDX

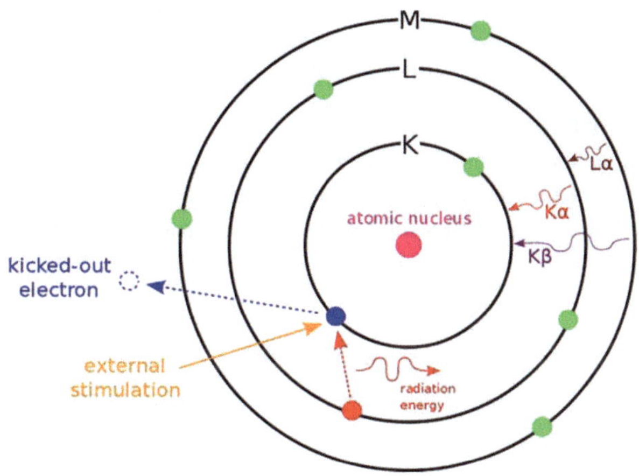

4.6 *Energy Dispersive X-ray Spectroscopy (EDS or EDX)*

Energy dispersive X-ray spectroscopy (EDS or EDX) is the analytical technique used to determine the elemental analysis, chemical characterization, and investigation of the sample. Also, the X-rays emitted by the matter showing the full quantitative of the sample composition are analyzed.

Working Principle

The ability of high-energy electromagnetic radiation (X-rays) is ejected as 'core' electrons from an atom, not in the outermost shell. The removal of these electrons from the system will leave behind a hole that can be filled in by a higher energy electron and release energy as it relaxes. The energy released during this relaxation process is unique to each element on the periodic table, and as such it can be used to bombard a sample with X-rays to identify which elements are present and the proportion in which they are present. Figure 9 illustrates the working principle of TGA.

4.7 X-Ray Fluorescence Spectrometers (XRF)

X-ray fluorescence (XRF) is a non-destructive analytical technique used to determine the elemental composition of materials chemistry of a sample by measuring the fluorescent X-ray emitted from a sample when it is excited by a primary X-ray source. The composition and layer thickness can be determined by XRF. X-ray fluorescence spectrometers analyze a variety of elements ranging from beryllium (Be) to uranium (U) with 100 wt% to sub-ppm levels of concentration.

Working Principle

The specimen is excited with the primary X-ray irradiation, and during the process, the electrons emitted from the inner electron shells are knocked (Fig. 10). The outer shell electron fills the voids emitting a fluorescence radiation that is distinctive in

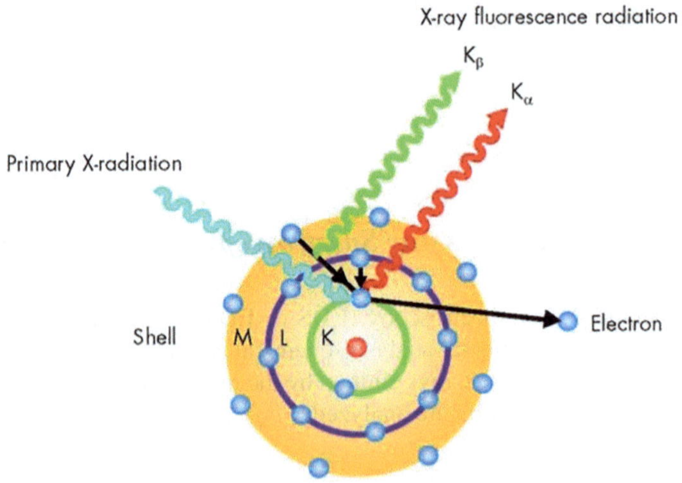

Fig. 10 Working principle of XRF

determining the energy distribution of a material. The detector senses the fluorescence radiation. The generation of the X-ray fluorescence radiation is shown in Fig. 10. When one electron from the K shell is knocked, the resulting void is filled by an electron from the L shell or the M shell. During the process, the K_α and K_β radiation is generated.

5 Results and Discussion

This section discusses the experimental outcome of the extraction of chitosan with the interpretation of data and analysis with the support of graphs and images.

5.1 *Study on Surface Morphology of Chitosan Using SEM*

The surface features of the extracted chitosan were analyzed using scanning electron microscopy, and the surface morphology is shown in Fig. 11. The SEM image indicates a scattered distribution of chitosan with an even spread and absence of any accumulation of particles, which endorses the successful synthesis of chitosan. The morphological characterization of chitosan powder exhibits the actual size and shape of the particles. The SEM image shows a rough and thick surface structure at

Fig. 11 SEM micrograph of chitosan powder

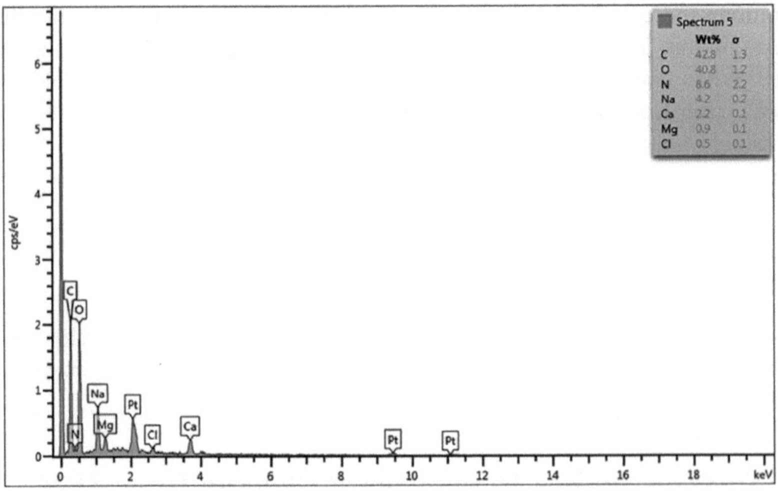

Fig. 12 EDX Spectra of chitosan powder

a magnification of 1100× and an excitation voltage of 15.0 kV, as illustrated in Fig. 11.

5.2 *Elemental Composition Analysis of Chitosan Using SEM EDX*

Figure 12 represents the energy-dispersive X-ray spectra (EDX) of the chitosan sample. The elemental composition analysis shows 43.3% C, 36.6% O_2, 16.1% Na, 2.5% Ca, 0.9% Mg, and 0.6% Al (Fig. 12). These analyses match with the literature (Zhang et al., 2018). All the samples tested for EDX analysis were coated with gold to prevent the accumulation of static electric fields during imaging.

5.3 *X-Ray Diffraction Analysis of Chitosan*

The XRD analysis of the extracted chitosan was performed at a scan rate of 1°/min with a diffraction angle varied from 2° to 40°. The phase and structural properties obtained from the XRD spectra are shown in Table 1. The XRD pattern confirms the crystalline nature of chitosan and the diffraction of the peak is visible at 2θ ranges between 10–15° and 25–30°. The maximum peak was observed at 1738.45–194.25 at a position of 2θ and equal to 9.1238–27.9082 as shown in Table 1. The results are matching with previous studies, and it is obvious that the samples retained the major crystalline structures. However, some additional peaks were observed, mainly due to the experimental conditions and nature of solvents used in the demineralization

Table 1 XRD data of chitosan powder

Pos. [°2Th.]	Height [cts]	FWHM Left [°2Th.]	d-spacing [Å]	Rel. Int. [%]
9.1238	1738.45	0.2007	9.69294	100.00
18.1688	488.38	0.2007	4.88278	28.09
18.6918	659.54	0.2007	4.74731	37.94
20.1816	1400.13	0.9368	4.40012	80.54
27.9082	194.25	0.1338	3.19699	11.17
29.4734	576.52	0.1171	3.03069	33.16
32.3711	293.15	0.1673	2.76570	16.86
33.9227	314.84	0.2007	2.64267	18.11
37.9620	165.03	0.2676	2.37026	9.49
39.4389	62.19	0.2007	2.28483	3.58
40.1463	70.93	0.4015	2.24619	4.08
44.8721	37.29	0.8029	2.02000	2.14
51.1413	21.56	0.8029	1.78613	1.24
52.5591	35.83	0.2676	1.74124	2.06
57.2427	14.36	0.8029	1.60941	0.83

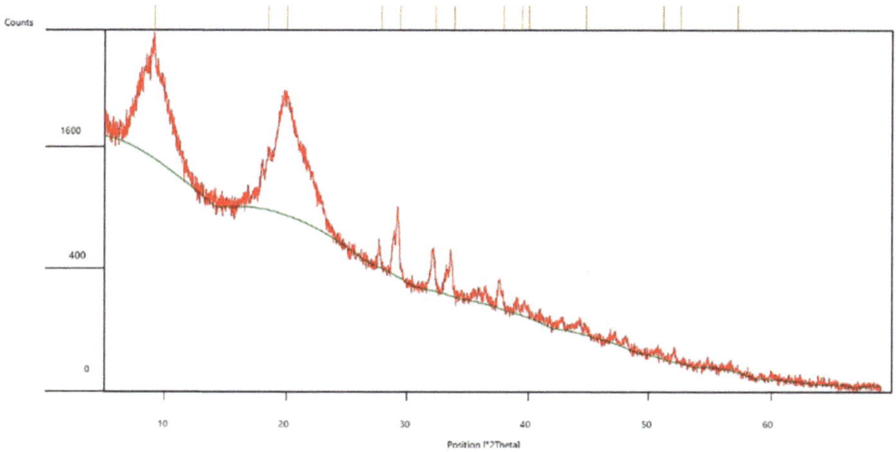

Fig. 13 XRD spectra of chitosan powder

and deproteinization stages. The reflection peak indexed at 9.31° confirmed the presence of extracted chitin and the chitosan peak was observed at 20°, which is in agreement with the previous studies (Rinaudo, 2006). The XRD spectra of chitosan powder are shown in Fig. 13.

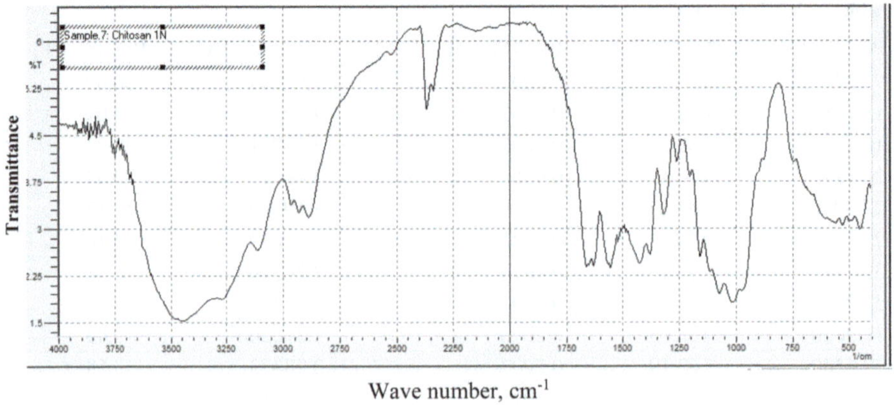

Fig. 14 FTIR spectra of chitosan powder

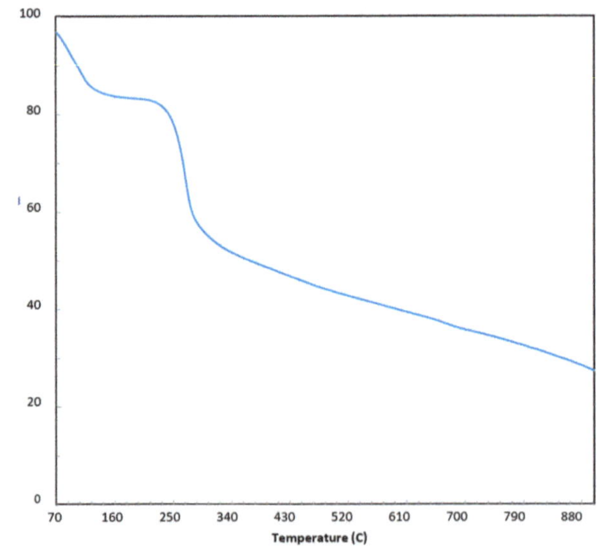

Fig. 15 TGA of extracted chitosan powder

5.4 *Fourier Transform Infrared Spectroscopic Analysis of Chitosan*

The chitin or chitosan samples were analyzed in an FTIR over a frequency range of 400–4000 cm^{-1} at a resolution of 4 cm^{-1}. The functional groups and bands present in the chitosan powder are shown in Fig. 14. The various characteristic peaks observed at wave number corresponding to 3469 cm^{-1} represent the O-H functional group, wave number corresponding to 2930 cm^{-1} refers to C-H, 1659 cm^{-1} refers to C = O, 1468 cm^{-1} refers to CH=H, 1380 cm^{-1} refers to CH2-OH, and 1580 cm^{-1} refers to N-H. The N-H and the O-H stretching bands of chitosan were observed

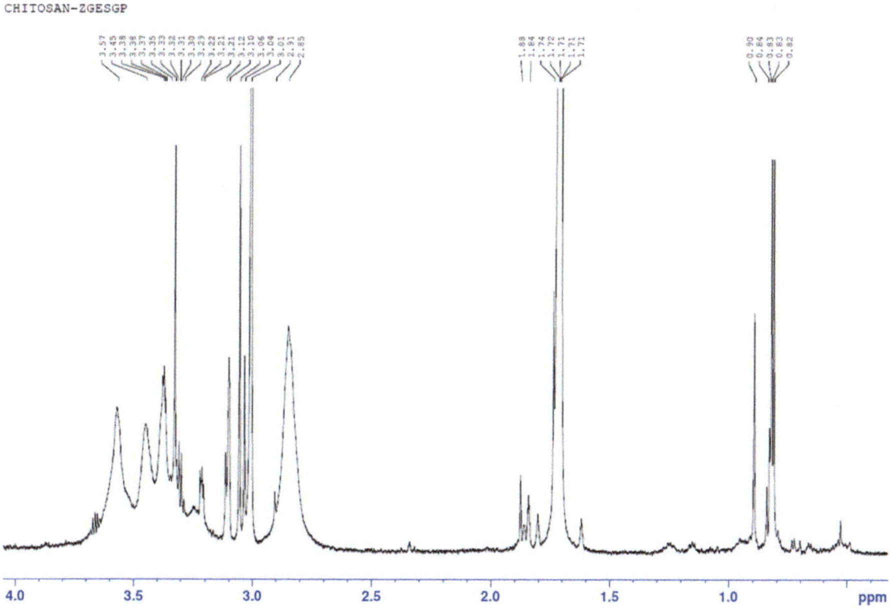

Fig. 16 ^{1}H NMR spectra of chitosan

between 800 and 1600 cm^{-1}. The FTIR results are in good agreement with the previous results found in the literature (Yen et al., 2009).

5.5 Thermo Gravimetric Analysis (TGA)

The thermal degradation temperature of chitosan was determined using TGA with two major degradation steps (Fig. 15). In Fig. 15, it was observed that the first stage of degradation of chitosan occurs between 30 and 90 °C with around 4% weight loss. The second stage was observed between 90 and 270 °C having 20% weight loss. This degradation can be due to the loss of moisture or water molecules from chitosan. The analysis is similar to that reported in the literature (Han et al., 2017).

5.6 Nuclear Magnetic Resonance

NMR is used to record the magnetic resonance spectra of the chitosan sample. Detailed information on the conformation, structure, and molecular motion of chitosan was detected using NMR. Chitosan samples were spun at 4000 Hz frequency and the 1H NMR spectra of chitosan were determined at 70 °C. The amine, amide, and nitrogen atoms are displayed in the spectra as indicated in Fig. 16. The chemical

Table 2 ^{1}H NMR spectral peaks of chitosan

Peak	ν(F1)	[ppm]	ν(F1)
1	176.2558	31,031.6668	73,154,104.5
2	19.6232	3454.8685	9,819,415.25
3	19.5113	3435.1673	17,787,105.25
4	19.3977	3415.1668	19,739,234.5
5	19.2852	3395.36	19,164,432.5
6	19.1734	3375.6765	10,188,643.5

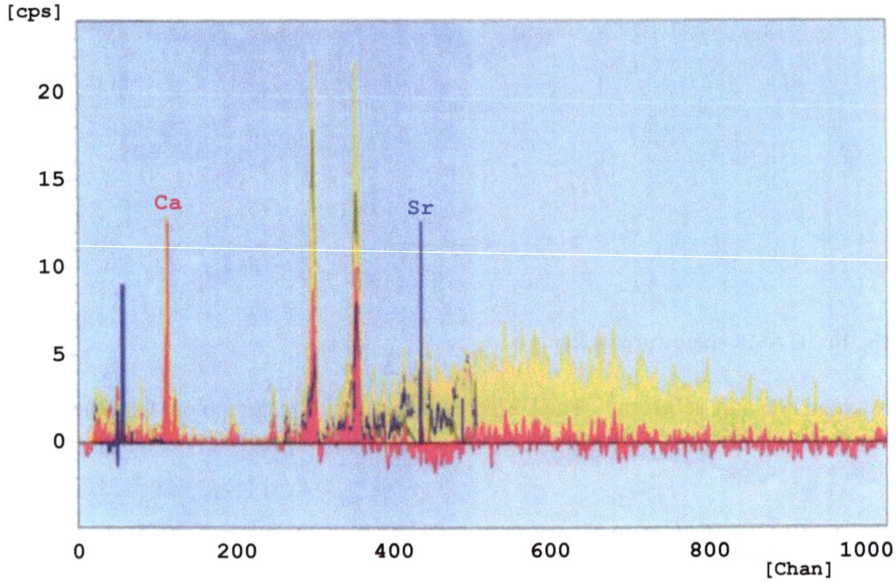

Fig. 17 XRF spectral analyses Chitin

shifts change with variations in allocations of sub-units of the polymer. The minor shift in peak positions is due to the nature of neighbouring subunits, the effect of temperature and solvents, and the spectral data (Table 2).

5.7 *X-Ray Fluorescence*

XRF is a non-destructive analytical technique used to determine the elemental composition of chitosan samples. XRF determines the chemistry of chitosan by measuring the fluorescent (or secondary) X-ray emitted from the sample when it is excited by a primary X-ray source. XRF was used to analyze the percentage of individual elements relative to the weight of the sample analyzed. Figure 17 shows the physical examination of the sample spectrum showing the prominent elements as the most towering peaks of calcium at 92.88%. Impurities present in the sample are shown as low and almost grounded peaks that are obtained after the deacetylation process. The amount of calcium present in the sample was 80.21%.

5.8 Application of Chitosan in Refinery Wastewater Treatment

The water scarcity along with escalating contamination of water bodies and reservoirs makes it unsuitable for drinking and other purposes. Recently, water resources have been contaminated due to fast-growing industrial development and inadequate sewage treatment facilities, etc. Environmental pollution is one of the major concerns due to the contamination of rivers, lakes, and oceans, depleting water quality and making it more toxic to the environment and humans. Human activity and rapid industrialization are primarily responsible for water pollution. The wastewater discharged from the various processing plants of refineries causes serious environmental pollution, which may lead to great threats to aquatic and terrestrial life. This is due to the presence of highly concentrated organic and inorganic effluents. Hence appropriate treatment of wastewater is highly recommended before its disposal. The selection of treatment method is based on the amount of organic and inorganic pollutants present in it and its discharge characteristics (Yen et al., 2009; Han et al., 2017). Conventional wastewater treatment methods are extremely energy-oriented and highly expensive processes and result in ecological and health issues. Thus, the development of a cost-effective and environmentally friendly water treatment technique is one of the most indispensable needs of the hour.

The water sector in the world is facing numerous challenges, which include water shortages, energy-intensive desalination requirements, increased domestic water consumption, unsustainable usage of groundwater in agriculture, and ineffective subsidies (Geetha et al., 2021; Khadija et al., 2021). There are different types of ongoing eco-friendly low-cost technology for water treatment that could reduce water pollution. As a Corporate Social Responsibility (CSR), each individual must conserve water and use environment-friendly biomaterials/biopolymers to minimize environmental pollution. The extracted chitosan powder was employed in the batch experimental studies for the treatment of refinery effluent.

The solution pH, stirring speed, stirring time, and dosage of chitosan are varied in the batch treatment of refinery wastewater and the optimum processing parameters are determined. The influence of change in refinery wastewater pH with the removal of contaminants was studied by varying the pH from 2.0 to 7.0. The treatment was performed by mixing 0.2 g of chitosan powder with 150 ml of refinery effluent under stirring for 45 min with a stirring speed of 25 rpm. The wastewater after treatment was tested for the determination of Total Dissolved Solids (TDS), Total Suspended Solids (TSS), Chemical Oxygen Demand (COD), Biological Oxygen Demand (BOD), and turbidity.

Mixing time plays a significant role in pollutant removal. The influence of variation of stirring time in the reduction of parameters was studied by varying the stirring time from 15 to 90 min. The pH of the effluent solution was maintained at 7.0 and a stirring speed was maintained at 25 rpm. The effect of variation of dosage of chitosan on the percentage removal of parameters is studied by altering the amount of chitosan from 0.1 to 0.5 g, with an optimized pH of the effluent and a mixing time. The influence of agitation speed on pollutant removal efficiency was investigated by changing the stirring speed from 25 to 125 rpm, keeping all other parameters at the optimized values.

The effect of variation of effluent solution on pollutant removal was studied by varying the pH from 2.0 to 7.0. The experiment was performed by mixing 0.2 g of extracted chitosan with 50 ml of the effluent solution and stirring for 45 min at a stirring speed of 25 rpm. The resulting mixture was tested for Total Dissolved Solids (TDS), Total Suspended Solids (TSS), Chemical Oxygen Demand (COD),

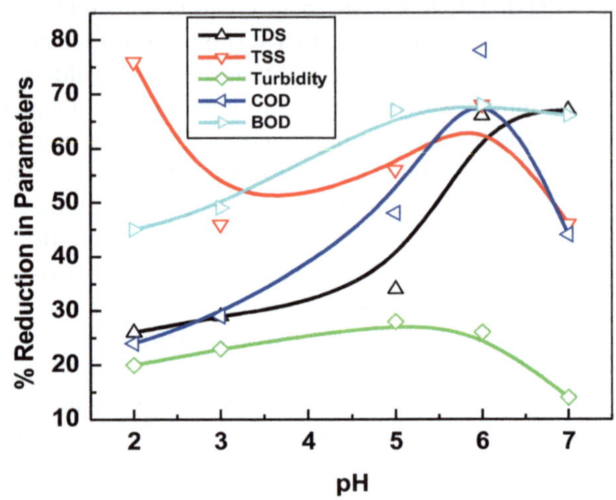

Fig. 18 Effect of variation of wastewater pH with parameter reductions

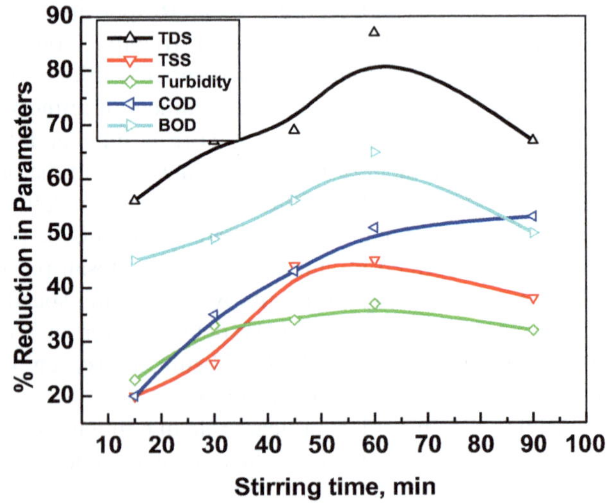

Fig. 19 Effect of variation of mixing time with parameter reductions

Biological Oxygen Demand (BOD), and turbidity. The best parameter reduction was obtained at an optimum pH of 7.0. The maximum reduction in parameters is obtained due to the coagulation and flocculation properties of chitosan. The surface charge of chitosan was high, and therefore, the adsorption capacity also increased resulting in the destabilization of particles in the waste water. This is because of the adsorption of excess polymer on the surface of the colloids. The charge transfer causes an electrostatic repulsion in the suspended solids leading to such type of behavior in the reduction of COD. The optimum pH was observed to be 6.0 with the highest COD reduction rate of 78%. Figure 18 represents the effect of variation of pH with parameter reductions.

As the contact time increased, the development of flocs increased and the flocculent started dispersing in the medium over a certain period of time. Period contact time enhanced the breakage of flocs into smaller ones, thereby retarding the flocculation rate. After treatment, the TDS, TSS, COD, BOD, and turbidity values were recorded. The best contact time for the maximum reduction of COD was observed at 90 min, with a percentage reduction corresponding to 53%. As stirring time increases, flocs formation increases and the flocculent disperses all over the medium over the stirring time at the same time, longer mixing will also break the flocs formed. This can sometimes lead to a reduction in the flocculation rate. Figure 19 illustrates the influence of contact time on parameter reductions. The percentage reduction in TDS was increased up to a contact time of 60 min, beyond which the percentage reduction shrinks.

The slow addition of chitosan powder into the refinery wastewater ensured an increase in the % reduction of TDS. The reduction in TDS was found to increase

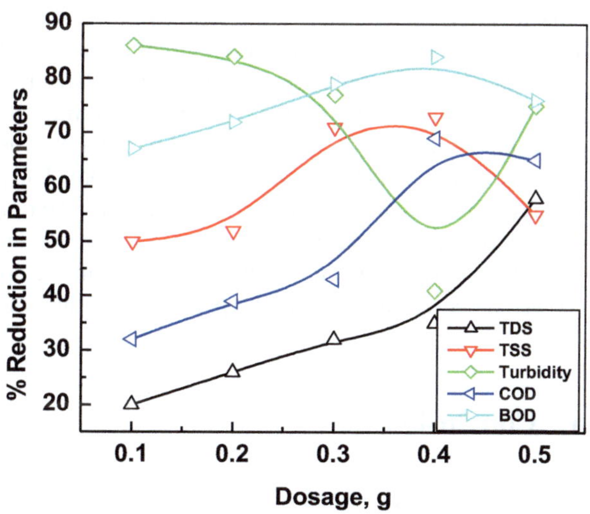

Fig. 20 Effect of variation of dosage of chitosan with parameter reductions

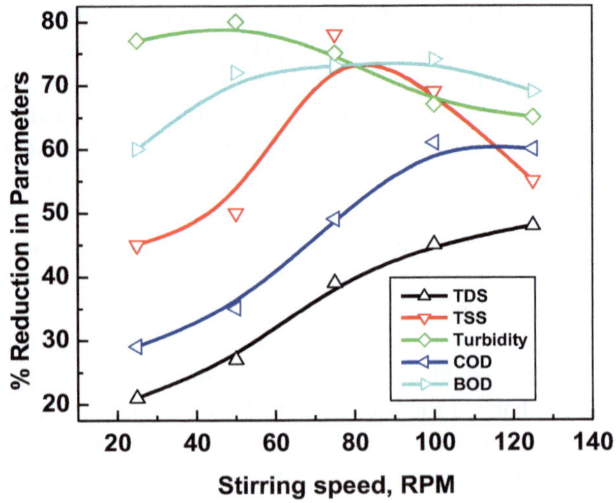

Fig. 21 Effect of variation of stirring speed with parameter reductions

steadily due to the availability of more adsorption sites on the surface of chitosan, which will enhance the deposition tendency of pollutants. This may be due to the strong intermolecular interaction between the amino groups present in chitosan and the pollutants present in the effluent. The percentage reduction in TSS exhibited inconsistent behaviour due to the formation of bigger flocs in the mixture upon changing the dosage of chitosan. As dosage increased, the turbidity reduction tendency was decreased. The positive charge present on the surface of chitosan along with the free amino groups will electrostatically interact with the negative charge of pollutants present in the wastewater, which results in turbidity reduction. The % reduction in COD steadily increased up to 0.4 g of chitosan and then decreased. This was due to the electrostatic repulsion between particles. The effects of variation in the dosage of chitosan with pollutant removal efficiency are shown in Fig. 20.

As the stirring speed increased, the particle size tended to decrease, thereby creating an enhanced surface area, which allowed improved adsorption efficiency. The percentage reduction in TDS increased with an increase in stirring speed. The optimum reduction in TDS was obtained at 100 rpm. The percentage reduction in COD increased with an increase in stirring speed with the range of values studied, whereas the turbidity decreased with an increase in stirring speed, as indicated in Fig. 21. The TSS value augmented with increased stirring speed up to 75 rpm, and then showed a decreasing trend in the percentage reduction of TSS.

6 Conclusion

In this chapter, the extraction of biopolymer chitosan from a crab shell using deproteinization and deacetylation processes was discussed. The surface morphology and chemical composition of chitosan were judged using various characterization techniques. The surface structural characterization, composition analysis, and functional groups of the extracted chitosan were characterized using scanning electron microscopy (SEM), Fourier transform infrared spectroscopy (FTIR), X-ray diffraction (XRD), X-ray fluorescence (XRF), nuclear magnetic resonance (NMR), and thermo gravimetric analysis (TGA). The extracted chitosan was successfully employed in the treatment of refinery wastewater treatment. The pollutant removal efficiency was assessed by varying the experimental parameters, and the optimum percentage reduction in parameters was determined. The study revealed that the extracted chitosan could effectively reduce the pollutants present in the refinery effluent and the parameter reductions were optimized. The extraction of chitosan from crab shells will serve as an incentive for the fish processing industries as it involves an environment-friendly approach for the extraction of chitin and chitosan and also to minimize marine pollution. The study demonstrates that the isolation of chitosan from crustacean shells would address the environmental pollution in marine sectors and also derive value-added products from the waste shells along with improving the water quality, creating clean environments, and hence protecting human health. It would also help to increase the country's per capita income, by exporting the extracted chitosan to suitable customers. This environmentally friendly approach for the green extraction of chitosan will be a better option for marine waste management without using harsh chemicals. This work is aligned with the United Nations Sustainable Development Goal (UNSD 2030 – Goal 6 Clean water and sanitation) and Oman Vision 2040 and provides a feasible solution with a clear emphasis on achieving sustainability in water demands.

References

Ahyat, N. M., Mohamad, F., Ahmad, A., & Azmi, A. A. (2017). Chitin and chitosan extraction from Portunus Pelagicus. *Malaysian Journal of Analytical Sciences, 21*(4), 770–777.

Al Sagheer, F. A., Al-Sughayer, M. A., Muslim, S., & Elsabee, M. Z. (2009). Extration and characterization of chitin and chitosan from marine sources in Arabian Gulf. *Carbohydrate Polymers, 77*(1), 410–419.

Bhatnagar, A., & Sillanpää, M. (2009). Applications of chitin-and chitosan-derivatives for the detoxification of water and wastewater—A short review. *Advances in Colloid and Interface Science, 152*(1–2), 26–38, 46, 83–86.

Dash, M., Chiellini, F., Ottenbrite, R. M., & Chiellini, E. (2011). Chitosan—A versatile synthetic polymer in biomedical applications. *Progress in Polymer Science, 36*(8), 981–1014.

Dong, C., Chen, W., & Liu, C. (2014). Flocculation of algal cells by amphoteric chitosan-based flocculant. *Bioresource Technology, 170*, 239–247.

Geetha, D., Khadija, S. A. B., Alaa, S. A. H., & Amira, S. R. K. (2021). Development of chitosan-TiO_2 thin film and its application for methylene blue dye degradation. *International Journal of Environmental Analytical Chemistry*. https://doi.org/10.1080/03067319.2021.1948540

Han, T., Martin, R., & Karina, Y. H. (2017). Occurrence and fate of emerging contaminants in municipal wastewater treatment plants from different geographical regions-A review. *Water Research*. https://doi.org/10.1016/j.watres.2017.12.029

Islam, M. M., Masum, S. M., Molla, M. A., Rahman, M. M., Shaikh, A., & Roy, S. K. (2011). Preparation of chitosan from shrimp shell and investigation of its properties. *International Journal of Basic and Applied Sciences, 11*(1), 116–130.

Kashyap, N., Kumar, N., & Kumar, M. N. V. R. (2005). Hydrogels for pharmaceutical and biomedical applications. *Critical Reviews in Therapeutic Drug Carrier Systems, 22*, 104–150.

Khadija, S. A. B., Geetha, D., Amira, S. R. K., Mohammed, A. S., Alaa, S. A. H., & Siham, S. K. S. (2021). Extraction of bio polymers from crustacean shells and its application in refinery wastewater treatment. *Walailak Journal of Science and Technology, 18*(5), 11543. https://doi.org/10.48048/wjst.2021.11543

Khoushab, F., & Yamabhai, M. (2010). Chitin research revisited. *Marine Drugs, 8*(7), 1988–2012.

Krajewska, B. (2005). Membrane-based processes performed with use of chitin/chitosan materials. *Separation and Purification Technology, 41*, 305–312.

Luo, Z., et al. (2013). Functional improvement of Saccharomyces cerevisiae to reduce volatile acidity in wine. *FEMS Yeast Research, 13*(5), 485–494.

Magwaza, S. T., Magwaza, L. S., Odindo, A. O., & Mditshwa, A. (2020). Hydroponic technology as a decentralised system for domestic wastewater treatment and vegetable production in urban agriculture: A review. *Science of the Total Environment, 698*, 134154.

Pedro, J. J. A., Candace, K. C., Menachem, E., Naomi, J. H., & Dino, V. (2018). Emerging opportunities for nanotechnology to enhance water security. *Nature Nanotechnology, 13*, 634–641. https://doi.org/10.1038/s.41565-018-04203-2

Pradip, K. D., Joydeep, D., & Tripathi, V. S. (2004). Chitin and chitosan: Chemistry, properties and applications. *Journal of Scientific & Industrial Research, 63*, 20–31.

Prashanth, K. V. H., & Tharanathan, R. N. (2007). Chitin/chitosan: Modifications and their unlimited application-An overview. *Trends in Food Science and Technology, 18*, 117–131.

Renault, F., Sancey, B., Badot, P. M., & Crini, G. (2009). Chitosan for coagulation/flocculation processes-aneco-friendly approach. *European Polymer Journal, 45*, 1337–1348.

Rinaudo, H. (2006). Chitin and chitosan: Properties and applications. *Progress in Polymer Science, 31*, 603–632.

Sowmya, S., Kumar, P. S., Chennazhi, K. P., Nair, S. V., & Tamura, H. (2011). Biocompatible β-chitin hydrogel/nanobioactive glass ceramic nanocomposite scaffolds for periodontal bone regeneration. *Artificial Organs, 25*(1), 1–11.

Yen, M. T., Yang, J. H., & Mau, J. L. (2009). Physico chemical characterization of chitin and chitosan from crab shells. *Carbohydrate Polymers, 75*, 15–21.

Zhang, J., Feng, M., Lu, X., Shi, C., Li, X., Xin, J., Yue, G., & Zhang, S. (2018). Base-free preparation of low molecular weight chitin from crab shell. *Carbohydrate Polymers*. https://doi.org/10.1016/j.carbpol.2018.02.019

Index

P. O. Isibor et al. (eds.), *Nanochitosan-Based Enhancement of Fisheries and Aquaculture*, https://doi.org/10.1007/978-3-031-52261-1

MIX
Papier aus verantwortungsvollen Quellen
Paper from responsible sources
FSC® C105338

If you have any concerns about our products,
you can contact us on
ProductSafety@springernature.com

In case Publisher is established outside the EU,
the EU authorized representative is:
Springer Nature Customer Service Center GmbH
Europaplatz 3, 69115 Heidelberg, Germany

Printed by Libri Plureos GmbH
in Hamburg, Germany